Young Sun, Early Earth
and the Origins of Life

D1376784

M. Gargaud · H. Martin · P. López-García
T. Montmerle · R. Pascal

Young Sun, Early Earth and the Origins of Life

Lessons for Astrobiology

Translated by Storm Dunlop

 Springer

Muriel Gargaud
CNRS-Université Bordeaux 1
Laboratoire d'Astrophysique de Bordeaux
Bordeaux, France

Hervé Martin
Université Blaise Pascal
Laboratoire Magmas et Volcans
Clermont-Ferrand, France

Purificación López-García
CNRS–Université Paris-Sud
Unité d'Ecologie, Systématique et Evolution
Paris, France

Thierry Montmerle
Institut d'Astrophysique de Paris,
Paris, France

Robert Pascal
CNRS–Université de Montpellier 2
Institut des Biomolécules Max Mousseron
Montpellier, France

ISBN 978-3-642-22551-2   e-ISBN 978-3-642-22552-9 (eBook)
DOI 10.1007/978-3-642-22552-9
Springer Heidelberg Dordrecht London New York

Library of Congress Control Number: 2012955750

© Springer-Verlag Berlin Heidelberg 2012
This work is subject to copyright. All rights are reserved by the Publisher, whether the whole or part of
the material is concerned, specifcally the rights of translation, reprinting, reuse of illustrations, recitation,
broadcasting, reproduction on microfilms or in any other physical way, and transmission or information
storage and retrieval, electronic adaptation, computersoftware, or by similar or dissimilar methodology
now known or here after developed. Exempted from this legal reservation are brief excerpts in connection
with reviews or scholarly analysis or material supplied specificall for the purpose of being entered and
executed on a computersystem, for exclusive use by the purchaser of the work. Duplication of this publication or parts there of is permitted only under the provisions of the Copyright Law of the Publishers
location, in its current version, and permission for use mustal ways be obtained from Springer. Permis-
sions for use may be obtained through RightsLink at the Copyright Clearance Center. Violations are liable
to prosecution under the respective Copyright Law.
The use of general descriptive names, registered names, trademarks, servicemarks, etc.in this publication
does not imply, even in the absence of a specific statement, that such names are exempt from the relevant
protective laws and regulations and there fore free for general use.
While the advice and information in this book are believed to be true and accurate at the date of publication, neither the authors nor the editors nor the publisher can accept any legal responsibility for any
errors or omissions that may be made. The publisher makes no warranty, express or implied, with respect
to the material contained here in.
Printed on acid-free paper
Springer is part of Springer Science+Business Media (www.springer.com)

Introduction

Questions relating to the origin of life on Earth, and its possible presence elsewhere in the Universe, have fascinated Man since antiquity, whether as a man of science, a philosopher, or quite simply as a man-in-the-street. Understanding how life appeared on Earth would be the culmination of a form of a very ancient quest for our origins. Such a culmination would also be a decisive advance in our knowledge of an extremely complex natural process, the sequence of which raises numerous questions. It is these questions, which are fascinating in themselves, and frequently still without any reply, that are at the heart of the current work.

To this day, we know of one single example of life: that of life on Earth. Defining life and its essential properties is not a simple task either for a biologist, who studies life, or for an epistemologist, as someone who seeks to understand the way in which Mankind forms a concept of the world around it. We are able, however, to recognize life as a state of complex matter that is evolving in a dynamical context. And as such, life does not, therefore, escape the laws of physics and chemistry. Far from it. It is based on physical and chemical mechanisms that take place in a specific geological and atmospheric environment: that of our planet. So, life is the result of a *natural* process, and that implies that it is possible that other life forms have appeared elsewhere in the universe, based on analogous physical and chemical foundations, yet must always be constrained by the universal laws of nature.

Exploring When and How Life Emerged

Is it really possible that the life that we know today on Earth resulted from a unique combination of circumstances in the universe? What was the relative significance of determinism (predictable events obeying natural laws), chance, and contingency (not predictable, historical sequences of events) in the emergence of living beings? Is life the result of a gradual and continual increase in complexity, or was there, at one specific moment, a sudden jump in complexity leading to the emergence of new properties – for example, through a combination of several elements that interacted? All these questions currently remain open, but it is very probable that examination of the scenarios capable of explaining the emergence of life on Earth could progressively provide us with some of the factors in the answer. But for all that, both the spirit and the approach of scientists who are interested in this problem are extremely varied, depending on the disciplines involved.

So, for example, astrophysicists seek to know if other objects in the Solar System could shelter life, or even if there exist, beyond the Solar System, "other Earths" or, at least, other "habitable" planets. They expect that the study of terrestrial life and its origins will reveal the conditions that were necessary for its development, and they focus their research on extraterrestrial objects where similar conditions are likely to be combined – such objects then being considered as potentially inhabitable. Chemists, however, try to understand the process of self-organization and the establishment of sequences of reactions leading to systems capable of evolving in the same way as living beings. They thus try to determine, how, on Earth, the passage from abiotic organic chemistry to biochemistry could have taken place. Geologists are interested in the history of the planet and the impact of life on its evolution, but, above all, they try to define as precisely as possible the environmental conditions that prevailed when life emerged and that favored its development. Finally, biologists seek to know how biological evolution started, such that it gave birth to an extraordinary diversity of organisms, with very different forms, sizes, and abilities, but which all possess common properties. Exploring *how* and *when* – two crucial questions – life emerged in this small corner of the universe that is the planet Earth, concerns any one of these just as much as any other, but the scientists are only able to answer within the limits of the field in which they are involved.

As regards the question "When?", we shall see that the first difficulty is linked with the question of time, that fourth dimension which is difficult to grasp, and where scientists of the different disciplines need to establish common conventions if they wish to speak of the same thing. In fact, time is measured differently by astrophysicists, for whom time advances in an absolute manner from an initial instant of reference (t_0) – which corresponds in this book to the start of the formation of the Sun, 4.57 billion years ago (4.57 Ga) – and by geologists and biologists, who measure time backwards, relative to the present. This characteristic is very real in this book where, for a given event, we progress from an absolute time scale expressed in billions of years running *towards the present day* from the reference point, t_0 (as will be the case in the early chapters, relating to the formation of the Solar System), to a relative scale expressed in billions of years, *towards the past* (or "before present", BP, and thus, by convention, before the year 1950) when, in our story, geology and then biology take over from astrophysics. Chemists are disoriented and worried by these notions of long time-scales, whether relative or absolute. In fact, for them, rather than the chronological moment when they occur, the important factor is the kinetics and thus the relative duration of chemical reactions which, to add to the difficulty, cannot be understood except in statistical terms (that is to say, over populations of molecules, in contrast to the individual random fate of a single molecule). Be that as it may, all these disciplines need to know to which common time scale they are referring when they attempt to reply to the question of "when did life appear?". We shall see in this book that, although this question remains without any precise answer, it is still possible to define a range of time during which the transition from inert to living matter occurred.

Replying to the question "How?" is more difficult and controversial. We shall never be able to obtain a definitive answer, because life is a historical (contingent) process, in other words it has evolved in an irreversible fashion over the course of time. At best we may hope to reconstruct a plausible scenario, compatible with the laws of physics as well as with the experimental data, and present-day and future observations. To this day there is no consensus about this problem – far from it! – and we sorely lack reliable, realistic data on the physical and chemical conditions that prevailed on the primitive Earth where life emerged. Consequently, numerous hypotheses, often mutually exclusive, have been suggested by researchers. Some of these, even if they do allow an explanation of the observations, could never be tested. Others, in contrast, are susceptible to being refuted one day if they do not agree with the constantly increasing body of observable data. Apart from these "structural" difficulties, there is a human factor, as shown by the fact that there exist opposing schools of thought, which sometimes rather dogmatically refuse to consider and analyze in detail any arguments that are not their own. However, it is essential to remain optimistic: such a situation tends to disappear as and when new, more reliable, data are acquired, and research into alternative pathways, often intermediate ones, enables more concrete scenarios to be proposed. We have chosen to give a broad and as neutral as possible view of these different models, preferring to put the emphasis on the existing data rather than to interpret them in a partisan manner.

A Novel Challenge: Getting Several Disciplines to Talk to One Another

The aim of this book is to present, in a chronological manner – or, at least, logically as a relative succession of events – the history of the origins of life on Earth and the conditions that allowed it to appear on our planet. The novel challenge is that for each of the time periods that form this chronology, we, as different specialists, will speak *together* to lift a corner of the veil; with the approach and questions appropriate for each original discipline. The image of the questions that it presents is therefore firmly multi-disciplinary. Astrophysics and geology will thus allow us to reconstruct the history of the formation of the Sun, of the Solar System, and of the Earth. Geology and chemistry, subject to certain constraints derived from observation that biology will impose, will deal with the occurrence of conditions required for complex chemistry and life to appear. Finally, biology will enable us to sketch the main features of evolution, and in particular to discuss the emergence of eukaryotic cells and their diversification, until the appearance of animals and terrestrial plants, which form the greatest part of the world visible to the human eye.

We have decided to stop this great tale at the Cambrian explosion, 540 million years ago, when the ancestors of the major animal lines that we see today made their appearance. At that time, biological evolution had been in progress for over 2 or even 3 billion years, and the multiplicity of directions that it was to follow subsequently – including the appearance of humans within a small phylogenetic line of descent among hundreds of others – is of lesser importance for our understanding of the origins and evolution of primordial life on Earth.

Lessons for Astrobiology?

The present book is based on a translation of the French original "Le Soleil, la Terre... la Vie" (the Sun, the Earth... Life), published in 2009 by Editions Belin (Paris). At the request of Springer, we have completed the original version by a new chapter on "Extrasolar planets", i.e., the hundreds of planets and planetary systems discovered since 1995 around normal stars other than the Sun. To stick to the original spirit of the book, we have put emphasis on the fascinating question of a particular sub-class called "habitable planets".

As the reader will see, the state-of-the-art in this field is still entirely astronomical, with no indication whatsoever of any "biological" evidence. In this context, is it really justified to use the terms "exobiology", or "astrobiology", which *stricto sensu* should etymologically mean "extraterrestrial biology" and "biology applied to astronomy", respectively? In his Preface to the Springer "Encyclopedia of Astrobiology" (2011), C. De Duve (awarded the Nobel-Prize for Physiology and Medicine in 1974), speaks of "the new discipline of exobiology-cum-bioastronomy-cum-astrobiology", implying that these three commonly found denominations are equivalent. In the context of the present book, however, we conclude that so far no evidence for life has been found elsewhere than on the Earth: neither in the Solar System, nor on planets around other stars – even if we suspect, and hope, that this evidence will come in the future. At this stage, astronomers, for their part, tend to support the term "bioastronomy"* (meaning, *astronomy applied to the search for life in the universe*, as "biophysics" or "biochemistry" etymologically are the fields of physics or chemistry applied to biological phenomena) as appropriate term to use for now.

But of course life is all about biology (the science of life!), so it is fair to say that all the disciplines combined in this book, to describe how we think life emerged on Earth, indeed can be considered as providing "Lessons for astrobiology" (the subtitle of this book), in the sense that the authors hope it can contribute to laying ground for the future – if, and when, a "biology" will be discovered in another world than the Earth.

---

* Indeed, a commission of the International Astronomical Union is called by this name (http://www.iau.org/science/scientific_bodies/commissions/51/).

Acknowledgements

This book is the result of a vast collective collaboration, in which more researchers have participated than the five authors mentioned on the cover.

The story started with a specialized CNRS* school organized in 2003 at Propriano (in Corsica, France) by Muriel Gargaud, and which was followed by two workshops organized in 2004 at Château Monlot-Capet at Saint-Émilion (in the Gironde) and at Château d'Abbadia (Académie des Sciences) at Hendaye (in the Pyrénées-Atlantiques). The aim of these meetings – financially underwritten by the Centre national de la recherche scientifique (CNRS), the Centre national d'études spatiales (CNES)**, the Conseil régional d'Aquitaine, the Université Bordeaux 1 and the Laboratoire d'astrophysique de Bordeaux – was to establish and discuss the chronology of the events that led to the appearance of life on Earth, between the formation of the Solar System, 4.57 billion years ago and the Cambrian explosion of lifeforms, 540 million years ago.

The work was put into concrete form in nine scientific articles in a special issue of the journal *Earth, Moon and Planets* (No. 98), entitled "From Suns to Life: a chronological approach to the origins of life on Earth", which appeared in 2006. It is those articles that form the basis for this work.

Twenty authors participated in writing the articles, and without them, this book would not exist. This is why we owe our warmest thanks to: Francis Albarède (École normale supérieure de Lyon), Jean-Charles Augereau (Laboratoire d'astrophysique de Grenoble), Laurent Boiteau (Département de chimie, Université de Montpellier), Marc Chaussidon (Centre de recherches pétrographiques et géochimiques, Nancy), Philippe Claeys (Vrije Universiteit Brussel, Belgium), Didier Despois (Laboratoire d'astrophysique de Bordeaux), Emmanuel Douzery (Institut des science d'évolution, Université de Montpellier), Patrick Forterre, (Institut de génétique et microbiologie, Université Paris-Sud, Orsay), Matthieu Gounelle (Muséum national d'histoire naturelle, Paris), Antonion Lazcano (Universidad Nacional Autónoma de México, Mexico), Bernard Marty (École nationale supérieure de géologie, Nancy), Marie-Christine Maurel (Institut Jacques-Monod, Université Paris 6), Alessandro Morbidelli (Observatoire de la Côte d'Azur, Nice), David Moreira (Unité d'écologie, systématique et évolution, Université Paris-Sud, Orsay), Juli Peretó (Insitut Cavanilles de Biodiversiat i Biologia Evolutiva, Universitat de València, Spain), Daniele Pinti (Université du Québec à Montréal, Canada), Daniel Prieur (Laboratoire de microbiologie des environnements extrêmes, Université de Bretagne occidentale, Brest), Jacques Reisse (Université libre de Bruxelles, Belgium), Franck Selsis (Laboratoire d'astrophysique de Bordeaux), et Mark Van Zuilen (Centre for Geobiology, Bergen University, Norway).

Finally, from its inception to the present, this project has benefited from the unfailing and enthusiastic support of Michel Viso, Director of the "Exobiology" programme at the Centre national d'études spatiales. We give him our warmest thanks.

Our deepest gratitude also goes to our editor at Editions Belin, whose role has been essential and who, by bringing the various sections together has been the true orchestrator of this work.

---

* CNRS = "Centre National de la Recherche Scientifique", the French national research agency.
** CNES is the French space agency.

Contents

The Formation of the Sun and Planets

Chapter 1

In the Beginning, There Was the Sun …

Age: 4.57 billion years

Place of birth: Unknown, but probably in a nebula similar to that of Orion.

Father: A molecular cloud, nowadays lost

Mother: Universal gravitation.

Gestation period: Ten thousand years.

Childhood: Very turbulent, even subject to tantrums, with the ejection of material and numerous consecutive, eruptive episodes in the presence of magnetic fields.

Descendants: Planets in the Solar System (giant planets in a few million years; terrestrial planets, including the Earth, in a few tens of millions of years).

◻ **The Orion Nebula seen in the near infrared, illuminated by the luminous "Trapezium cluster" of massive stars (seen in the center of the picture).** It was probably within a nebula of this type, inside a stellar association, that the Sun was born. (Credit ESO)

M. Gargaud et al., *Young Sun, Early Earth and the Origins of Life*,
DOI 10.1007/978-3-642-22552-9_1, © Springer-Verlag Berlin Heidelberg 2012

1

On a fine, Moonless, winter's night, out in the country, look up at the sky. The Milky Way, our Galaxy, is revealed in all its splendor, all light and shade. The light is, of course, that of stars and bright nebulae, such as the celebrated Orion Nebula; the shade is that of interstellar clouds, or rather minute grains of dust that they contain (a few hundred microns across at the most) and which hide the stars in the background (■ Fig. 1.1). In fact, these clouds primarily consist of transparent molecular gas: molecular hydrogen H_2, consisting of two atoms of hydrogen (the simplest element and the most abundant in the universe) and a minute quantity of more complex molecules, most carbon-based, which produce multiple spectral lines that may be detected by means of radiotelescopes. It is because of this that these interstellar clouds are nowadays known as molecular clouds. Using simple binoculars it is possible to see the close interleaving of nebulosity and the dark material that surrounds them. This dark material does, in fact, consist of grains of dust, illuminated either from the front or from behind (as for example in the Horsehead Nebula, ■ Fig. 1.1). But we also know, from the detection of radio emission from the molecules (in the sub-millimeter to centimeter range), that the illuminated portions represent only a small fraction of the total mass of these clouds: these are, in reality far more extensive than the nebulosity visible to the naked eye or in a telescope that is sensitive to the same wavelengths as the human eye (from about 0.4 to 0.7 μm, ■ Fig. 1.2).

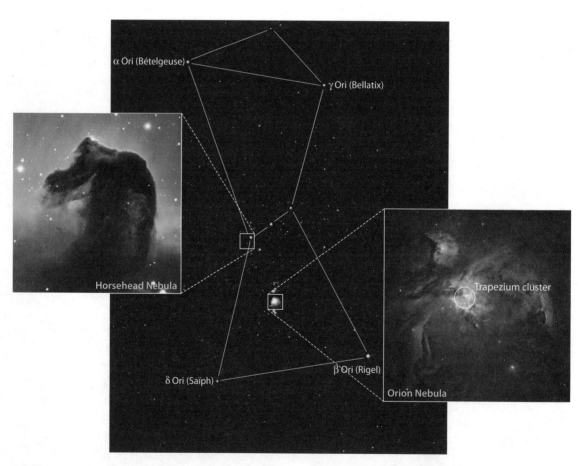

■ **Fig. 1.1 The constellation of Orion and the Orion Nebula as observed with a small telescope.** Inset: the Horsehead Nebula and the Orion Nebula as observed by the *Very Large Telescope* (ESO, Chile) and by the *Hubble Space Telescope* (NASA/ESA). Like all bright nebulae, the Orion Nebula is genuine nursery for stars, where numerous bodies are forming from surrounding clouds of gas and dust (molecular clouds). Close to bright nebulae, these clouds are illuminated by the intense light from stars. In the Horsehead Nebula, the silhouette of a horse corresponds to a molecular cloud seen like a shadow puppet: it masks the light from a bright nebula lying in the background.

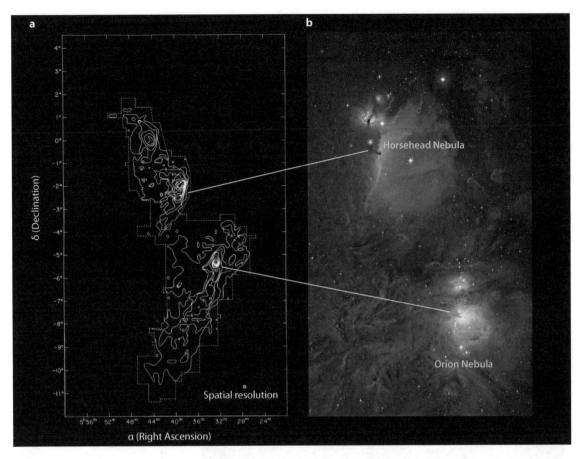

◘ **Fig. 1.2 The molecular clouds in the constellation of Orion seen (a) at the wavelength of the ¹²CO molecule (ordinary carbon monoxide) and (b) at visible wavelengths.** The most clearly visible condensations (the most dense) are associated with visible nebulosity. The illuminated portions of the molecular clouds (appearing as the bright nebulae) therefore represent just a small fraction of their mass and overall extent. The small circle at bottom right in diagram (a) indicates the resolution of the orbiting telescope used, here a few minutes of arc. (a: after Columbia University, New York, United States.)

In fact, all the bright nebulae in the sky are stellar nurseries: a whole multitude of stars is constantly forming there, under our eyes, from molecular clouds, rather like a cigar, where the nebulosity is the glowing tip, and which is slowly being consumed. Gravitation, one of the Universe's fundamental forces, is at work. Within molecular clouds it leads to the collapse of gas and dust, thus giving rise to "associations" of hundreds of thousands of stars over a time-span of only a few hundred thousand to a few million years. The most massive of these may amount to up to 100 times the mass of the Sun (denoted $M_\odot$), whereas the least massive go down as far as 0.1 $M_\odot$, or even less (as with brown dwarfs, stars where thermonuclear reactions will never take place).

Observation of bright nebulae such as Orion enable us to catalogue and classify the stars that illuminate them as a function of their luminosity and their temperature. This classification, whose origin dates back to the beginning of the 20th century, is known as the Hertzsprung-Russell Diagram (◘ Box 1.1). This is a fundamental diagram, which, thanks to the theory of stellar evolution, enables us to deduce the mass and age of stars, as well as their future state (◘ Box 1.1). It shows, for example, that the four brightest stars in the Orion Nebula (the Trapezium, ◘ Fig. 1.1) are also the most massive (up to 45 times the mass of the Sun) and between 2 and 3 million years old.

Thanks to images from the Hubble Space Telescope, we have also been able to discover that the stars in the Orion Nebula are, in the majority of cases, surrounded by dark circum-

◻ Box 1.1   The Masses and Ages of Stars

The fundamental tool for the study of the evolution of stars is the Hertzsprung-Russell Diagram, more commonly known as the "HR Diagram." This empirical diagram, devised at the beginning of the 20th century when the source of energy in stars was still unknown, may take several forms. Basically, it relates the color of a star and its absolute magnitude, which is often expressed in terms of luminosity, L_*, and surface temperature (known as effective temperature, T_{eff}). This assumes an accurate knowledge of the properties of its atmosphere (for example, its spectral type) and its distance from the Earth. For the great majority of stars, that conversion is well-understood.

In an HR diagram, stars appear in well-defined groups. One of the major topics of astrophysics has been – and remains – to understand the physical significance of such a diagram. For most stars, the key is to be found in the successive nuclear reactions that produce different phases in stellar evolution (◻ Box 1.2). Thanks to theoretical models, it is possible to express the HR diagram as a function of just two parameters: the mass of the star M_*, and its age t_*: for every star of luminosity L and temperature T_{eff} there is a corresponding pair of parameters M_*, and t_*. The validity of the models is confirmed by the use of binary stars, because they have the same age, while Kepler's laws allow the mass of the two components to be determined. This explains, for example, why the band on which most stars (including the Sun) are found, known as the Main Sequence, corresponds to the phase during which hydrogen is transformed into helium, and where the luminosity is proportional to the cube of the mass (◻ Fig. 1.8).

However, in the case of young stars, nuclear models are not applicable, because the temperature at their center is not sufficient to initiate the reactions that convert hydrogen into helium: these stars are said to be in the "pre-Main Sequence" phase. In this case, only gravity is acting: such "pre-Main Sequence stars" (in particular young stars known as T Tauri stars) shine because they are contracting, until the point when the central temperature becomes high enough to initiate the conversion of hydrogen into helium. This contraction may also be modeled, finding a curve (M_*, t_*) corresponding to the (L_*, T_{eff}) pair that is observed (◻ Fig. 1.3), that may be displayed as a function of time. The time for the star to evolve until it arrives at the Main Sequence is thus found to be of the order of a few million years (for stars of $M_* >$ 10–20 $M_\odot$) to several tens of million years (for M_* about the mass of the Sun, ◻ Fig. 1.3), which may be compared, for example, with the Sun's lifetime on the Main Sequence, which is about 10 billion years (◻ Fig. 1.8 in the ◻ Box 1.2).

Having thus been able (via theoretical models) to determine the mass of stars in different regions of the Galaxy, it appears that stellar masses are distributed according to a power law that is more-or-less universal: by mass interval, the number of stars $N(M_*)$ of mass M_* is $M_*^{1.35}$ (◻ Fig. 1.4), for $M_* > 0.5\ M_\odot$. This law flattens out

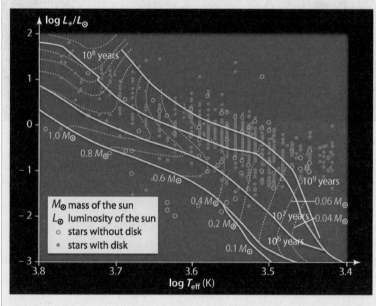

◻ **Fig. 1.3 The Hertzsprung-Russell (HR) diagram of young stars associated with the Orion Nebula.** All the stars in the sky may be placed on a similar diagram, which is first constructed from directly observed data: the magnitude (apparent) is plotted as the ordinate, and color (the difference in magnitude in two different colors) as the abscissa. Various combinations are possible, but one derives the intrinsic, current, physical values for the stars, with luminosity (L_*) in ordinate, and surface temperature (known as the "effective temperature," T_{eff} in abscissa), from the distance and from a model stellar atmosphere, respectively. Then, one calculates stellar evolution models (according to the type of star), which, for a given mass, enables evolutionary tracks to be plotted. These are so many "clocks," which enable us to know about the past and future of the stars, and where each point is associated with a luminosity and an effective temperature. The evolutionary tracks may be used to construct a graph, where the reference lines in the "vertical" direction indicate a given mass, and those "inclined" downwards are isochrones (that is points with the same age). This enables conversion of a (L_*, T_{eff}) pair into a (mass, age) pair for every star. Thanks to this diagram, one can see that stars in Orion between 40 $M_\odot$ (top left, beyond the range shown here), and 0.1 $M_\odot$, have ages between 100 000 years and a few million years. (After L. Hillenbrand, 2006.)

at lower masses, but the point to note is that in a star-formation region, the presence of a few massive, and thus bright stars, is accompanied by the presence of thousands of stars with smaller masses, like the Sun ...

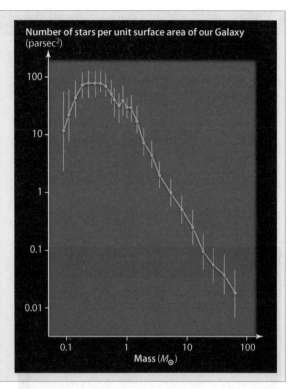

� Fig. 1.4 The initial stellar mass-function. This function, first established by E. Salpeter in 1955, determines the mass-distribution of all stars at the time of their birth, or, in other words, the number of stars formed (per unit surface area of our Galaxy, here square parsecs) as a function of that mass. It is determined from an empirical count of stars whose distance from Earth has been established, as well as their mass and age, thanks to the HR Diagram and evolutionary models that allow them to be traced back to their birth. Between 0.5 $M_\odot$ and 100 $M_\odot$, the stellar mass distribution approximately follows a power law (a straight line on a logarithmic scale, as here). From the figure it may be seen, for example, that for every star of mass 100 $M_\odot$ that forms, about 10 000 stars of mass 0.5 $M_\odot$ (a typical mass for "T Tauri" stars) are born. Below 0.5 $M_\odot$, the number of stars falls: about 10 times fewer stars of mass 0.1 $M_\odot$ are formed than stars of mass 0.5 $M_\odot$. Understanding the origin of this function is one of the great challenges of theories of stellar formation. (After J. Scalo, 1986.)

stellar disks, with radii of several times that of the Solar System and consisting, like the molecular clouds themselves, of molecular gas and of dust grains (■ Fig. 1.9; ■ Box 1.4). In this chapter we shall see why astrophysicists describe these disks as being protoplanetary; numerous arguments – whether theoretical ones or derived from observation – have actually convinced them that planets are born inside such structures. This is one of the reasons (but there are others) why it is now thought that the Sun and the Solar System were born in a similar environment, as was once proposed, on purely qualitative grounds, by Emmanuel Kant and Pierre-Simon Laplace at the end of the 18th century.

There is thus a close, natural link between what we observe today of the formation of stars and protoplanetary disks on the one hand, and what we have learnt *in situ* (notably thanks to meteorites) about the formation of our own planetary system, which we shall describe later in detail. This will lead us to cover the time from the beginning of the collapse of the interstellar cloud in which our star, the Sun, formed, until the birth of the planets orbiting around it. We shall discover that this fundamental (not to say founding) episode, as far as we, as humans, are concerned, was astonishingly brief on the geological time-scale, and yet action-packed.

It took just some ten thousand years for the proto-Sun to initiate its development from a molecular cloud. Later, in a few million years (not more than ten!), the progressive agglomeration of dust grains and, simultaneously, the concentration of gas at several points in the disk gave birth to the giant planets (Jupiter, Saturn, Uranus, and Neptune) and to the meteorites (which are the debris resulting from the incessant collisions between the small protoplanetary bodies in the nascent Solar System). Finally, much more slowly, it would be the turn of the rocky planets (Mercury, Venus, Mars, and ... the Earth) to form from the remaining material.

How may these events be dated in relation to the situation in the Solar System at the present day? Is it possible to establish an absolute chronology? By using radioactive-dating methods (*see* ▶ Chapter 2, and ■ Box 2.2), we have been able to measure, extremely precisely, the age of the most primitive meteorites, known as carbonaceous chondrites. The most famous example is that of an enormous meteorite (weighing 250 tonnes!) fragments

◻ Box 1.2   Stellar Destinies

The life of a star is entirely one of a fight against gravitation. This force, which brought about its birth from a molecular cloud, will subject it to greater and greater compression, induce radical changes in its internal structure, and, in the ultimate, decide its eventual fate. This explains why the parameter that determines the structure and evolution of a star throughout its life is its mass. But gravitation, one of the three fundamental forces of nature, is always attractive, which is to say, acts in one direction only. This means that the compression of the star is converted into the liberation of energy, produced within its interior and transferred to the ex-

terior: this is its luminosity. But this transfer may occur in several different ways, depending on the "opacity" of the atoms forming the star. The hottest regions are radiative (only radiation can thread its way through the material), and the cooler regions are convective (the radiation "drives" the material into turbulent motion). It is this, finally, that determines the temperature of the stellar surface (known as the photosphere).

Luminosity and temperature are the two parameters in the Hertzsprung-Russell Diagram that are accessible to observation (◻ Box 1.1). They enable us, through theoretical models, to know the nature and evolutionary stage of a star at any given time. These models lead to the consideration of two principal stages of evolution: a non-nuclear phase, when the star is young and it releases only gravitational energy, becoming progressively hotter (as in the T Tauri stars, for example; ◻ Fig. 1.3), and a thermonuclear phase, in which the star releases nuclear energy over a much longer time-scale. This latter phase itself consists of two stages: first, the Main Sequence, where the source of energy is basically the conversion of hydrogen into helium, which starts when the star's core temperature reaches 15 million degrees (as with the Sun at present); and then with what is called "post-Main-Sequence" evolution, where multiple nuclear reactions, of greater and greater complexity, rapidly succeed one another, forming heavier and heavier elements (carbon, oxygen, silicon, etc.) – a process known as nucleosynthesis. At the same time, the structure of the star is drastically altered. The outer layers expand more and more; they are thus less subject to gravitation, but more affected by radiation, which causes an intense mass-loss (◻ Fig. 1.5). In stars with the mass of the Sun, and up to about 8 $M_\odot$, grains of dust condense and form a spectacular, expanding envelope around the star: these are the planetary nebulae (◻ Fig. 1.6). More massive stars, which are hotter, are the source of very intense and rapid stellar winds.

The very last stages of stellar evolution depend even more strictly on the mass than the earlier ones, and are marked by very different end products. Below 8 $M_\odot$, the shell of dust surrounds a tiny white dwarf, no larger than the Earth, but which is very hot and dense: it will take billions of years to cool. Above 8 $M_\odot$, the thermonuclear reactions run away, and the star explodes as a supernova (◻ Fig. 1.7). In a few hours, the supernova releases as much energy as the whole Universe! Below about 25 $M_\odot$, the remnant is a neutron star, even more compact (about the size of Paris, which measures 10 x 18 km) and denser than a white dwarf; its rapid rotation means that it may be observed as a pulsar for millions of years. Above 25 $M_\odot$, the material collapses onto itself

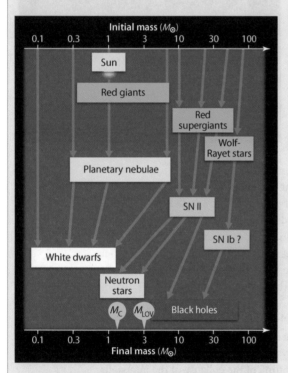

◻ **Fig. 1.5 Stellar destinies.** The evolution of a star is determined at its birth by the mass that it acquires at the time of its formation. This mass evolves over the course of time, because its internal structure evolves under the influence of the thermonuclear reactions that provide its energy and which (except for the least massive stars, which evolve extremely slowly) cause different mass-loss phenomena to occur. This loss of mass is frequently observed and has resulted in a whole menagerie of spectacular objects, like planetary nebulae or supernovae (SN). The figure shows the wide variety of "stellar remnants" as a function of their initial mass. "Compact objects" (neutron stars and black holes), whose mass is determined by quantum physics (the "Chandrasekhar mass" M_C for neutron stars), or by general relativity (the "Landau-Oppenheimer-Volkoff mass" M_{LOV} for stellar black holes), are the stellar evolutionary remnants of massive stars ($M_* > 8\ M_\odot$). They are not visible directly, but are detectable by the very energetic radiation that they emit. For massive stars, the Wolf-Rayet phase immediately precedes the explosion as a supernova.

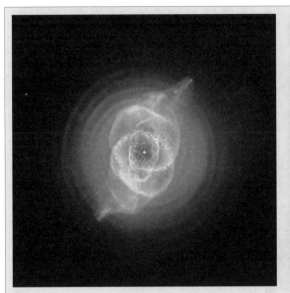

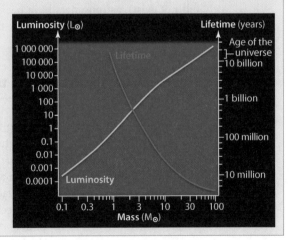

☐ **Fig. 1.6 The "Cat's Eye" planetary nebula (NGC 6543).** The overall shape like a spiral shell is evidence of a intense mass-loss in the form of dust. This mass-loss occurred in several successive episodes, which were modified by the star's rotation (a bit like a rotating garden sprinkler). This shell extends to some hundreds of astronomical units. The Sun will experience a similar fate and the whole Solar System will be engulfed ... in five billion years.

☐ **Fig. 1.7 A famous supernova remnant: the Crab Nebula.** This is the debris of a massive star that exploded in 1054. According to Chinese astronomers, the "guest star" – to use their term – remained visible for several months. The filaments are the very rapidly expanding ejecta (several thousand km s⁻¹), of the star itself. The halo surrounding the central regions is caused by what is known as "synchrotron radiation," from extremely energetic electrons, accelerated by the shockwaves from the filaments.

and forms a black hole, from which light cannot escape. Gravity has finally gained the upper hand ...

What is important to remember in the context of the formation of the Solar System is that stars more massive than 25 $M_\odot$ have a lifetime of 10 million years at most (☐ Fig. 1.8). They therefore explode where they

were born, so to speak, and their dispersed material is thus capable of "contaminating" the stars of lesser mass that are surrounded by protoplanetary disks, as for example with the one that existed at the beginning of the Solar System's formation.

☐ **Fig. 1.8 The luminosity and the lifetimes of stars are a function of their initial mass.** The "ascending" curve (in yellow) corresponds to the luminosity (the left-hand ordinate), and the "descending" curve (in red) to the lifetime (the right-hand ordinate), with the mass given in the abscissa. There is a very strong relationship between the two curves: the luminosity is simply the energy expended by the star in counteracting gravity. The more massive it is, the more energy it needs to expand, and the shorter its lifetime. There is a factor of a thousand in the range of masses for stars. The corresponding range in luminosity varies by a factor of one billion and, conversely, the age decreases by a factor of 10000. The least massive stars are "eternal" (their lifetimes are greater than the age of the Universe), whereas the most massive stars die by exploding as supernovae at the same location as they were born.

of which fell in 1968 not far from the village of Allende in Mexico. It has been possible to determine its age to within one million years: 4.5685 billion years (Ga) ± 0.5 million years (Ma). By convention, this age is considered to be that of the Solar System (designated t_0 throughout the rest of this book). It will be understood that this dating does not take account of the time necessary for the formation of the Sun itself, and which is thus added to

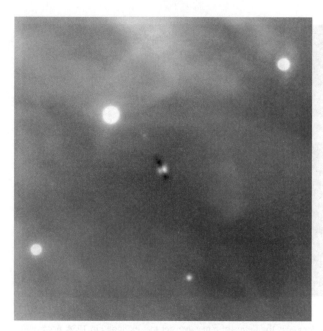

□ **Fig. 1.9 A circumstellar disk in the Orion Nebula.** By far the majority of stars in this nebula are low-mass stars surrounded by dark disks, consisting, like molecular clouds, of gas and dust grains. These are "circumstellar disks," with radii that are several times that of the Solar System. Astrophysicists are nowadays convinced that such structures give birth to planets.

these 4.5685 billion years. In practice, that time being very short in comparison with the age of the meteorites, we shall simply say that our trip back to the origin of the Solar System takes us back in time to 4.57 billion years, in round figures[1] (in the rest of the book, we shall simply write "at 4.57 Ga"), to the heart of a vast, anonymous interstellar cloud of molecules and dust grains ...

The Sun's Protoplanetary Infancy

Thanks to the observation of numerous regions of star formation, and helped by theoretical models (the oldest of which date back to the 1960s, at a time when even pocket calculators did not exist), astronomers have been able to reconstruct the different stages in the birth and early lifetime of stars of the same type as the Sun – which are those that primarily interest us in this book.

The Paradoxes Surrounding the Birth of the Sun

Ever since the 18th century, following the writings of Kant (*Allgemeine Naturgeschichte und Theorie des Himmels* – "Universal Natural History and Theory of Heaven," 1755) and of Laplace (*Exposition du Système du Monde* – "Description of the System of the World," 1796), the prevailing idea has been that the Sun and the planets, whose orbits all lie in the same plane, were born at the same time. To Laplace, the Sun formed from a cloud that was collapsing gravitationally and was in rotation. Because it contracted, this cloud turned more rapidly, like a skater who folds their arms (both obey the same physical law of the conservation of angular momentum, a crucial idea to which we shall return shortly). According to Laplace, the young Sun, because of its fast initial rotation, flattened, ejecting through the effects of centrifugal force a sort of equatorial disk that cooled, giving birth to the planets. This scenario allowed the Sun to slow its original speed of rotation at the same time as it contracted to its current

1. This figure may be compared with the age of the Universe: 13.7 billion years. Contrary to a widely held belief, the Sun did not, therefore, form directly from the "Big Bang", but independently, and 9 billion years later.

size, to the benefit of the planets, which were assumed to have formed from the material in the disk.

In this scenario (which was purely qualitative), Laplace, who was an excelled physicist, recognized the necessity of decreasing the Sun's kinetic moment, so that it could contract, by transferring, via the disk, the difference in kinetic moment to the planets as they formed. In fact, to give a figure, Jupiter currently accounts, by itself (because it is very far from the Sun, even though only one thousandth of its mass), for 99 percent of the Solar System's kinetic moment, including the Sun. (The problem is that we do not know the "initial value" of the angular momentum of the protoplanetary disk, 4.57 billion years ago.)

Nowadays, the background remains the same, but the play is performed differently, notably with an essential, contemporary actor that Laplace could not know about: the magnetic field. Before becoming acquainted with the new actor in the drama, let us start from the astronomical observations carried out nowadays. In the infrared and millimetric regions (i.e., at wavelengths that go from a few micrometers to a few millimeters), telescopes are able to detect molecules and dust grains, including those "inside" the depths of molecular clouds, because the latter are more and more transparent to radiation, the more the wavelength increases. (This is the principle behind infrared binoculars, which enable one to see through fog.) It is therefore possible to observe strange phenomena in the very heart of these clouds, which would be completely inaccessible to wavelengths in the visible region, and which allow us to understand the true nature of a protostar.

Intuitively, Laplace had the right idea: everything starts from a large-scale condensation[2] of gas and dust, in the heart of a cold, dense, molecular cloud (between 10 and 50 K, depending on the region), that is rotating – very slowly – on its own. It has been found that such condensations (known as protostellar condensations) which form the basic structure of the cloud (■ Fig. 1.10), are almost exactly in gravitational equilibrium. If no external perturbation occurs, the clouds do not collapse. However, detailed observations, utilizing the Doppler effect on the spectral lines of the molecular gas, show that these condensations do collapse in about 10 000 years. They then become very extended protostellar envelopes (10 000 au across; where 1 AU = 1 astronomical unit = the Earth–Sun distance = 150 million km), sheltering within them stellar embryos, which grow at their expense. A protostar is thus a hybrid body, consisting of an envelope and an embryo. To this day, the exact cause of the initiation of the collapse is still unidentified (it is not necessarily unique), but there is no lack of working hypotheses, notably in the vicinity of massive stars. The most popular currently are the effects of shockwaves caused either by the powerful winds create through the pressure of their UV radiation, or by the explosion of supernovae when they reach the end of their lives (■ Boxes 1.1 and 1.2).

Once it has started, this collapse continues, more rapidly within the inner regions of the envelope, which are denser. But, for many reasons (the inherent rotation of the original molecular cloud on its own, accompanied by galactic rotation, turbulent motion, etc.), this envelope is itself rotating, and its inner regions rotate faster than its outer ones. As a consequence, and again as foreseen by Laplace, the central portions flatten out to form a dense disk, onto which the rest of the envelope will "rain down" at large distances from the center (■ Fig. 1.11). This disk is to play a crucial role in stellar formation.

In fact, it is now known that the presence of such a disk enables us to explain, though a complex magnetic mechanism, the most spectacular phenomenon accompanying the formation of stars, at the very early protostar stage, namely a significant loss of mass in the form of winds or jets, which are strongly collimated and emitted in opposite directions. (They are, for

---

2. This term, widely used in astronomy, means "concentration", or "a region of high density", that is observed "instantaneously" (relative to the time-scales discussed here). It does not have the dynamical connotation of rapid collapse, as used by planetologists or geologists.

1

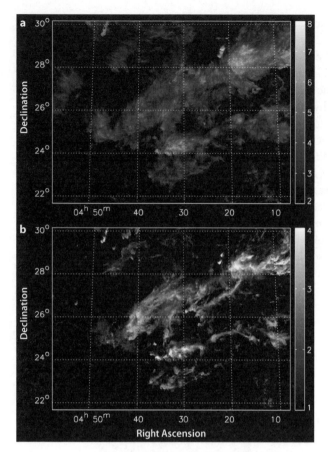

◩ **Fig. 1.10 A map of the molecular clouds in the constellation of Taurus. (a)** plots of the ^{12}CO molecule and **(b)** of the ^{13}CO molecule at a wavelength of 2.6 mm. Deep within molecular clouds one discovers a number of condensations, i.e., regions of higher density. As a result of external perturbations that are still not perfectly understood, the latter may begin a process of gravitational collapse. This is the first act in the formation of a star. As with ◩ Fig. 1.2, the ^{12}CO molecule (plot a) traces the large-scale structure of the clouds, whereas in plot b, the ^{13}CO molecule (where ordinary ^{12}C is replaced by the ^{13}C isotope), traces the dense "skeleton" of the molecular cloud, in effect revealing the pre-stellar condensations. (Diagram: *Five College Radio Astronomy Observatory*, Amherst, United States.)

this reason, known as "bipolar jets".) These jets, which are of fundamental importance, were discovered in 1980. Visible at practically all wavelengths (◩ Fig. 1.12), they "punch through" the molecular clouds and commonly reach lengths of 50 000 AU, or even far more (up to a light-year).

The existence of bipolar jets implies that here Laplace's scenario has to be modified, because not only does the (proto)Sun eject material from its poles (and not from its equator), but because it rotates far too slowly to expel an equatorial disk as a result of centrifugal force. Quite the contrary, it seems that stars of the solar type form *from a disk,* which has itself been born within a slowly collapsing envelope and in which, under the influence of gravitation, material flows from the envelope (around the periphery) towards the stellar embryo (in the center). More specifically, as can be shown solely by spectroscopic methods (through the Doppler effect), the material in the disk (gas and dust) "falls" from its inner region onto the embryonic star, and causes it to grow progressively until it becomes a star in its own right: this is known as accretion. (Again, the term accretion is here used in a different sense to that used by planetologists and geologists. The latter speak of accretion in the sense of "aggregation," in describing the phenomena that, much later, give rise to planets within the protoplanetary disk.) However, the quantity of material ejected (as seen on a large scale) represents just a small fraction of the material (as seen by spectroscopy) that is accreted by the stellar embryo from the disk (between 10 and 30 percent according to observation).

This paradox of nature, whereby to form a star, it must simultaneously gain and lose material, is known as the "accretion-ejection" phenomenon. The "engine" responsible for this phenomenon is, as we shall see, completely inaccessible to observation, and has to be analyzed in an indirect manner.

Fig. 1.11 The structure of a protostar. A protostar is a hybrid body, consisting of two parts: an envelope and an embryo star. The central portion of the envelope, which is denser and in rapid rotation, flattens and forms a disk, onto which the remaining material "rains" at a considerable distance from the center. The growth of the extremely young star is accompanied by another spectacular phenomenon: the ejection of material in the form of bipolar jets. This is an extraordinary natural paradox: the birth of a star implies that it both gains and loses material simultaneously. (au: astronomical unit.)

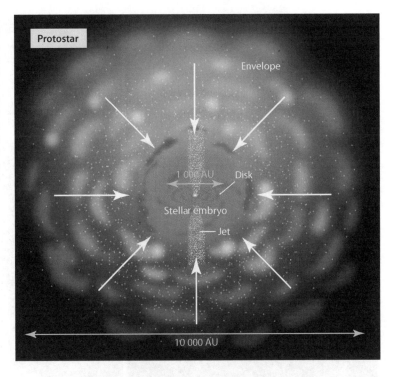

There remains the notorious angular momentum, a decrease in which is indispensable for the embryonic star to grow: the latter must brake its rotation more rapidly than its mass increases, to counteract the opposing effect of centrifugal force. The discovery of jets, which eject material to a great distance, solves part of the problem, but it nevertheless implies that some of the material in the disk must also move outwards – i.e., in the opposite direction to the accretion! The question has not really been resolved at present; it is, in fact, extremely complex, especially if one takes account (as Laplace did) of the formation of planets and, equally, of the transfer of material from the disk to the jets, which involves magnetic fields, as we shall see later.

Despite these lacunae in the theories, it is possible to observe that the accretion phenomenon takes place at the rate of about one solar mass every 100 000 years, which illustrates the (relatively) rapid transition from a stellar embryo to a "true" star like the Sun. The circumstellar disk itself, persists for much longer. The planets will form within it, but later, and much more slowly. Among other effects, this progressively decreases the rate of accretion (there is less material available) and eventually stops the ejection.

The Primitive Solar Nebula Revealed by Young Stars

We said at the beginning of this chapter that a star like the Sun would not have been born alone. In fact, observation of molecular clouds typically shows hundreds of stars being formed, each with its bipolar jet, rather as if they wanted to destroy the medium in which they were born (Fig. 1.12c). Indeed, these jets tend to "blow away" the material of the cloud that surrounds them, thus revealing the central star, which is then observable in the visible region for the first time. From being a protostar, the nascent star becomes what is known as a T Tauri star, after the name of the first star of this type to be discovered (in 1945).

At the T Tauri stage, the envelope that surrounded the embryonic star has almost disappeared in favor of the latter, which has thus been able to "grow" and reach its final mass. However, there is still a circumstellar disk through which material is flowing. If the latter is

1

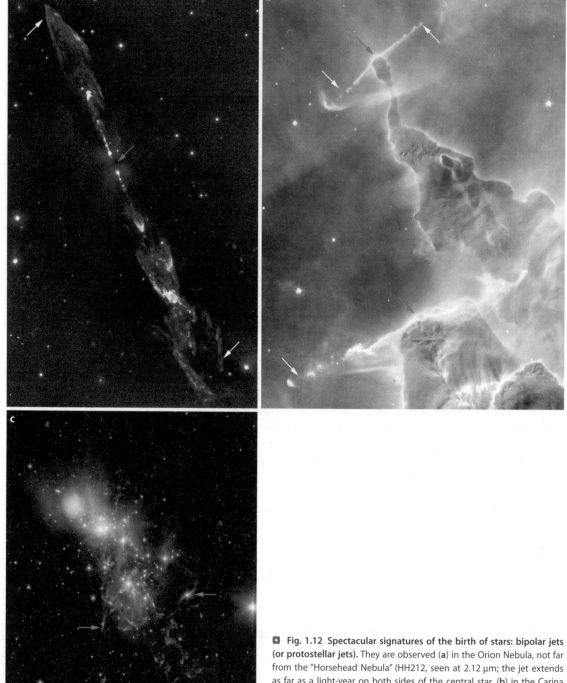

■ **Fig. 1.12 Spectacular signatures of the birth of stars: bipolar jets (or protostellar jets).** They are observed (**a**) in the Orion Nebula, not far from the "Horsehead Nebula" (HH212, seen at 2.12 μm; the jet extends as far as a light-year on both sides of the central star, (**b**) in the Carina Nebula (in the visible), and (**c**) in the cluster NGC 1333 (in the infrared). A complex magnetic mechanism accounts for the formation of these jets (■ Fig. 1.18). The red arrows indicate the position of the central star, and the white arrows indicate the ends of the jets. (Numerous other jets are visible in image c.) (a: M. Mc Caughrean *et al.* ESO/VLT.)

relatively low in mass when compared with the central star (a maximum of 0.01 $M_\odot$, i.e., one tenth to one hundredth of the star's mass), it represents a crucial transition stage: on the one hand it continues to provide material to the star through a very specific process that we shall describe shortly and, on the other hand, it begins to gradually take on a structure that will give birth to planets (0.01 $M_\odot$ represents ten times the mass of Jupiter) (■ Fig. 1.13).

Observation together with the theoretical understanding of the structure and evolution of these circumstellar disks (now synonymous with "protoplanetary disks" or "proplyds")[3], thus give us some of the keys to the analysis of the first stages of the formation of our own Solar System. In other words, what we currently know about the protoplanetary disks around young stars, helps us to gain a better picture of the primitive nebula within which the Solar System was born. The other keys are to be found within the Solar System itself, where meteorites and comets allow us to reconstruct *in situ* the principal stages of formation 4.57 billion years ago (*see later*).

It is time to give some figures enabling us to visualize the structure of a protoplanetary disk, using the size of the current Solar System, that is 50 AU (■ Fig. 1.14), as a comparison. First, the overall size: circumstellar disks generally have immense radii (200 to 500 AU), and thus 4 to 10 times the radius of the Solar System today, although "small" relative to the length of the jets (which may reach thousands of AU). Currently, the best instruments available do not enable us to see these disks with a spatial resolution better than 20 AU, which more or less corresponds to the orbit of Uranus around the Sun. Patience! It will probably have become possible in a few years time!

Towards the center, various indirect arguments show that these disks cannot be in contact with the star. They exhibit a circular cavity: a bit like old 33 rpm long-playing records. The estimated radius of this "hole" is around 0.1 AU, or about 5 times the radius of the central star[4] (or the size of a pin-head relative to the same 33-rpm record) (■ Fig. 1.14). Theoretically, the presence of such a central cavity is to be expected. Indeed, the temperature of the disk increases the closer one gets to the star, and at a certain limiting distance, it rises above the sublimation temperature of the dust grains (that is, the transformation from the solid to the gaseous state), namely around 1500 K. Below this distance, the disk should therefore be transparent. Equally, the density increases towards the center of the disk. However, most of the mass is concentrated in the outer, cooler regions. For example, an "ice line" is defined beyond which the water ice that covers the grains of dust is no longer vaporized by the radiation from the central star. For a T Tauri star with the mass of the Sun, this boundary lies at about 4 AU. We shall see that this ice line plays an important part in the formation of the giant planets. But before that, let us provide ourselves with a very powerful zoom lens and study the central regions of the disk (the "pin-head") in more detail.

---

3. Throughout this chapter, the disks lying around young stars play an essential role. We have initially described them as "circumstellar", and later, we shall speak of "accretion" disks. These are the *same* disks, but discussed in different contexts. We call them "circumstellar disks" when they are simply observed around stars; "protoplanetary disks" when the discussion is explicitly related to planetary formation; and "accretion disks" when we are talking about the transfer of material from the disk to the central star. Occasionally, we shall use the simple term "disk", when there is no ambiguity.

4. At the T Tauri stage, the star has a radius of about 3 solar radii, or 2 million km. It is slowly contracting under the influence of gravitation, which provides its luminous energy. This phase lasts about 100 million years, at the end of which nuclear reactions begin at its core, which arrest the contraction. From this moment onwards, the star will shine through the same thermonuclear mechanisms at the Sun today (converting hydrogen to helium, ■ Box 1.2).

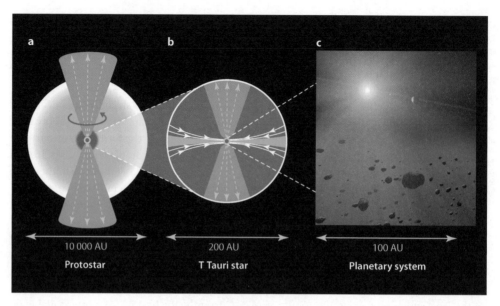

☐ **Fig. 1.13 From a protostar to a T Tauri star and planets.** At the protostar stage (**a**), the material in the envelope surrounding the growing star "falls" on the latter, which is thus able to "grow." At the T Tauri stage (**b**), this envelope has almost completely disappeared, and the young star has almost attained its definitive mass. It is still surrounded by a circumstellar disk, with one tenth to one hundredth of its mass. It is in this disk that planets will progressively form (**c**). Whether it is a protostar or a T Tauri star, the star that is forming ejects material from its poles in the form of bipolar jets. (AU: astronomical unit.)

The Mysterious Central Cavity

Observations show that the circumstellar disk is in Keplerian rotation around the star, that is to say that the material within it should, at any point, be orbiting as if it were a planet (there is an exact balance between the force of gravity and the centrifugal force). In principle, the particles of gas and dust each have their own orbit, but in reality they are constantly colliding with one another (though Brownian motion, turbulence, etc.), which causes them to loose energy. The overall result is that they have a tendency to drift in the direction of the center. Overall, the particles have a spiral motion towards the star, becoming faster the closer they are to it, and it is precisely this mechanism, known as "accretion," that causes the star to grow (☐ Fig. 1.14). To use another image, the disk resembles a gigantic sink, and the minute outlet is the star.

In this context, the circumstellar disk may be considered an accretion disk. The story would be (relatively) simple if we were to stop there, but there are also the conspicuous bipolar jets. How can we explain this coexistence between the increase in matter (the accretion) and the mass-loss (the ejection)?

Failing the ability to observe phenomena directly at such small distances as 0.1 AU (or one-thousandth of the radius of the disk), we are reduced to theoretical considerations, which are still the subject of major debate. But even if there are important differences between the models, these do agree on at least two arguments: (i) the respective influence of two universal forces that have already been mentioned, the gravitation attraction exerted by the star on the one hand, and centrifugal force on the other; (ii) the predominant role of the magnetic field in the cavity separating the star from the inner edge of the disk. If the first argument seems to be perfectly natural within an astronomical context (*vide* Laplace!), the second seems arbitrary, or at least, rather far from the subject. We shall see that this is not so: it is based on the property that magnetic fields possess of being able to channel material that is electrically charged.

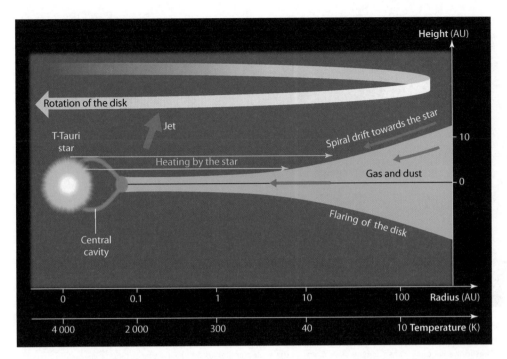

⬚ Fig. 1.14 Simplified cross-section of a circumstellar disk. In this disk, the particles of gas and dust are in Keplerian rotation around the star (curved arrow), but they are also constantly colliding with one another, which causes them to lose energy. The result is that they drift in the direction of the center, spiraling in towards the star and moving more rapidly the closer they get. It is precisely this accretion mechanism that causes the star to grow. The disk is not concentrated in the equatorial plane but, on the contrary, tends to "flare out" with increasing distance from the central star and as the local gravitational field decreases. The structure of the disk (notably its extent in depth) depends in detail on the amount of heating by radiation from the star. In the inner regions ($R < 1$ AU), the grains of dust themselves are vaporized. So the disk is not directly in contact with the star, it is separated from it by a minute central cavity. The outer regions ($R > 1$ AU, except the areas that are away from the plane of the disk, which are directly illuminated, and thus heated, by the star) are thus governed by the cold dust, together with ices (in the equatorial plane) and molecules.

It has been observed for a long time that the Sun and similar types of stars have magnetic fields. Their existence is a direct consequence of the convection motion that stirs their outer layers (one speaks of the "dynamo effect"). The intensity of these magnetic fields may be measured directly (through the Zeeman effect) on the Sun and on certain nearby stars: it varies from several 100 to 1000 Gauss (or several 0.01 to 0.1 Tesla; for comparison the terrestrial magnetic field has an intensity or, more correctly, a magnetic flux density of some 10^{-5} T). More generally, observations at X-ray wavelengths of thousands of solar-type stars (notably protostars and young stars of the T Tauri type, ⬚ Fig. 1.15) show that the latter are the site of eruptions that are quite similar to those that affect the surface of the Sun (⬚ Fig. 1.16), i.e., they are of magnetic origin, despite being 1000 to 10 000 times as powerful. (It is, among other factors, the great intensity of these stellar eruptions that allow us to see them at great distances, in the Orion Nebula, for example, or even much farther away in our Galaxy.) Calculations show that these eruptions are the result of magnetic fields with strengths that are comparable with those that have been measured directly on the sample – still very small – of nearby stars.

The existence of magnetic fields at the surface of T Tauri stars is thus firmly established by observations. That being so, can we explain the "accretion-ejection" phenomenon within the context of the theory governing interactions between particles and magnetic fields ("magnetohydrodynamics" or MHD)?

Without going into details (which basically concern the topology of the magnetic field and which may differ significantly from model to model), the pattern that is generally accepted is as follows. To simplify matters, we assume that the central star is surrounded by a

☐ **Fig. 1.15 The Orion Nebula imaged in the near infrared (*left*) and at X-ray wavelengths by the *Chandra* Observatory satellite (*right*).** This "nursery" has formed about 2000 solar-type stars. Comparison of the two images shows that each of these stars is the site of intense X-ray radiation. Study of this emission enables us to establish that gigantic magnetic eruptions are taking place, which irradiate the protoplanetary disks around these stars.

vast magnetosphere, which to a first approximation is a dipole (a North magnetic pole and a South magnetic pole), which rotates as a unit, carried by the star: it is described to be in "solid body rotation" (as is the case with the Earth's magnetic field). The radius of this magnetosphere (the space enclosed by the lines of force that link the two poles, again like that of the terrestrial magnetosphere, ☐ Fig. 1.17) is determined by the balance between the magnetic pressure[5] and the gas pressure in the accretion disk. It may be shown that this radius is practically equal to the radius of co-rotation, that is, at which the material is in a Keplerian orbit with the same angular velocity as the star.[6]

By virtue of the equality between the radius of the magnetosphere and the co-rotation radius, the radius of the central cavity is equal to the co-rotation radius. That means, in particular, that the magnetosphere of a young star and the inner boundary of its accretion disk rotate at the same velocity. Beyond that point, the disk itself remains magnetized and is "threaded" by magnetic-field lines that are open on both sides. Anchored in the disk, these lines wind round one another in a helical fashion along the polar axis (☐ Fig. 1.18).

As just defined, the radius of the central cavity therefore plays a double role that is simultaneously magnetic and gravitational. It corresponds to the location of the bifurcation between, on the one hand, the closed magnetic trajectories that allow direct passage between the inner boundary of the disk and the surface of the star and, on the other hand, open magnetic trajectories along which material may escape for good. In fact, the material coming from the disk is electrically charged and is thus strongly coupled to the magnetic field: it will flow along the magnetic-field lines[7]. Because of this, the flux of material splits into two: part

---

5. The magnetic field exerts an additional pressure when a gas is electrically charged (when it is said to be ionized), which is the case in the neighbourhood of almost all stars.

6. For the Earth, this radius is 36 000 km, which is that of the geostationary Keplerian orbit, in which most telecommunications satellites are placed.

7. In reality, the particles describe trajectories in the form of a tight spiral with its axis aligned along the magnetic-field lines, and the radius of which is known as the "Larmor radius". Here we simply say that the particles follow the field lines.

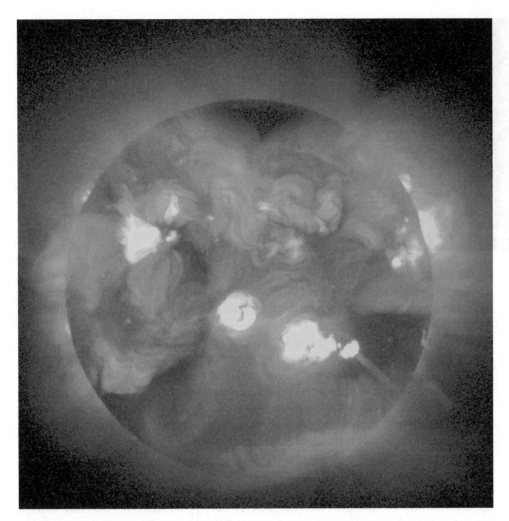

Fig. 1.16 The Sun observed at X-ray wavelengths. On this false-color image, the brightest regions are also the hottest, typically at several million degrees. There is an exact correspondence between these regions and the well-known "sunspots." The gas radiating at X-ray wavelengths (a "plasma," consisting of atoms that are fully ionized) is confined by large magnetic loops, visible here in the form of filaments anchored to the surface of the Sun. The energy responsible for the heating of the plasma is of magnetic origin: when two magnetic loops of opposite polarity come into contact as a result of the motion of the surface of the Sun (and, by extension, of the surface of stars of the solar type), it produces a violent eruption, a sort of "short-circuit" which suddenly releases the magnetic energy and heats the surrounding gas to the temperatures observed, Subsequently, the gas cools: the phase during which there is X-ray emission lasts only a few hours, before recurring elsewhere. (Image obtained by the Japanese satellite *Yohkoh*.)

will follow the magnetosphere's closed field lines and thus fall onto the star (this is known as magnetospheric accretion: gravitation predominates), and the other part will escape, still following magnetic field lines, but this time those that are open to space (this results in ejection: centrifugal force predominates) (**◻** Fig. 1.18).

The decisive advantage of these models is that of directly linking accretion and ejection though a single phenomenon, namely the spiral flow of material in the disk, from the envelope towards the inner regions. They are in quite good agreement with observation, and they predict that a fraction of the material will fall onto the star (the exact value, between 70 and 90 percent, depends on details of the models) and that the remainder (10 to 30 percent) is ejected. Magnetospheric accretion is nowadays considered to be the secret governing the growth of young stars. This phenomenon is relatively well-understood and specific observations have allowed us to measure certain physical quantities such as the accretion rate, the velocity of the falling gas, and even to set constraints on the topology of the magnetosphere.

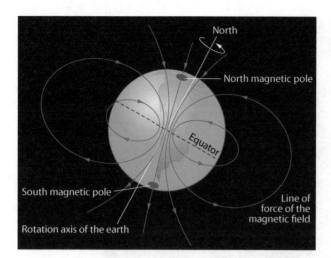

□ Fig. 1.17 **The structure of the terrestrial magneto-sphere.** As an approximation, the large-scale magnetic field produced by a young T Tauri star has a structure comparable with the Earth's magnetic field. It is dipolar and the associated magnetosphere (the space enclosed by the magnetic lines of force that link the two poles) rotates as a whole, carried by the star.

On the other hand, it has still not been possible to identify the precise mechanism that initiates ejection, that is, how a portion of the material can "take off" from the disk, in the neighborhood of the central cavity for centrifugal force to act effectively and accelerate it along the open magnetic-field lines.

The central cavity is thus the seat of a complex magnetic mechanism. Might it be possible for us to find, today, in our Solar System, any signature of this (obviously universal) mechanism, and which operates for several million years, from the protostellar stage to the T Tauri stage (at which we see the first stages of planetary formation occur in the disk, as described later)? Yes, thanks to players that are perhaps unexpected in this context: meteorites, and to be more precise, the oldest of them, the carbonaceous chondrites, which may be considered direct witnesses of the very first stages of the formation of the Solar System, when the Sun was a young T Tauri star.

The link is that some of these meteorites contain inclusions that carry the traces of the existence of radioactive nuclei with short half-lives (that is, that disintegrate in about one million years or less), which implies that they must have been produced simultaneously with the meteorites themselves, at the very beginning of the Solar System's formation. As explained in the □ Box 1.3, debate is rife today, trying to understand the origin of these elements (notably ^{7}Be, ^{26}Al, and ^{60}Fe). One of the most "unifying" hypotheses is that, except for ^{60}Fe, they were all produced 4.57 Ga ago, near the central cavity through nuclear interactions that are one consequence of the accretion-ejection phenomenon that we have just described.

In this model, the radioactive nuclei are created through intense bombardment of the dust grains arriving at the inner border of the accretion disk. The bombardment consists of energetic particles (protons, and helium nuclei) that are accelerated by the powerful magnetic eruptions occurring on the surface of the central star, as observed today on the Sun. Although a portion of these grains that are thus "polluted" is fated to disappear by falling onto the star through the accretion process, the remainder is carried out of the disk by the blast from the jet, and then falls back ballistically (because these grains, being electrically neutral, are not bound to the magnetic field) in a few years, into what is now the asteroid belt (between 2 and 4 AU from the Sun). These irradiated grains are then incorporated into the first planetesimals from which the planets will form (*see later*).

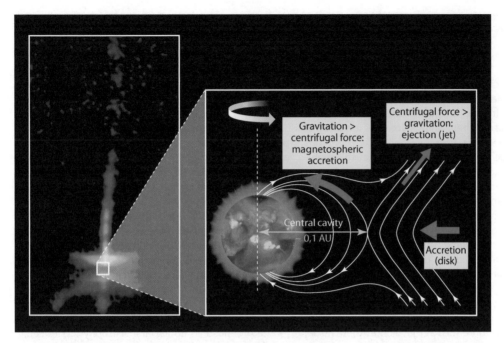

◻ Fig. 1.18 A close-up view of the magnetic structure of the central cavity of a protoplanetary disk. The radius of the central cavity corresponds to the location of a bifurcation between the closed magnetic trajectories that allow direct passage from the inner edge of the disk and the surface of the star, and the magnetic trajectories that are open to space. Electrically charged matter from the disk, is channeled along these field lines. Part follows the magnetosphere's closed lines and falls onto the star (giving magnetospheric accretion, where gravitation predominates), and the other part escapes along the field lines that are open to space (giving ejection, and the origin of the bipolar jets (centrifugal force predominates).

Ultimately, it is within "primitive" meteorites – which resulted from collisions between these early planetesimals[8] – and within refractory inclusions that are present, unaltered, today that the radioactive elements reside. The latter decay over a million years leaving just "daughter nuclei" as witnesses. For example, ^{26}Al decays into ^{26}Mg, an element of which an excess was discovered (in 1969) in the Allende meteorite. In this scenario, it is thus possible to find, in the Solar System today, direct fossil evidence of the irradiation of its disk by the young Sun. The value found is comparable with that deduced from the X-ray emission currently observed in T Tauri stars.

◻ Box 1.3 explains the rival theories, but it seems clear that certain radioactive nuclei with short half-lives are well and truly the consequence of the Sun's magnetic activity at the T Tauri stage, and of the accretion-ejection mechanism. (It is certainly the case with ^{7}Be, with a half-life of only two months, which inevitably implies that it was formed "on the spot"). However, others may have been produced by different events, notably, as we have said, ^{60}Fe, whose origin would be linked to the occurrence of a nearby supernova that appeared right at the start of the formation of the Solar System. So at least two "polluting" mechanisms may have existed simultaneously, one internal, and generic to all stars (irradiation by the central star), and the other external, and exceptional (a supernova). The result is that the situation remains extremely controversial at present.

---

8. At this extremely early stage, the first planetesimals are essentially homogeneous, and their debris (primitive meteorites, primarily those classed as "carbonaceous chondrites") are therefore the most ancient bodies in the Solar System, very similar to simple rocks (silicates with traces of carbon and of metals). More recent meteorites carry the traces of later "differentiation" in the parent bodies (with increasing complexity of their mineral composition as planetary formation progressed).

1

◘ Box 1.3   Did a Supernova Explode at the Time the Sun Was Born?

In 1969, a meteorite weighing 250 tonnes fell near the small village of Allende in Mexico (◘ Fig. 1.19). It was to provide specialists with unexpected clues about the birth of the Solar System.

This meteorite was a carbonaceous chondrite, and these, from the geochemical point of view, are some of the least evolved objects in the Solar System, and thus the most ancient. The particular feature of this meteorite – and which remains exceptional even today – is that it contained refractory inclusions (that is ones that have resisted the high temperatures prevailing in the primitive solar nebula), and which are rich in nuclei resulting from the decay of several radioactive elements. The first to be discovered was ^{26}Al, whose daughter is ^{26}Mg and where the stable isotope is ^{27}Al. Other elements have been detected (in order of mass: ^{7}Be, ^{10}Be, ^{41}Ca, ^{53}Mn, and ^{60}Fe). The significant common factor in these radioactive isotopes is that their half-lives are, at most, around one million years. In other words, to be present in a meteorite such as the Allende, these nuclei must have been incorporated within it very quickly, less than a million years after the formation of the Solar System. By now they have completely disappeared and, for this reason, are referred to as "extinct radioactivities."

The Allende meteorite and the few similar meteorites that have been recovered since, therefore represent true "Rosetta Stones." The contain radioactive archives of the very earliest times in the Solar System, but in extremely low quantities (for example, $^{26}Al/^{27}Al \sim 5 \times 10^{-5}$ and $^{40}Ca/^{41}Ca \sim 10^{-8}$). Their interpretation is subject to major controversy, in that, as we shall see, nuclear physics demonstrates that the two extreme nuclei, ^{7}Be and ^{60}Fe, necessarily have very different origins.

Put more precisely, there are, *a priori*, three possible paths for the origin of elements in the Universe: (i) nucleosynthesis by thermonuclear reactions (such as that of helium at the time of the Big Bang), or of heavier elements, up to iron, in stars; (ii) "explosive" nucleosynthesis in supernovae (which is the origin of elements heavier than iron); and (iii) spallation reactions, that is, "cold" collision-induced nuclear reactions between very energetic nuclei (for example those of cosmic rays) and the ambient material (gas, grains of dust, etc.). This last type of reaction explains, in particular, the abundance of the light elements found in nature (lithium, beryllium, and boron).

The discovery of aluminium-26 in the Allende meteorite produced an upheaval in astrophysics. Indeed, given that it is produced by various processes in stellar nucleosynthesis and by spallation, but far more efficiently by explosive nucleosynthesis, its existence was at first interpreted as proof that the Solar System was formed *because of* the explosion of a nearby supernova, with the implication that therefore the birth of such a system would be rare! This hypothesis has subsequently been progressively abandoned, but it has recently returned in force, and is currently again the subject of heated discussion.

First of all, we now make the distinction between the *presence* of a supernova and any cause-and-effect relationship with its formation. On the one hand, it is nowadays accepted, both thanks to observations and to theoretical considerations, that stars like the Sun do not require a supernova to form. On the other hand, we now know more than 750 confirmed extra-solar planets (or exoplanets), as well as over 1300 candidates awaiting confirmation. In fact, we have good reason to think that planetary formation is widespread, if not universal: according to the most recent estimates, every star in our galaxy (except the most massive ones, which are much less numerous anyway) is orbited by at least one planet! (*See* ▶ Chapter 8.) Yet, it is also well established astronomically that most stars are born in associations with several hundred, or even thousands, of members, including massive stars whose fate is to explode as supernovae after a few million years (◘ Box 1.2). In fact, we do actually know of a certain number of supernova remnants in star-formation regions (◘ Fig. 1.20).

So, one might say, where is the problem? Since the 1980s, two other astronomical discoveries concerning young solar-type stars (the T Tauri stars) have produced a different approach to the origin of extinct radioactivi-

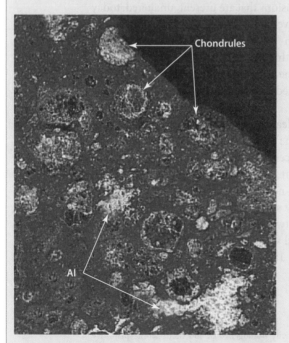

Chondrules

Al

◘ **Fig. 1.19 A fragment of the Allende meteorite.** It was the discovery of this meteorite that allowed the age of the Solar System to be dated to 4.5685 Ga ± 0.5 Ma. (Photo: M. Chaussidon.)

Fig. 1.20 The supernova remnant IC443 (nicknamed "the Medusa"). It was produced by the explosion of a massive star (☐ Box 1.2) which was undoubtedly born in the nearby star-formation region (the bright nebula on the left, IC444). The two bright stars, γ and η Gem, lie in the foreground (left and right, respectively).

ties. The first of these is that, during the first millions of years of their life, all the T Tauri stars are surrounded by circumstellar disks. The second discovery is that solar-type stars are the site of extremely powerful eruptions of magnetic origin, visible at X-ray wavelengths by satellites (☐ Fig. 1.15). Compared with those that the Sun experiences today (☐ Fig. 1.16), these eruptions are one thousand to ten thousand times as powerful, and one hundred times as frequent! This implies that, as with our star, these eruptions are accompanied by a proportionally intense production of energetic particles (mainly protons and helium nuclei). The latter must irradiate the protoplanetary disk and initiate spallation nuclear reactions, primarily with the dust grains. The models show that ^{26}Al is found among the reaction products, as well as *almost* all of the other nuclei from extinct radioactivity, and in proportions comparable with those observed in the Allende meteorite and its like. This is particularly the case with ^{7}Be, which, because of it extremely short half-life (52 days) could not, in any case, have been produced by a supernova.

Because of its overall consistency with measurements made on meteorites and because it is "generic" (it should operate around all T Tauri stars), this mechanism is very appealing. But it cannot explain a very significant extinct radioactive nucleus: ^{60}Fe. This isotope (where the $^{60}Fe/^{56}Fe$ ratio has recently been measured to be about 10^{-8}) contains 4 more neutrons than the stable isotope ^{56}Fe and it is impossible for it to be synthesized other than by explosive nucleosynthesis – in other words, in a supernova! Immediately, several authors re-examined the probability that the Solar System was "contaminated" (with ^{60}Fe, with ^{26}Al, and also with other elements) by a supernova. This contamination should occur within such a short time-window (according to the meteorite data, the Sun's protoplanetary disk would have had to be contaminated in less than one million years) that, in reality, the probability is extremely low (less than 1 percent, according to certain estimates). But then we could invoke a supernova from an earlier generation in the same association. In this case, however, it would be necessary to be "spot on" for the contamination not to be stronger than observed and should occur sufficiently early.

Be that as it may, in a strange turn of events, almost forty years after the discovery of aluminium-26 in the Allende meteorite, the same question is thus posed – although for very different reasons – is the Solar System exceptional?

From Disks to Planets

As we have seen earlier, because of the simple fact that the disk "empties," first by the infall of material onto the central star, and second by ejection along the rotational axis of the system on both sides, after a certain time, a young star should appear "naked," that is, without a circumstellar disk. This is basically what is observed by making use of the fact that circumstellar disks all have a spectroscopic signature. The presence of dust within them creates a strong emission in the middle and far infrared regions (> 2 µm to 200 µm and even beyond, depending on the telescopes' abilities, whether ground-based or in space), which is superimposed on the spectrum of the central star (◘ Fig. 1.21). Logically, the absence of excess infrared radiation implies the absence of a disk but, at the same time, it does not allow us to describe a star as "young," because, in the vast majority of cases, stars are not surrounded by disks!

The Lifetimes and Evolution of Circumstellar Disks

To get over this dilemma, the method that allowed the discovery of "diskless," but undoubtedly young, stars is based on the detection of the X-ray emission produced by young stars in general (*see earlier*). Indeed, observations at X-ray wavelengths of nearby star-forming regions, such as those lying in the constellations of Taurus or Chamaeleon have, since the 1990s shown that alongside what may be called the "classical" T Tauri stars, already catalogued and with infrared excesses, there was a comparable number of young stars with the same "X-ray

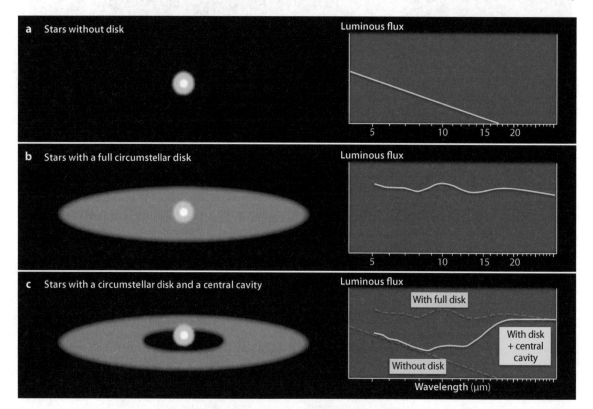

◘ **Fig. 1.21 Infrared spectra of three types of stars:** (**a**) a star without a disk, (**b**) with a full circumstellar disk, and (**c**) with a circumstellar disk and a central cavity. The presence of dust within circumstellar disks creates strong excess emission in the medium and long infrared regions (> 2 µm to 200 µm and longer). It is this spectroscopic signature that enables the detection of circumstellar disks. (After NASA/Spitzer.)

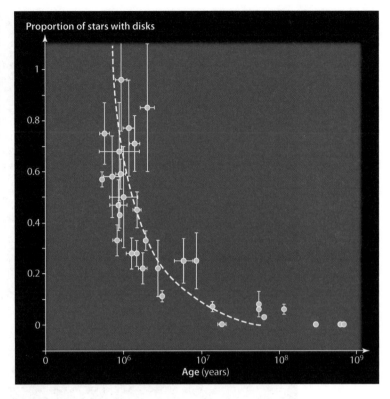

☐ **Fig. 1.22 Percentage of T Tauri stars of a given age, which, within a cluster, are surrounded by a disk, as a function of the age of the individual stars.** Each point corresponds to a cluster of young stars that has been studied. At an age of 2 Ma, 80 percent of the stars in a cluster are surrounded by disks, whereas at an age of 10 Ma, the disks have practically disappeared. From this we may determine that the half-life of circumstellar disks is approximately 3 Ma.

properties," but without an infrared excess. These stars without disks (called "naked T Tauri stars" by some workers), had not previously been identified as "young stars."

This result had two important consequences: on the one hand, the number of T Tauri stars doubled at a stroke; on the other, the location of these stars, with and without disks, on the Hertzsprung-Russell diagram, which gives an estimate of their ages (☐ Box 1.1), enabled the duration of the lifetimes of disks to be measured for the first time (in the statistical sense, for several hundreds of stars). This revealed that there were a small number of very young T Tauri stars (say, 1 Ma) without disks, and a small minority of "old" T Tauri stars (say, 10 Ma) still surrounded by disks, with the number of stars with disks decreasing rapidly at the end of a few million years. More recently, the study of clusters of T Tauri stars in various star-formation regions has confirmed this tendency: at an age of 2 Ma, 80 percent of stars in a cluster are surrounded by disks, whereas at an age of 10 Ma, the disks have practically disappeared. Taken as a whole, the observational data may be represented by a decreasing exponential law, which enables us to express the overall result in the form of a characteristic half-life for circumstellar disks: this time is about 3 Ma (☐ Fig. 1.22).

What is the cause of the disappearance of circumstellar disks in such a short time? The first explanation that comes to mind is that, as mentioned at the beginning of this section, the material has disappeared because it has been accreted by the central star, or ejected. But observations show that this is not the case: the accretion rate during the T Tauri stage is not high enough (it would require several tens of millions of years to "empty" a disk in this way). So we need to invoke a more radical process. But what?

A wind or UV photons from the central star itself could "blow away" a circumstellar disk, but theoretical estimates show that this mechanism is not very effective. So the cause of the disappearance of disks might be external: when they are located near massive stars, and thus exposed to strong external UV radiation, calculations show that they may "evaporate" over a period of about a million years. However, the vast majority of disks do not exist in such a situation, which reduces the overall efficiency of such a mechanism. In the end, it

1

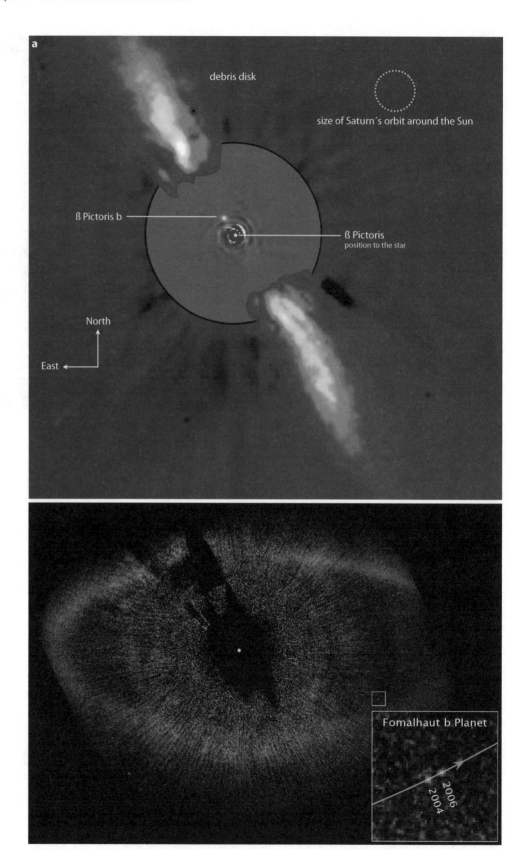

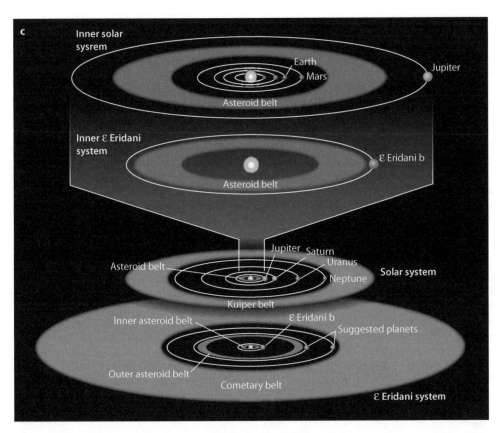

Fig. 1.23 Some planets and planetary systems recently observed around young stars (direct imagery): (a) the planet β Pictoris b, orbiting the star β Pictoris within the inner region of the debris disk that surrounds the latter. The "waves" that appear in the central region of the image, and amongst which the new planet is visible, are an artifact caused by the remnants of the normal optical diffraction disk (the "Airy disk"), after major correction to compensate for the brightness of the central star. The colored regions on both sides correspond to the circumstellar disk in orbit around the star, but seen in profile; **(b)** the planet Fomalhaut b, observed in 2004 and 2006 in the debris disk around Fomalhaut (α Piscis Austrinus). It was the displacement of the spot of light that revealed the existence of this planet; **(c)** representation of the planetary system around ε Eridani. Its structure is compared with that of the Solar System. To interpret the observations, it is necessary to assume the presence of two asteroid belts and two remote planets, not directly visible. In the case of ε Eridani, the discovery of exoplanets preceded that of the debris disk, whereas for β Pictoris **(a)** and Fomalhaut **(b)**, a debris disk was observed before the planet was detected. (a: A.-M. Lagrange and D. Ehrenreich, Institut d'Astrophysique et de Planétologie de Grenoble and ESO/VLT; b: P. Kalas, University of California at Berkeley and NASA/HST.)

is generally agreed that the phenomenon that is most efficient overall is ... the formation of planets.

Explaining a formation by a disappearance may *a priori* seem paradoxical, but it is necessary to understand that the disappearance of disks to observation may only be apparent, and does not necessarily imply the disappearance of material. For example – and we shall return to this point – small pebbles a few centimeters or tens of centimeters in size and, *a fortiori*, bodies that are meters, tens of meters or hundreds of meters in size are, in effect, undetectable, whether by spectroscopy or direct imagery. Not a chance: the laws of physical optics prevent us from witnessing the actual beginnings of planetary formation!

Before tackling this subject in detail, let us enquire whether observation can provide any indications of the growth of the dust grains or of the existence of "small bodies" (in the planetology sense: planetesimals or comets, for example) within circumstellar disks. The answer is "Yes." In the case of microscopic particles, such as dust grains, we may make use of the fact that, in physical optics, these primarily emit at wavelengths comparable with their size: notably in the near infrared (between about 2 and 10 μm). In the case of young disks (around

1

▣ Box 1.4　Dust and the Mineralogy of Circumstellar Disks

The importance of cosmic dust for our own history may be summarized in a single sentence: No dust, no planets! From the very start of this chapter, we have mentioned the existence of the essential, and universal, component of the interstellar medium, and which we find, not only in the Milky Way, but also in all galaxies, to the farthest distances that we can observe them. And it is the interstellar medium that provides the dust seen in circumstellar disks, and which will lead to the formation of planets in a few million years.

We know the properties of interstellar dust reasonably well. By mass, it represents only about 1 percent of that of the gas, and it is found that this ratio is uniform (at least on a large scale) in galaxies, which implies a strong link between the gas and the dust. Besides, this link is so strong that it is on the surface of dust grains that many molecules form, of which the most important is molecular hydrogen, H_2. From various sources of information, and assuming for simplicity that they are spherical (*see later*), we find that their sizes lie between ~0.01 μm and 10 μm, in other words about those of the particles suspended in cigarette smoke. They essentially consist of a core of carbon (generally graphite), surrounded by silicates. But when they occur in dense, cold environments (< 10 K), protected from UV radiation from stars, particularly in molecular clouds, they also include a mantle of ice (water ice, and also carbon-monoxide ice CO). This mantle evaporates as soon as the temperature is high enough (~100 K), in the vicinity of a forming star, for example.

This description becomes more complicated if one looks at a microscopic scale. For example, the grains are probably not solid, but porous. Measurements of interstellar polarization suggest that they generally have an elongated shape, like a rugby ball, rather than a football. We should also mention that there is a population of grains or aggregates of polycyclic aromatic hydrocarbon (PAH) grains, which are strong emitters of narrow spectral lines when they lie in a region of intense UV radiation (near massive stars, for example). These aggregates are well-known to chemists, and are found in abundance on Earth, such as in combustion products (soots). These are extremely large organic molecules, generally flat or in sheets, consisting of up to 50 or 60 carbon atoms organized into hexagonal benzene cores, with a certain number of peripheral bonds occupied by hydrogen atoms and others by heavier atoms.

Interstellar grains thus form a complex material, difficult to study "at a distance," and at the boundary between the physics of solids and organic chemistry. Where do they originate? To find out, it is necessary to know the origin of the carbon, their principal component. The answer is provided simultaneously by observations and by the theory of stellar nucleosynthesis. To summarize, we now know that it is stars that, via a

series of thermonuclear reactions taking place within their cores, form almost all of the elements in nature, in particular carbon, oxygen, nitrogen, and silicon. (The sole exceptions are the "light" elements lithium, beryllium, and boron, which are created by collisions between cosmic rays and the interstellar medium.) At some time after having remained on the main sequence of the Hertzsprung-Russell diagram (converting hydrogen into helium), stars like the Sun, and of an intermediate mass ($M_* \sim$ 0.3–8 $M_\odot$), enter a stage of colossal expansion, but where their surface temperature is relatively cold (~3000 K): these are the "red giants." It is during the course of this expansion that large quantities of dust and carbon molecules are produced. These are then dispersed into the interstellar medium during the course of the spectacular planetary-nebula phase[1] (▣ Fig. 1.6). Certain grains, called "presolar grains" recovered from meteorites carry very clear signature of heavy atoms (isotopes of C, N, and O, in particular), or molecules (such as SiC), synthesized by red giants well before the formation of the Solar System, and now undoubtedly far from their birthplace.

We need to view things in this context in considering the problem of dust in circumstellar disks, here interpreted in terms of "protoplanetary disks" (*see* note 3). Even before it was possible to observe such disks by direct imaging, we had an idea of the radial structure (through their infrared emission – between 1 and 100 μm – that is, at submillimetric and millimetric wavelengths). If we assume that the dust grains emit like black bodies at a given temperature, it is possible to reconstruct the temperature distribution within the disks from their continuous spectrum (known as the "spectral energy distribution"). This method was the first to give results, because it does not require spectral resolution: comparison of photometric measurements through different, sufficiently narrow-band filters (passing between 1 and 5 μm, for example) is sufficient to indicate that the temperature decreases with distance from the star, passing from 1500 K at a few stellar radii to 100 K or less, beyond a few AU.

This spatial temperature distribution is essential: there is an "ice line" beyond which grains are "cold" and covered with a mantle of water ice. (In the Solar System, this "frontier" between warmth and cold is found at about 4 AU from the Sun.) As far as planetary formation is concerned, or at least its initial conditions, this leads to the recognition of two zones: that of the rocky

---

1. This historical description is extremely inappropriate, because the phenomenon has nothing whatsoever to do with planets or their formation. It harks back to the 19ᵗʰ century, at a period when telescopes only revealed tiny, fuzzy, luminous patches, which were mistaken for planets.

planets near the star, inside the ice line, and that of the giant planets, outside it.

It is possible to go further in analyzing disks before planetary formation: (i) by obtaining images at several wavelengths and by using the fact that the peak thermal emission from the grains lies at a wavelength that is of the same order as their size; (ii) by using spectrographs that will, this time, be able to determine their composition and structure.

The first approach has recently drawn attention to a key phenomenon in planetary formation, namely the sedimentation of grains towards the equatorial plane: for example, thanks for images of the disk of GG Tau, a well-studied T Tauri star, with the Hubble Space Telescope, it has been possible to reveal greater emission from grains with a typical size of 1.5 μm along the equatorial plane, compared with the emission from grains, 0.5 μm in size, lying at a greater distance above and below that plane. In other words, the "larger" grains have fallen onto the disk, just like dust onto furniture ...

The second approach enables the composition of the grains or, in other words, their mineralogy, to be determined. Thanks to the infrared satellites ISO (ESA, 1995–1998) and Spitzer (NASA, launched in 2003, and still functioning), there is access (between 1 and 30 μm) to emission lines characteristic of key solid materials such as water ice and CO ice, and silicates. Let us consider the latter, which are also an essential component of terrestrial rocks. As shown in ◘ Fig. 1.24, their spectral signature has been obtained from the disks of young stars of various masses, from brown dwarfs ($M_* < 0.1\ M_\odot$), to "Herbig stars" ($M_* \sim 2\ M_\odot$), and compared to spectra

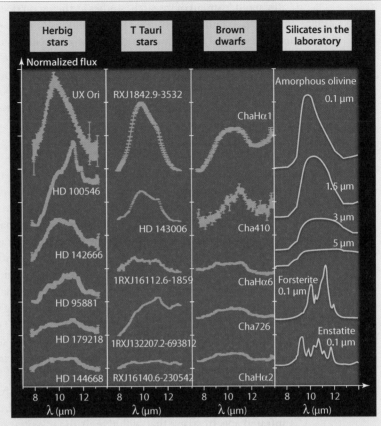

◘ **Fig. 1.24** Infrared spectra observed by the Spitzer satellite of the circumstellar disks of young stars: Herbig stars, with intermediate masses ($M_* > 2\ M_\odot$); T Tauri stars ("young Suns," with masses between 0.5 and 2 $M_\odot$); brown dwarfs of very low mass ($M_* < 0.1\ M_\odot$). On the right, the spectra of micron- or submicron-sized silicates, obtained in the laboratory, ranging from amorphous (poorly differentiated spectra: a broad line with a maximum around 10 μm), to a purely crystalline form (spectra containing numerous narrow lines). The amorphous form, the only one present in the interstellar medium, dominates in "poorly evolved" disks (around very young stars). The spectral resolution is insufficient to differentiate individual lines in the more evolved disks shown here, but it may be demonstrated that the proportion of crystalline silicates increases with age. At present, the reason for this is unknown, but the tendency approaches the terrestrial composition. (After A. Natta *et al.*, 2007.)

obtained in the laboratory. It has also been established that, the more evolved the disks (the older the stars), the more they approach terrestrial mineralogy (at least in terms of silicates), reinforcing the idea that such disks might eventually give rise to rocky planets.

T Tauri stars), by the use of sophisticated models to predict the infrared emission from the grains, it has been possible to reveal a concentration of larger particles, some micrometers in size, in the equatorial plane (◘ Box 1.4). In what is known as sedimentation, the grains slowly fall onto the disk – rather like dust on furniture – as they grow, while still remaining in orbit around the central star. Such a process is would undoubtedly make it easier for planetary formation within the disk to start.

In another strand of research, studies are being carried out concerning stars that are much older than T Tauri stars (a few tens of millions of years old), where the original disks have disappeared, and in the hope of discovering indications of planetary systems that are forming, or have already formed. At present such studies have been few in number, but they have enabled images to be obtained of disks consisting of micron-sized dust that are extensive and of low mass (around 10^{-6} $M_\odot$), commonly known as debris disks. The best-studied example is that around the star β Pictoris, where the disk was discovered in 1986. (It was the first of the type.) Early on, the variations in spectrum from this star's atmosphere were interpreted as being linked to the impact of comets onto the star, caused by gravitational perturbations created by at least one invisible giant planet, orbiting close to the star. This theoretical prediction was wonderfully confirmed in 2008, thanks to direct, high-resolution imaging, which enabled observation of a planet (β Pictoris b), orbiting the star within the inner region of the debris disk (◘ Fig. 1.23a).

This was not an isolated case, and thanks to direct imaging, progress has been rapid. Several nearby stars have been observed by using coronagraphic methods (that is, by using an occulting disk to hide the light from the star itself), and this has enabled the detection of deformation of the debris disks (annular and spiral structures, gaps, etc.), which are interpreted as caused by one or more bodies of planetary mass. Although most of these bodies still remain invisible, despite everything we are beginning to be able to observe some of them, such as that orbiting Fomalhaut (α Piscis Austrinus, ◘ Fig. 1.23b). The opposite may apply, namely the detection of a planet, or even a planetary system – by indirect means (*see later*, and ► Chapter 8) – preceding the discovery of the associated debris disk (as in the case of ε Eridani, ◘ Fig. 1.23c).

Another piece of the puzzle is the "mineralogy" of circumstellar disks or, in other words, the chemical composition, in the geological sense, of their dust grains. Here, once again it is infrared spectroscopy that will give us the essential means of replying to this question (◘ Box 1.4). Every time that they have been analyzed, the chemical composition of the disks surrounding T Tauri stars has proved to be comparable to that of bodies in the Solar System, such as meteorites or comets. This composition therefore seems to be universal: it presumably applies to all planetary systems, or at least those associated with solar-type stars. What has been detected are molecules or complex aggregates: crystalline silicates (fosterite, enstatite, and even diopside – all minerals rich in calcium and magnesium), polycyclic aromatic hydrocarbons, etc. Olivine, an important silicate in the terrestrial crust, is also present. Although this is not definitive proof, we may deduce from this that circumstellar disks can form rocky planets with the same composition as those in the Solar System.

To summarize, we currently have numerous indications that enable us to conclude that circumstellar disks are definitely the stage for the early phases of planetary formation, including the terrestrial, "rocky," type, and that they are genuinely "protoplanetary."

The Major Stumbling Block: The Growth of the Grains of Dust

In re-reading the last section, an attentive reader will realize that "something's wrong." In fact, in disks surrounding T Tauri stars, we are able to observe, directly, grains of a few micrometers to a few millimeters in diameter, and by in direct imaging, in debris disks, bodies several kilometers in size, or even by direct imaging, bodies of planetary size. But what is actually going on between the two? This still mysterious and unobservable transition between sub-centimeter particles (the dust grains) and bodies kilometers in size (the planetesimals) is a real *terra incognita*, which people are nowadays trying to elucidate, simultaneously, by experiment and theory.

In the laboratory, micron-sized grains, differing in their structure (porous or solid) and in their composition (or both) are forced to collide over a whole range of velocities that are representative of those operating in disks and in low-eccentricity Keplerian orbits (from a few meters per second to a few tens of meters per second). The principal question is to draw up a balance sheet for these collisions: do the grains have a tendency to stick together (to agglomerate) or to rebound or fracture? The results are ambiguous. Slow collisions favor accretion. Porous grains clump together more easily if they undergo head-on collisions, but they may be destroyed by smaller, faster grains. However, between 5 and 50 m.s^{-1}, current experiments are not conclusive. Under astrophysical conditions like those in disks, it is probable that the processes of accretion and destruction are both in operation, and that all we see is the net balance, favoring the growth of grains.

For larger grains (as large as pebbles, for example) and, *a fortiori*, for bodies that are tens or hundreds of meters in size, it is impossible to carry out laboratory experiments, all the less because it would be necessary to force the simultaneous interaction of a whole range of grains or smaller bodies. This collision regime has been studied by way of numerical simulations. The result: if the bodies are monolithic, it seems that destruction overwhelms consolidation! The problem is that we are completely ignorant of the internal structure of the colliding bodies and it is possible that, despite their large size, they are not yet truly in a solid state: we might instead be dealing with poorly bound, or porous, spheres, which is likely to change the result of collisions (*see later*: the formation of giant planets).

Another theoretical path that has been explored is that of a "forced" collective agglomeration (meaning that it is not just subject to the vagaries of collisions), and only influenced by local gravitational effects. This assumes that a clump of material may form, sufficiently isolated from the disk for a gravitational well to arise, and cause the said material within its vicinity to collapse on itself. This situation appears to be theoretically possible within a turbulent disk, which naturally contains whirls (known as "vortices") at various scales. This idea of a vortex of material being implicated in the formation of planets is rather old (Laplace mentions it) and it had been revived by numerous authors attempting to explain the (false) "Bode's Law," including during the 20th century[9]. With regard to forced agglomeration, equally, the surrounding gas must be taken into account: it exerts friction on the grains, which has a tendency to brake them and force them to fall towards the central star, rather than clustering together at a point. Numerical simulations of this process have been carried out. Here again, the results are ambiguous, in so far as the conclusions change if one extends calculations in two dimensions (that is, in just the plane of the disk) to calculations in three dimensions, which are obviously far more complex.

The consequence of all this is that today, in the absence of any indications or constraints from observation, the crucial transition from millimeters to kilometers – a factor of one thousand in size, but one thousand million in volume! – remains what is undoubtedly the least understood problem of planetary formation, and the one that is the most difficult to resolve.

From Planetary Embryos to Planetary Systems

Reluctantly leaving aside the problem of the transition between dust grains millimeters in size to planetesimals kilometers across, we find ourselves confronted by another difficulty: that of "manufacturing," starting with these planetesimals, a whole planetary system, or at

---

9. "Bode's Law", proposed in the 18th century, states empirically that the radii of the orbits of the planets in the Solar System form a geometrical progression. On the one hand, however, it applies rather poorly to the distant planets in the Solar System and, on the other, does not apply at all to exoplanetary systems. It is therefore of no help in explaining the formation of planets.

least giant planets, whose size amounts to several tens of thousands of kilometers, and for this to occur in a few million years at the most (to satisfy the constraint imposed by the lifetime of the disks). A key to understanding this process is, of course, the Solar System itself, but we must keep in mind that it is by no means representative of what we know today about extra-solar planetary systems, and their much-discussed "exoplanets" (*see* ▶ Chap. 8).

First, let us note some useful data. The Solar System includes two very distinct regions: the region of what are known as the terrestrial (rocky) planets – in other words that are in a solid state, at least as far as their outer layers are concerned. These are Mercury, Venus, Earth, and Mars: small planets whose radii lie, roughly speaking, between 0.5 and 1 Earth radii (R_E), with densities between 4 and 5 g.cm^{-3}, and distances from the Sun between 0.5 and 1.5 AU. Then there is a transition zone, lying between Mars (1.5 AU) and Jupiter (5.2 AU), empty of planets, but filled with millions of asteroids of all sizes. (The largest, Ceres, a dwarf planet, has a diameter of 950 km.) This is the asteroid belt. Beyond that we have the beginning of the region of the giant planets (Jupiter, Saturn, Uranus, and Neptune), which are extremely different from their rocky cousins. They are gaseous spheres, primarily consisting of molecular hydrogen, H_2. They are vast (4 to 12 R_E), low in density (0.7 for Saturn – which would thus "float" in water – to 1.8 g.cm^{-3}), and lying far from the Sun, between 5.2 and 30 AU. Beyond that, there is the Kuiper-Edgeworth belt, another asteroid belt, which is distant and poorly known, which extends between 30 and 55 AU, to which Pluto, with its low mass (1/500 Earth mass) belongs following a decision of the International Astronomical Union in 2006.

As for planets orbiting stars other than the Sun (exoplanets), currently over 750 have been discovered, and about 1300 others are "candidates" awaiting confirmation; the list grows from day to day. Over 120 multiple systems are known, containing up to 6 planets. In the vast majority of cases they have been discovered by indirect methods, primarily through the very weak gravitational perturbations they create on the star itself, and which betray their presence, and also when they are caught passing in front of their parent star ("planetary transits"). (More details are given in Chapter 8.)

What conclusions can be drawn from this harvest? The key word that occurs repeatedly, as far as planetary systems are concerned, is "diversity." In other words, our own Solar System is by no means typical. On the contrary, it is thought nowadays that it is just one chance "realization" (as it is called in the numerical simulations that we shall return to later), among a large number, perhaps even an infinite number, of realizations that are, *a priori*, possible.

Let us now return to the formation of planetary systems, taking the most general possible overall view, without, however, forgetting the need to explain, as a minimum, the formation of the Solar System. Thanks to theoretical models and heavy computations (*N*-body simulations, where *N* is of the order of a few thousands), we can distinguish three phases, which take place in parallel, but on timescales that may be very different.

Between 1 and 10 Million Years: The Formation of Planetary Embryos

Starting at a size of about one kilometer, the nature of the interactions between the bodies changes radically. Gravitation plays no role between grains of dust, whereas it becomes the dominant one between planetesimals, where it profoundly influences the outcome of the competition between accretion and destruction. The most spectacular process is the "snowball effect" (exponential growth), starting with the agglomeration of small solid bodies, and which leads to the formation of true planetary embryos. This snowball effect is all the more effective if the bodies involved are covered with water ice, as is the case beyond what is known as the ice line (which lies at 4 AU for a solar-type star; *see above*). Under such conditions they "stick" better. Modeling predicts that within the ice line, bodies of the mass of the Moon may form in less than a million years, and that beyond it, bodies of several Earth masses may form in a few million years. This is satisfactory when considered in terms of the lifetimes of the disks. The process is rendered all the more effective by the fact that the low relative velocity of small-sized bodies (a few meters per second, *see earlier*) decreases the likelihood of their encountering one another and of thus forming other, larger-sized bodies. When it comes to a planetary system in formation, bodies that are already large grow faster than the smaller ones. This is what is known as oligarchic accretion.

Up to 10 Million Years: The Formation of Giant Planets

Here again, we can distinguish three phases: (i) the formation, beyond a few AU, of a planetary embryo (the core) of some ten Earth masses (M_E), starting with small bodies and initially rapidly, in the process described in the last paragraph; (ii) then, much more slowly, a phase of progressive accretion of gas from the disk, which, in a few million years, increases the mass of the planet to 20–30 M_E; (iii) a collapse of the "atmosphere" on itself, which has its origins in another "snowball" effect, this time produced by the accumulation of gas: gravity increases at the surface of the planet, which accelerates the accretion, which increases the mass and thus the gravity, and so on. As a result, a mass of a few hundred Earth masses (Jupiter = 318 M_E) may be attained in 10 000 years (☐ Fig. 1.25)!

Satisfying though it may appear, this scenario raises significant questions. Moving, as it does, relative to the gas in the disk, the planet will be braked and thus will drift slowly, spiraling towards the star (this is "type-I migration"). In addition, if the gas accretion is sufficiently efficient, the planet, as it forms, will rapidly collect material from the disk in its neighborhood. This will create a gap, which will confine it and bind it gravitationally to the disk

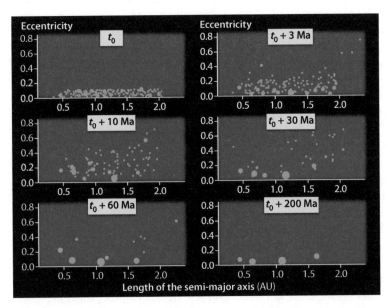

☐ **Fig. 1.25 Numerical simulations of the formation of the terrestrial planets in the Solar System from a disk of planetary embryos.** The figure shows the results of numerical simulations, in the form of points which each correspond to a body in formation (the size of each point is proportional to its mass). The distance from the Sun is given on the abscissa (in astronomical units, AU). Each body has an associated elliptical orbit, the eccentricity of which, *e*, is given on the ordinate (*e* = 0 for a perfect circle; *e* = 1 for an infinitely elongated orbit). Most of the initial mass of the embryos is assumed to have been gathered in a circular annulus lying between 0.5 and 2.0 AU. Following collisions and gravitational deflections, certain bodies tend to grow, and the eccentricity of their orbits is little changed (*e* ≈ 0), whereas the smallest tend to be more and more dispersed (with great eccentricities), and then leave the system for good. In this simulation, a planetary system fairly similar to the inner Solar System (within the asteroid belt) forms at the end of 200 Ma. (After Chambers, 2001.)

1

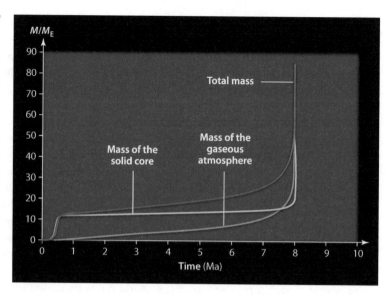

■ **Fig. 1.26 The "snowball" formation of Jupiter.** The abscissa represents time (in Ma) and the ordinate the mass accumulated, in Earth masses. Initially, a solid core forms from small bodies (yellow curve), and then a gaseous atmosphere progressively accumulates from the disk (green curve). The increase in the overall mass is shown by the red curve. At the end of a few million years, the mechanism suddenly surges under the influence of gravity: the atmosphere collapses and the forming planet attracts gas from the disk more and more strongly, but this process ceases for lack of "fuel" (with the creation of a gap in the protoplanetary disk, ■ Fig. 1.28). (After Pollack *et al.*, 1986.)

(■ Fig. 1.26). As the latter is flowing towards the star, it will inevitably drag the planet with it ("type-II migration" towards the star) ...

These different forms of migration were, in fact, predicted *before* the discovery of "hot Jupiters." So the presence of these giant planets in the immediate proximity to their stars was not really a surprise to the theoreticians. But there still remains one small detail to be explained: how does the migration stop? This point is still obscure. Among other suggestions, one model takes explicit account of the central cavity, which would prevent the planet from "jumping" onto the star, and would trap it, once for all, in corotation with it. But for this to occur, the said cavity must exist, i.e., the "accretion-ejection" mechanism must still be operating at this late stage. Yet the accretion rate decreases with time. So there would be a race against the clock, with an uncertain outcome – the planet would need to arrive at the corotation radius before the central cavity disappears. In other models, planets are well and truly "swallowed up" by the central star, and it is only the last to arrive, when no more of the disk survives, that have a chance of seeing their migration cease. By contrast, in our own Solar System, the giant planets have certainly not undergone any significant migration that brought them closer to the Sun. It is fair to say that the problem of the migration of planets and how they come to a stop is still unresolved.

Up to 100 Million Years: The Formation of the Rocky Planets

Numerical simulations show that once the giant planets have formed (and a large part of the mass of the initial disk has been thus consumed), there remain bodies of intermediate mass, sort of "failed" planetary embryos, with sizes comparable with those of the Moon (0.013 M_E) or Mars (0.11 M_E). Under the influence of mutual gravitational interactions, the orbits cross, which creates chaotic instabilities and collisions. On the one hand, the latter favor a slow accretion, and on the other, force the ejection from the developing planetary system of bodies that have neither grown by accretion nor been destroyed. Bodies of a terrestrial mass are thus formed in 100 million years, within a radius of less than 2 AU.

As regards the formation of the Earth, this timescale turns out to be a bit too long, relative to the one that is obtained from a method of radioactive dating based on an analysis of the hafnium/tungsten ratio (^{182}Hf/^{182}W). This method is able to measure the time required for our planet's differentiation, that is, for the internal structure of core and crust to develop. This time is about 40 million years (*see* ▶ Chapter 2). A possible explanation is that the differentiation might have started within the planetesimals even before these assembled into the

The last stage in the formation of a giant planet involves a snowball effect, with accretion of the gas becoming more and more effective. The planet rapidly removes the disk material in its vicinity, thus creating a gap, which channels it, and drags it with the disk, towards the star. In addition, as the forming planet moves relative to the gas in the disk, it is braked and thus slowly drifts towards the star. How does this migration towards the central body come to a stop? Several models attempt to explain this, but the matter still remains obscure.

Earth: again we encounter the problem that has already been mentioned regarding our lack of understanding of the internal structure of the bodies that are involved in collisions. This question is also linked to that of the formation of the Moon, which is now generally believed to be the result of a gigantic impact between the 90-per-cent-formed Earth and a body the size of Mars (**Box 1.5**). But, as explained in the following chapter, other purely geological theories have also been suggested to resolve this apparent contradiction between astronomy and geology.

Box 1.5   The Formation of the Moon

The formation of the Solar System is a gigantic game of cosmic billiards governed by gravity: at great distances and collectively, under the gravitational influence of the Sun; and at close distances and individually by the influence of the bodies that are forming, in line with their increase in mass. We have an illustration of this in the celestial body that is closest to us: the Moon.

There is a broad consensus nowadays that attributes the formation of our satellite to a gigantic impact, occurring between the Earth as it was forming and another body comparable in size to that of Mars (so half the size and one tenth of the mass). As numerical simulations have shown, such collisions are not rare at this stage of the Solar System's evolution, where numerous bodies of planetary mass (asteroids, etc.) have been formed. It is even thought nowadays that the formation of the terrestrial (rocky) planets takes place through the occurrence of several giant impacts.

Nevertheless, the Earth-Moon system is unique in the present-day Solar System. First, because the Moon is a particularly large body relative to the planet that it is orbiting, even though it is far less massive. (In round figures, we have $R_{Moon} \sim 1/2\,R_{Mars} = 1/4\,M_E$; $M_{Moon} = 0.1\,M_{Mars} = 0.01\,M_E$; density $\varrho_{Moon} \sim \varrho_{Mars} \sim 1/2\,\varrho_E$.) Second, the Earth is the most massive of the rocky planets and the only one to have a satellite: Mars has two satellites, Deimos and Phobos, but these are only small, shapeless asteroids that the planet has captured, or, according to recent results, large boulders that have been blasted away from Mars itself by a catastrophic collision. In this sense, the formation of the Earth and the formation of the Moon are inextricably linked.

However that may be, the parameters of this decisive impact are being sought through numerical calculation. The (fundamental) difference between a classical game of billiards and our game of cosmic billiards it that in the former case, physicists speak of "elastic" collisions. (The bodies, which are hard, exchange energy at the instant they collide, and subsequently separate symmetrically with respect to their common center of gravity.) In the latter case, the collision occurs between "soft" bodies and is going to fracture them – at least partially. That causes a major dissipation of energy, which may go so far as to vaporize material, a portion

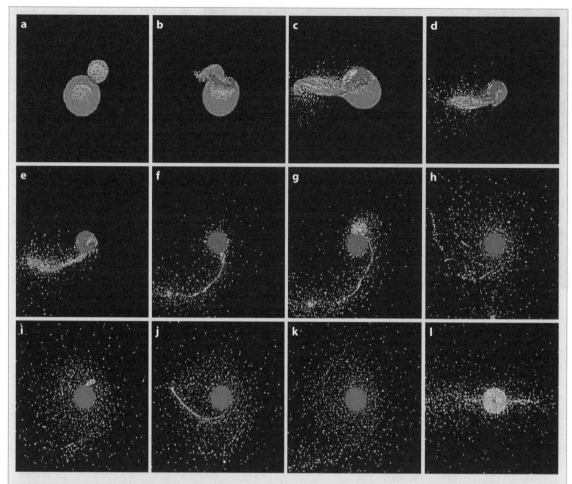

▣ Fig. 1.28 A numerical simulation of the giant impact that led to the formation of the Moon, carried out by R. Canup and his collaborators (2004). A body similar to Mars, collides with the proto-Earth at a velocity comparable with the mutual escape velocity (10 km/s), necessary for at least a portion of the ejected material to remain in orbit around the Earth. The various panels are separated by a few tens of minutes (**a** to **c**), and by a few hours (**d** to **k**). Panel l corresponds to panel k seen edge-on. In this simulation the formation of the protolunar disk (from **a** to **l**) takes place in 27 hours. The colors correspond to temperature, from the coldest (blue) to the hottest (red). (Images: R. Carup.)

of which may then escape from the system, while the remainder falls back into orbit around the most massive of the remnants.

In physical terms, the initial parameters to be integrated into the numerical calculation are, at first, relatively intuitive: (i) the mass of the two bodies (total at least equal to the sum of the current masses of the Earth and the Moon); (ii) the mutual separation between their respective trajectories (this is what is known as the "impact parameter": a collision occurs if it is less than the sum of the radii of the two bodies); (iii) their relative velocity. The last two parameters are constrained by the current kinetic moment of the Earth-Moon system, which should be equal to that existing at the instant of collision, at least with the hypothesis (verifiable *a posteriori*) that little material escapes from the system. The relative velocity is obviously a decisive parameter: if it is high, the collision might be capable of pulverizing one of the bodies, or even both; if it is low (around a few km.s⁻¹), the collision will be "gentle" and the damage will be primarily caused to the less massive body. (Some of the material forming it might fall onto the other body, and another portion might even remain in orbit around the latter.) In addition, in the case of a grazing collision, some of the kinetic moment will create rapid rotation, or "throw" (to revert to the billiards analogy). Models show that the initial length of the "terrestrial day" must have been about 4 hours. This result tends to support the idea of such a collision, and in this case, the quantity of material torn from the more massive body is not necessarily negligible, which will, in fact, lead to a mixture of the two materials!

Calculating the exchange of material between the two bodies is obviously the most sensitive part of the problem. But there is a key to getting around it: the analysis of the abundance of iron in the Moon. The idea is that the collision took place at a sufficiently late stage for both the Earth and the "impactor" to already have a partially formed iron core, such that the abundance of this element in their respective mantles was less than its initial value. Under these conditions, the whole is constrained by two quantities that are now known: (i) the proportion of iron in the Earth (about 30 percent, and primarily in the core); and (ii) the proportion of iron in the Moon, which is estimated, on the basis of measurements by the Clementine lunar probe, to be just 10 percent, whereas the composition of all the bodies in the Solar System should, initially, have been identical.

Numerical calculations that give rise to a good mixture show that the phenomenon lasted less than 30 hours (☐ Fig. 1.28). During this lapse of time, after the initial impact, the Moon accreted via the formation of a very hot (about 10 000 K), protolunar disk in orbit around the Earth. The solid particles of which it consisted subsequently took several centuries to cool and condense. Once this had happened, the Moon formed in a year, almost instantaneously in comparison with the preceding phase. In total, a small fraction of the disk (10–20 percent) originated in the terrestrial mantle, and the remainder primarily came from the mantle of the impactor, the balance being vaporized and leaving the system: this is the reason suggested for the Moon's relative poverty in iron.

One of the unknowns in the problem is the period at which the impact took place or, putting it in other terms, the mass that the proto-Earth had acquired at the time of the collision. Taking account of the relatively low mutual velocity of the two bodies (less than the escape velocity of the system), little of the material would have been lost, and the Earth's final mass would be practically equal to the sum of the masses. Two possibilities would satisfy the constraints imposed by the collision: (i) an early giant impact (the Earth 60 percent formed, mass of the impactor equal to 30 percent of the terrestrial mass, the remaining 10 percent being progressively accreted subsequently) and (ii) a late giant impact (the Earth 90 percent formed, mass of the impactor equal to the remaining 10 percent, i.e., the mass of Mars). The detailed composition of the Moon enables us to decide in favor of the second scenario. In fact, to determine the formation date of our satellite, we use both for lunar rocks and for the Earth (as explained in ▶ Chapter 2), the hafnium/tungsten isotopic chronometer, $^{182}Hf/^{182}W$ (~ 1.1×10^{-4} initially, with a half-life of 9 Ma), as well as the $^{182}W/^{184}W$ ratio, which is approximately 1 and does not vary over the course of time. It has recently been shown that the lunar and terrestrial mantles have identical ratios of elements, which implies that there was no ^{182}Hf at the time the Moon was formed. This conclusion implies that the Moon formed, at the latest, at $t_0 + 60$ Ma.

This work, which favors a very late formation of the Moon, reinforces the results of numerical simulations of the giant impact and, more generally, those for the formation of the Solar System. They do not seem, however, to be in complete agreement with the age of the Earth as determined independently from the $^{182}Hf/^{182}W$ chronometer. This indicates that our planet formed at about $t_0 + 40$ Ma (see ▶ Chapter 2). However, it is possible that there was some alteration to the Earth's differentiation after the impact took place, which would modify the interpretation of the ratio: hafnium being lithophilic, that is, with a tendency to remain in the rocky mantle, and tungsten siderophilic, that is, tending to become incorporated into the ferrous core.

However the case may be, at present no one queries this scenario of the Moon's formation (and, consequently, the last stage in the Earth's formation) through a late impact. But our detailed understanding of the mechanism and its consequences are certainly not perfect … In particular, a new problem has arisen very recently, with the discovery of water in localized lunar melt inclusions, which implies that some parts of the lunar interior contain as much water as Earth's upper mantle. While the general scenario will probably remain, to what extent this discovery challenges it remains to be seen.

Chapter 2

Formation and Early Infancy of the Earth

Between 4.568 and 4.4 Ga: An Uninhabitable Planet?

Immediately after its formation, the surface of the Earth was particularly inhospitable: no continents, no liquid water, but instead, a magma ocean subjected to intense meteoritic bombardment. The period between 4.57 and 4.4 billion years before the present (4.57–4.4 Ga BP) nevertheless saw the progressive establishment of the environmental elements that, later, would allow both the prebiotic chemistry to begin, and the emergence of life.

🔲 An artist's impression of the magma ocean that covered the very early Earth.

M. Gargaud et al., *Young Sun, Early Earth and the Origins of Life*,
DOI 10.1007/978-3-642-22552-9_2, © Springer-Verlag Berlin Heidelberg 2012

2

The Earth is a living planet: its surface is perpetually remodeled by plate tectonics as well as by the surface mechanisms of alteration and erosion. As a result, unlike "dead" bodies such as the Moon, where all the stages of their primitive history have been preserved, practically all the traces of the first 500 million years (Ma) of the Earth's history have been obliterated. We know almost nothing about either the mineralogical and petrographic composition, or the surface structure of the primitive Earth.

The oldest rocks currently known outcrop in Canada. They are the Acasta gneisses, dated to 4.031 Ga, and the Nuvvuagittuq greenstone belt, whose age is greater or equal of 4.0 Ga (*see* ▶ Chap. 6). Only a few minerals, zircon crystals ($ZrSiO_4$), recently discovered at Jack Hills in Western Australia, are older (4.404 Ga, *see* ▶ Chap. 3), but the rocks that contained them have been eroded away and have completely disappeared long ago. The period of time between the Solar System's "t_0" (4.568 Ga, *see* ▶ Chap. 1 and below) and the origin of the Acasta gneisses (4.031 Ga), has been called the Hadean aeon. This chapter will concentrate on the first 170 million years of the Hadean, up to the time when the Jack-Hills zircons crystallized, thus corroborating the presence on our planet of a continental crust and liquid water (*see* ▶ Chap. 3). In the absence of tangible markers such as rocks or minerals, we shall try to recon-

---

◼ Box 2.1   The Age of the Earth

Determining the age of the Earth has always been one of Mankind's preoccupations. The first attempts during the Middle Ages, were based on biblical texts and on the lifetimes of successive patriarchs. They led to the conclusion that our planet had been formed 4000 years B.C. – James Ussher (1581–1656) even calculated that the Earth had been created on 23 October 4004 B.C. (but without specifying the time of day!) It was only in the 18th and 19th centuries that scientists such as Georges-Louis Leclerc, Comte de Buffon (1707–1788) or Lord Kelvin (1824–1907) tried a physical approach. Assuming that the Earth formed as a red-hot body and by calculating the time required for it to cool, Kelvin concluded that our planet was between 20 and 400 Ma years old. At the same period, basing their view on the thickness of sedimentary deposits, geologists dated the formation of the Earth as between 3 Ma and 3.5 Ga. The real revolution did not appear until the 20th century with the establishment of dating methods based on the radioactive decay of certain natural elements such as uranium (◼ Box 2.2).

These so-called "absolute" dating methods are capable of attaining an accuracy better than 1 Ma on ages over 4500 Ma. The problem is then one of knowing exactly what event is being dated. In the case of a granite body, for example, is the age obtained that of the melting event that generated the granitic magma? That at which magma was emplaced? Or its age of crystallization? In fact, the isotopic clock will trigger when the object to be dated ceases to exchange isotopes with its environment (gas, fluids, other rocks, etc.), that is, when it starts to behave as an isolated system. Such exchanges operate through chemical diffusion, a mechanism whose rate (and thus efficiency) is strongly temperature dependent. What is known as the closure temperature for an isotopic system is the temperature below which the rate of diffusion of the isotopes under consideration becomes too low to allow redistribution (homogenization) of the parent and daughter isotopes. In fact, most of the ages measured by long-period isotopic systems are crystallization ages or cooling ages. In other words, what we are measuring is the time that has passed since the system dropped below its closure temperature.

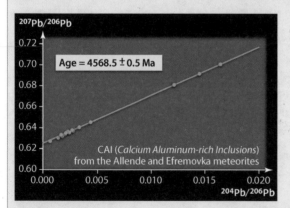

◼ **Fig. 2.1** ($^{207}Pb/^{206}Pb$) *versus* ($^{204}Pb/^{206}Pb$) isochron diagram for CAI (Calcium Aluminum-rich Inclusions) extracted from the Efremovka and Allende carbonaceous chondrites (◼ Box 2.2 and ◼ Fig. 2.6 for the description of the dating method). In such a diagram, the age is proportional to the slope of the straight-line isochron. Here, the calculated age, is 4568.5 ± 0.5 Ma; by convention, this age is considered to be the starting point, "time zero" (t_0), for the formation of the Solar System. (After Bouvier *et al.*, 2007.)

In the case of the age of the Earth, the problem is complicated by the fact that, as we have seen in Chap-

struct the early Hadean history of our planet from the data given by global geochemistry, the study of meteorites, and comparative planetology.

Before starting on this history, it seems useful to define as clearly as possible the chronological landmarks that will be used. By convention, astronomers and geologists are in agreement with regard to an initial time (t_0) that they have adopted as being the age of the Solar System. We note that this age of 4.568 Ga, has been measured from refractory inclusions (the CAI, *Calcium Aluminum-rich Inclusions*) inside carbonaceous chondrites; the CAI are, at present, the most ancient, dated materials in the Solar System, which condensed at high temperatures (> 1800 K) from the gases in the primitive planetary nebula (*see* ▶ Chap. 1 and the box below). It is, however, necessary to remember that all the stages of stellar formation occurred before t_0.

It was only after t_0 that the planets began to take shape, through successive collisions between small bodies. These collisions extended over a period of several million years in the case of the giant planets and probably over several tens of millions of years for the Earth. This phase in the construction of the planets, often called the accretion phase by geologists, was not therefore a punctual event, but rather lasted a long period of time, whence the difficulty in

---

ter 1, its formation was not some instantaneous phenomenon, but was, on the contrary, spread over several tens of millions of years.

In a rather arbitrary manner, purely by convention, the age of the carbonaceous chondrites is ascribed to the Earth. These meteorites have, in fact, not undergone any differentiation, and possess the same composition as the solar photosphere: They are primitive meteorites formed early in the Solar System's history. Moreover, in general these meteorites are of small size (when compared to the planets) such that they may be considered to have cooled extremely rapidly. These carbonaceous chondrites do not just consist of chondrules in a matrix, but also contain refractory inclusions that are rich in calcium and aluminum, known as CAI (Calcium Aluminum-rich Inclusions). These inclusions are thus older than the chondrules and the matrix (because they condensed at higher temperatures). The CAI are among the oldest materials currently known in the Solar System. Extremely precise measurements using lead isotopes have enabled the chondrules to be dated to 4564.7 ± 0.6 Ma and the CAI to 4568.5 ± 0.5 Ma (◨ Fig. 2.1). Thus, 3.8 Ma elapsed between CAI condensation and chondrule formation. This difference is in perfect agreement with the relative chronology of the different elements in carbonaceous chondrites, which shows that the CAI are older than the chondrules. An approach based on the now extinct radioactive isotope ^{26}Al (◨ Box 2.2), confirms these results, because it gives an age difference of about 3 Ma.

It is this age of 4568 Ma that is considered to be "time zero" (t_0) for our Solar System (and thus for the Earth). It is, in fact, that of the first solid particles that condensed at high temperatures from the gases in the protoplanetary nebula. It is essential, however, to remember that this reference t_0 does not actually correspond to the age at which the Sun was formed, nor to that of the

Earth. Indeed, terrestrial accretion did actually occur through collisions between small bodies that formed after the condensation of the CAI and the chondrule formation, a process which could have required several tens of millions of years. It was only after this accretion had finished that the Earth could begin its life as an "autonomous" planet (*see* ▶ Chapter 1). In addition, all the stages of stellar formation (in other words, the birth of the Sun) occurred before the condensation of the CAI, that is, before t_0.

So the chronology based on extinct radioactivity (^{182}Hf, ^{26}Al, ^{53}Mn, ^{60}Fe, ^{41}Ca, ^7Be, etc…) does, in fact, refer to an event that precedes t_0. Indeed, these isotopes could not have arisen except as the result of nuclear reactions earlier or, at the best, contemporary with the formation of the CAI (◨ Box 1.3). We may recall that two mechanisms have currently been proposed to account for their presence in the most primitive meteorites:

– the explosion of a supernova, the shockwave induced the gravitational collapse of a presolar molecular cloud. With this hypothesis, a portion of the supernova's products is injected into the presolar cloud. This mechanism primarily concerns ^{60}Fe (which is extremely rich in neutrons, and thus, by necessity the result of explosive nucleosynthesis) and ^{26}Al. In this scenario, the time between the explosion and the formation of the CAI, (t_0), would be, at most, about one million years;

– powerful X-radiation of the protoplanetary disk by the extremely violent eruptions of the young Sun in its T Tauri phase. This mechanism includes all the isotopes mentioned above, with the exception of ^{60}Fe. In this case, the CAI and (t_0) would be more-or-less contemporary with the explosion.

giving a precise age for the planet's accretion. Be that as it may, it was only after this accretion had finished that the Earth could begin its life as an independent planet, and its geological history began.

The Rapid Differentiation of a Metallic Nucleus: The Core

Unlike the formation of the continental crust, which as we shall discover later in the book (*see* ▶ Chaps. 6 and 7), has been extracted progressively from the mantle throughout the last 4.4 Ga, the segregation of the core (consisting of an iron/nickel alloy) and of the mantle (consisting of silicates), as shown in Chap. 3 (◻ Fig. 3.2), was a far more rapid and dramatic event. The age of this differentiation has recently been measured using the extinct radioactivity of hafnium, ^{182}Hf (◻ Box 2.2). In fact, ^{182}Hf decays into tungsten ^{182}W, with a half-life of 9 Ma, which implies that all the ^{182}Hf would have disappeared less than 60 Ma after the beginning of the condensation of the Solar System (t_0). Consequently, the ^{182}Hf/^{182}W pair is a chronological marker particularly well adapted to the earliest stages of the Solar System. In addition, it is particularly well suited to the study of the mantle/core differentiation, because Hf and W have contrasting geochemical behaviors: W is a siderophile element, which means that it concentrates within a metallic phase, such as the core, whereas Hf is lithophilic, and concentrates within a silicate phase such as the mantle. If the mantle/core separation took place more than 60 Ma after the Solar System condensation began, it would have taken place when all the ^{182}Hf had already been transformed into ^{182}W. Under such conditions, the ^{182}W should now be found solely within the core. However, recent measurements have revealed an excess of ^{182}W in the mantle. This means that ^{182}Hf still existed when the mantle and core separated, and that this element, being lithophilic, remained within the silicates. It subsequently decayed to produce the excess ^{182}W measured today in rocks from the mantle. We may therefore conclude that the mantle/core differentiation occurred less than 60 Ma after the start of the Solar System's condensation. More precise calculations based on the amount of the ^{182}W in excess in the Earth mantle has enabled the determination of the age of this differentiation as being between 11 and 50 Ma, with an average value of 30 Ma, if the event was unique and global (◻ Fig. 2.2).

What mechanisms governed the mantle/core separation? Although the general principle of the process is simple – the metallic phase is far denser that the silicate phase and, as a result, the force of gravity will attract it towards the center of the planet – the details remain complex and largely speculative. Most researchers agree on the fact that the metal was in liquid form when it migrated to the center of the planet. But although some believe that the metallic liquid percolated between solid silicate grains, others consider that the sedimentation took place from two immiscible liquid phases (molten metal and silicates) within the terrestrial magma ocean (*see later*).

The type of planetary body in which this differentiation occurred is also subject to debate. According to certain researchers, the process took place at high pressure, at the end of the accretion phase, in an Earth whose size was close to the one it currently has. In this theory, the ages averaging 30 Ma measured by the ^{182}Hf/^{182}W system disagree with the recent data from theoretical modeling of the accretion of the terrestrial planets, which all indicate that the latter formed in several tens of millions of years, probably between 50 and 100 Ma (*see* ▶ Chap. 1). According to others, the differentiation began earlier, at low pressure and in the interior of planetesimals. It may be noted that in the context of this second theory, the Earth's core would be the result of the agglomeration of the cores of several planetesimals, which would imply that a second migration phase would have brought these "mini-cores" together in the center of the Earth.

Currently, the Earth's core consists of two parts: a solid inner core surrounded by a liquid outer core. As the Earth is slowly cooling, the solid core is growing at the expense of the liquid outer core.

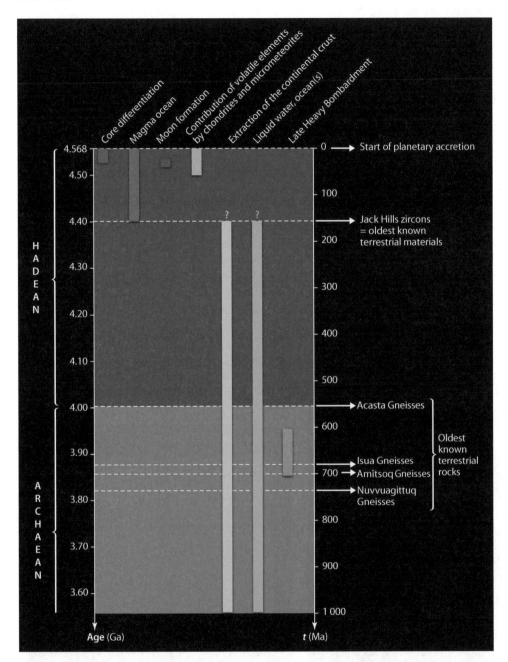

▫ **Fig. 2.2 Simplified geological time scale showing a possible succession of the different geological events that took place during the first billion years of the Earth's history.** The separation into a metallic core and a silicate mantle was rapid and dramatic in comparison with the growth of the continental crust, which has taken place in a continuous manner since 4.4 Ga, at least (*see* ▶ Chap. 3). Moreover, several arguments indicate that the mantle of the very young Earth underwent an episode of general melting. In other words, the surface of our planet was completely molten, forming what geologists term a magma ocean (which persisted until 4.4 Ga at the latest). Between t_0 + 10 Ma (the minimum age for the differentiation of the core) and t_0 + 70 Ma, carbonaceous chondrites and micrometeorites contributed a significant proportion of terrestrial water, such that at 4.4 Ga, liquid water was stable on the surface of our planet (*see* ▶ Chap. 3). The meteoritic bombardment, a cataclysmic event that affected the Earth about 3.9 Ga ago, will be discussed in Chap. 5.

2

◘ Box 2.2   Geochronology

The isotopes of an element are atoms that have the same number of protons (and electrons), but a different number of neutrons. Stellar nucleosynthesis gives rise to several stable or unstable isotopes, the latter decaying over the course of time into another element. For example, rubidium has, in total, 24 isotopes of which one of the most abundant is $^{87}_{37}$Rb, which has 37 protons and 50 neutrons. $^{85}_{37}$Rb is the stable form of rubidium; it has 37 protons and 48 neutrons, giving a total of 85 nucleons. $^{87}_{37}$Rb is unstable and decays into $^{87}_{38}$Sr (which is stable) according to the following reaction:

$$^{87}_{37}\text{Rb} \longrightarrow {}^{87}_{38}\text{Sr} + e^- \tag{2.1}$$

This reaction, which emits an electron (e⁻), corresponds to what is known as β⁻-decay. There are other types of radioactive decay (α, β⁺, and ε), but in every case the rate of decay is totally independent of the nature and history of the object being considered. For example, the time required for the decay of half of the atoms of $^{87}_{37}$Rb that are present in the system into $^{87}_{38}$Sr will always be 48.81 Ga: this period of time is also known as the half-life ($T_{1/2}$) (◘ Fig. 2.3).

In a more general case, if we consider a nuclear reaction of the form:

$$P \rightarrow D$$

where P = number of parent atoms ($^{87}_{37}$Rb in our example) and D = the number of daughter atoms ($^{87}_{38}$Sr in our

example), then the proportion of parent atoms that decay per time unit t is a constant:

$$\frac{dP}{Pdt} = -\lambda \tag{2.2}$$

where λ is the decay constant and corresponds to the probability of a decay event per time unit. The half-life may thus also be expressed as a function of λ:

$$T_{1/2} = \frac{\ln 2}{\lambda} \tag{2.3}$$

$T_{1/2}$ being expressed in years and λ in yr⁻¹.

If, at a time t_0, a system (such as a rock) contains a number P_0 of parent atoms, then the integration of equation (2.2) may be expressed as follows:

$$P = P_0 e^{-\lambda t} \Rightarrow P_0 = P e^{\lambda t} \tag{2.4}$$

In the same way, at time t_0, the system may also contain some amounts D_0 of daughter atoms (◘ Fig. 2.4), which means that at time t, the total number of daughter atoms D will be:

$$D = D_0 + D^* \tag{2.5}$$

D^* being the number of daughter atoms produced through decay since t_0; D^* is thus equal to the number of parent atoms that have disappeared:

$$D^* = P_0 - P \tag{2.6}$$

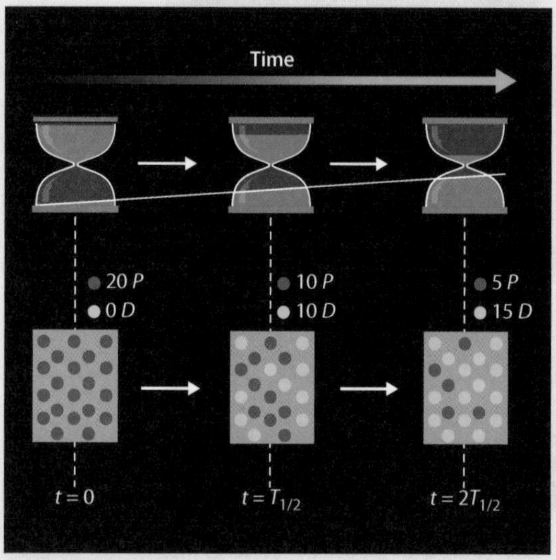

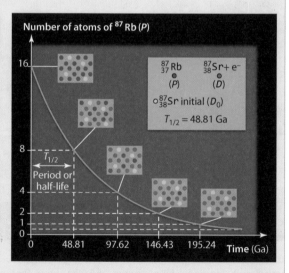

◘ **Fig. 2.3 Diagram showing the temporal evolution of the number of parent atoms P and daughter atoms D in a decaying radioactive system.** The half-life $T_{1/2}$ is the time required for half of the parent atoms that are present to decay. At the end of $t = T_{1/2}$, half of the initial 20 parent atoms have decayed, so only 10 remain. At $t = 2T_{1/2}$, just half of the parent atoms present at $t = T_{1/2}$ remain, i.e., 5 parent atoms, etc. The half-life $t = T_{1/2}$ is a characteristic constant of any given radioactive isotope.

◘ **Fig. 2.4 Diagram showing the evolution of the number of parent atoms ($^{87}_{37}$Rb; maroon circles) and daughter atoms ($^{87}_{38}$Sr; red circles) versus time.** The daughter atoms derived from the decay of $^{87}_{37}$Rb. The yellow circles represent atoms of $^{87}_{38}$Sr initially present in the system (at time t_0). As these atoms were stable, their number remains constant over the course of time. The number of parent atoms decreases in accordance with an exponential law: $P = P_0 e^{-\lambda t}$.

Parent nuclide = P	Daughter nuclide = D	Half-life = $T_{1/2}$
^{7}Be	^{7}Li	53.1 days
^{228}Th	^{224}Ra	1.91 years
^{226}Ra	^{222}Rn	1602 years
^{14}C	^{14}N	5730 years
^{59}Ni	^{59}Co	76 000 years
^{41}Ca	^{41}K	100 000 years
^{36}Cl	^{36}Ar	301 000 years
^{26}Al	^{26}Mg	707 000 years
^{60}Fe	^{60}Ni	1.5 Ma
^{10}Be	^{10}B	1.51 Ma
^{182}Hf	^{182}W	9.0 Ma
^{129}I	^{129}Xe	15.7 Ma
^{53}Mn	^{53}Cr	37.1 Ma
^{146}Sm	^{142}Nd	103 Ma
^{235}U	^{207}Pb	704 Ma
^{40}K	^{40}Ar	1.31 Ga
^{238}U	^{206}Pb	4.47 Ga
^{232}Th	^{208}Pb	14.0 Ga
^{176}Lu	^{176}Hf	35.9 Ga
^{187}Re	^{187}Os	42.3 Ga
^{87}Rb	^{87}Sr	48.81 Ga
^{147}Sm	^{143}Nd	106 Ga

◘ **Table 2.1** Half-life $T_{1/2}$ of the main isotopic systems used in geochronology.

◘ **Fig. 2.5 Zircon crystal extracted from a gneiss (SV11) from the Sete Voltas massif in Brazil (a) and the ^{206}Pb/^{238}U against ^{207}Pb/^{235}U diagram that enabled it to be dated (b).** The *Concordia* curve (orange) is calibrated in Ga; the *Discordia* straight line is shown in blue. One of the zircon crystals analyzed (red dot) lies on the *Concordia* curve whereas the others (blue dots) have partly lost their lead such that they plot on a *Discordia* straight line. The intersection of the *Discordia* straight line with the *Concordia* curve at top right gives the age of the zircon population (3.394 ± 0.005 Ga). (Dating after Martin *et al.*, 1997; photo: H. Martin.)

element being considered (◘ Table 2.1). A distinction is made between elements that have a short half-life (< 500 Ma; short-period radioactivity) and those where the half-life far exceeds 500 Ma (long-period radioactivity).

Combining equations (2.4), (2.5) and (2.6) gives:

$$D = D_0 + P_0 - P$$

$$D = D_0 + Pe^{\lambda t} - P$$

$$D = D_0 + P(e^{\lambda t} - 1) \qquad (2.7)$$

The age t of the system may therefore be obtained as follows:

$$t = \frac{1}{\lambda} \ln\left(1 + \frac{D - D_0}{P}\right) \qquad (2.8)$$

Although the parameters P and D are measured directly from the rock or mineral, D_0 itself is unknown, so the equation (2.8) cannot be used directly. The solution employed is different depending on the radioactive

1. Long-period Radioactivity

Here, two cases should be considered:
- D_0 is negligible relative to D ($D_0 \ll D$)

This is basically the case with the uranium-lead systems ^{235}U $\rightarrow$ ^{207}Pb and ^{238}U $\rightarrow$ ^{206}Pb used for dating the zircons. Because during their crystallization, natural zircons are practically unable to incorporate lead into their crystalline lattice, the initial quantities of ^{207}Pb and ^{206}Pb are utterly negligible relative to the quantity of lead produced by the decay of the uranium. Equation (2.8) may thus be written:

$$t = \frac{1}{\lambda} \ln\left(1 + \frac{D}{P}\right) \quad \text{that is: } t = \frac{1}{\lambda_{235_U}} \ln\left(1 + \frac{^{207}Pb}{^{235}U}\right)$$

or $t = \dfrac{1}{\lambda_{238_U}} \ln\left(1 + \dfrac{206_{Pb}}{238_U}\right)$.

Classically, the two isotopes ^{235}U and ^{238}U are plotted in a $^{206}Pb/^{238}U$ versus $^{207}Pb/^{235}U$ diagram (◘ Fig. 2.5). In such a diagram, the identical ages, independently calculated from the two isotopic systems lie along a curve known as *Concordia*. If a zircon crystal has evolved as a closed system (without any loss of lead), then its isotopic composition will lie exactly on the *Concordia* curve, which enables its age to be determined directly (the zircon is said to be concordant). However, it may happen that a zircon crystal loses lead over the course of its geological history (during a metamorphic event, for example); then the two isotopic systems give different (discordant) ages. In this case, the population of studied zircons lies on a line known as *Discordia* and located below the *Concordia* curve. The intersection of the *Discordia* and *Concordia* curves at top right (called upper intercept) then gives the crystallization age of the zircon (◘ Fig. 2.5).

– D_0 is not negligible relative to D

In the case where D_0 is not negligible relative to D, a computation method based on the principle of isotopic homogenization is used. For example, during melting or crystallization processes, the chemical elements may have different behavior with respect to the solid and liquid phases; in contrast, the isotopes of a single element do not undergo any fractionation. The ratio of two isotopes of a specific nuclide remains constant. In this case, at a time t, equation (2.7) may be written as:

$$\left(\dfrac{D}{D'}\right)_t = \left(\dfrac{D}{D'}\right)_0 + \left(\dfrac{P}{D'}\right)_t (e^{\lambda t} - 1) \qquad (2.9)$$

where D' represents the number of atoms of a stable isotope of D. In a closed system, D' remains constant.

Taking as an example the rubidium-strontium (Rb-Sr) system and taking ^{86}Sr as the stable isotope, equation (2.9) may be written:

$$\left(\dfrac{^{87}Sr}{^{86}Sr}\right)_t = \left(\dfrac{^{87}Sr}{^{86}Sr}\right)_0 + \left(\dfrac{^{87}Rb}{^{86}Sr}\right)_t (e^{\lambda_{87_{Rb}} t} - 1)$$

Several cogenetic samples (generated, for example, from the same magma) may have different $(^{87}Rb/^{86}Sr)_t$ ratios (because of magma fractionation; ◘ Box 6.1), and identical $(^{87}Sr/^{86}Sr)_0$ ratios. In a $(^{87}Sr/^{86}Sr)_t$ versus $(^{87}Rb/^{86}Sr)_t$ diagram, the isotopic composition of each sample evolves following a linear law (◘ Fig. 2.6),

defining a straight line called an *isochron*, whose slope a is equal to $e^{\lambda_{87_{Rb}} t} - 1$. This enables determination of t:

$$t = \dfrac{1}{\lambda_{87_{Rb}}} \ln a$$

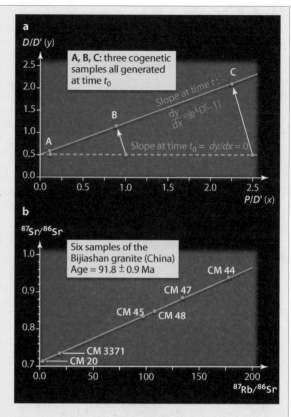

◘ **Fig. 2.6 a. (D/D') versus (P/D')diagram:** the age of a set of samples is proportional to the slope of the straight-line isochron, (see equation 2.9); **b. Example of a $^{87}Sr/^{86}Sr$ versus $^{87}Rb/^{86}Sr$ diagram:** the straight-line isochron obtained from six granite samples from the Bijiashan Massif in China gives an age of 91.8 ± 0.9 Ma. (Isochron for the Bijiashan granite after Martin *et al.*, 1994.)

In order to determine as accurately as possible the equation of the isochron, several samples must be analyzed.

2 Short-period Radioactivity or Extinct Radioactivity

The basic principle is that the nucleosynthesis of these radionuclides was caused by a supernova explosion at the very beginning of the formation of the Solar System, 4.568 Ga ago, and that they were homogeneously distributed within the solar nebula. This latter hypothesis is fundamental and indispensable, because it allows the assumption that different objects (meteorites, for example), which, *a priori*, were generated in different places in the Solar System, have identical initial isotopic ratios. As explained in the ◘ Box 1.3, there is currently a debate about their origin, which is doubtless to be found in some "accommodation" between the two

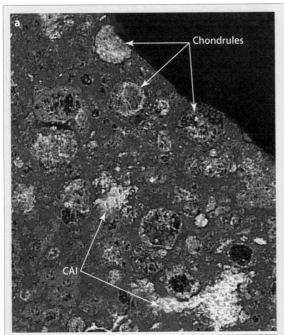

Given their short half-lives, all the parent atoms have now completely decayed into daughter atoms ($P = 0$ and $D^* = P_0$). Equation (2.9) may thus be written in the following form:

$$\left(\frac{D}{D'}\right)_t = \left(\frac{D}{D'}\right)_0 + \left(\frac{P}{D'}\right)_0 \qquad (2.10)$$

To be able to use this equation as a chronometer, it is necessary to input a stable isotope P_2 of the parent, short-lived radionuclide P:

$$\left(\frac{D}{D'}\right)_t = \left(\frac{D}{D'}\right)_0 + \left(\frac{P}{P_2}\right)_0 \cdot \left(\frac{P_2}{D'}\right) \qquad (2.11)$$

When this equation (2.11) is applied to the $^{26}\text{Al} \rightarrow {}^{26}\text{Mg}$ system (where $T_{1/2} = 707\,000$ years), and where ^{27}Al is taken as the stable isotope of aluminum (P_2), this becomes:

$$\left(\frac{^{26}\text{Mg}}{^{24}\text{Mg}}\right)_t = \left(\frac{^{26}\text{Mg}}{^{24}\text{Mg}}\right)_0 + \left(\frac{^{26}\text{Al}}{^{27}\text{Al}}\right)_0 \cdot \left(\frac{^{27}\text{Al}}{^{24}\text{Mg}}\right)$$

In this short-period system, the isotopic composition of the parent nuclide ^{26}Al will vary very swiftly: the $^{26}\text{Al}/^{27}\text{Al}$ ratio dwindles to half in 707 000 years. As a consequence, the temporal accuracy of such a chronometer is excellent. If we know the value of this ratio in the solar nebula at t_0, then this chronometer may be used to calculate the age t of a sample. Indeed, we have:

$$\left(\frac{^{26}\text{Al}}{^{27}\text{Al}}\right)_t = \left(\frac{^{26}\text{Al}}{^{27}\text{Al}}\right)_0 e^{-\lambda \cdot t}$$

In fact, extinct radioactivity is most frequently used to measure the age difference between two samples generated from a single reservoir, at times t_1 and t_2 ($\Delta t = t_1 - t_2$). This difference in age may thus be expressed in the following manner (■ Fig. 2.7):

$$\frac{(^{26}\text{Al}/^{27}\text{Al})_{t_1}}{(^{26}\text{Al}/^{27}\text{Al})_{t_2}} = e^{-\lambda \cdot \Delta t}$$

$$\Delta t = \frac{1}{-\lambda} \cdot \ln\left(\frac{(^{26}\text{Al}/^{27}\text{Al})_{t_1}}{(^{26}\text{Al}/^{27}\text{Al})_{t_2}}\right) \qquad (2.12)$$

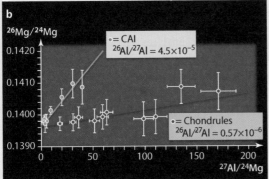

■ **Fig. 2.7 a. The Allende meteorite (a carbonaceous chondrite):** note the distribution of chondrules and the CAI (Calcium-Aluminum-rich Inclusions; photo. M. Chaussidon). **b. Isochron diagram of $^{26}\text{Mg}/^{24}\text{Mg}$ versus $^{27}\text{Al}/^{24}\text{Mg}$:** the slope of the regression lines is equal to the ratio ($^{26}\text{Al}/^{27}\text{Al}$), which is itself dependent on time. The Δt calculated from equation (2.12) is 2.02 Ma. (After Chaussidon, 2005.)

mechanisms: X-ray radiation within the solar nebula by eruptions from the young star in its pre-T Tauri phase, and "pollution" by nuclei produced by the explosive nucleosynthesis in a nearby supernova.

The formation of the metal core could have had several consequences for the overall evolution of the Earth. First, the differentiation between metal and silicates liberated a huge amount of gravitational energy, which, if it was released in a single episode, would have been capable of raising the overall temperature of the planet by about 1500 K, thus contributing in a substantial manner to the melting of the silicate rocks and the formation of a magma ocean (*see later*). Second, the differentiation of the core could have played a part in the birth of the Earth's magnetic field. It is this second point on which we will now focus the discussion.

Opening a Protective Umbrella:
the Birth of the Earth's Magnetic Field

The formation of a metallic core is one of the conditions necessary for the appearance of a magnetic field. In fact, the movements affecting the liquid core (they result both from the phenomenon of convection and from the Coriolis forces that are linked to the rotation of the Earth), induce electrical currents which produce a magnetic field: this is known as the geodynamo. This geodynamo is self-sustaining : the magnetic field itself produces induced electrical currents, producing a magnetic field, etc.

However, there is nothing to prove that the appearance of the terrestrial magnetic field was concomitant with the core/mantle differentiation, nor even that it followed shortly after. In other words, no one knows precisely when the magnetic field appeared! Some minerals can record, in the form of remanent magnetism, the presence and the characteristics of a magnetic field that existed at the time they crystallized. Unfortunately, if they are heated (through a metamorphic process, for example), they can lose this magnetic memory, and record the magnetic field present at the time of reheating. At present, the oldest trace of a magnetic field preserved by minerals has been discovered in 3.6 Ga-old rocks in South-Africa; however, in any case, in the absence of rocks older than 4.0 Ga, the search for a fossil magnetic field is difficult to apply to the Hadean. Nevertheless, preliminary studies carried out on the Australian zircons from Jack Hills (4.4 Ga) appear to indicate that these oldest terrestrial materials, recorded the marks of a magnetic field contemporary with their formation. If these results are confirmed, that would indicate that the magnetic field appeared very early in the Earth's history, between 4.568 and 4.4 Ga. Moreover, analyses carried out on the martian meteorite ALH84001 have been able to establish that it possesses remanent magnetism that was acquired between the time of its formation (between 4.5 and 4.4 Ga) and 4.0 Ga. So, it seems that a geodynamo has been active extremely early on the planet Mars, which suggests that the magnetic fields of the terrestrial planets were established shortly after the separation of their mantles and cores.

The time when the Earth's magnetic field arose is significant, both for the history of life and for that of the atmosphere. Indeed, on the Earth, the magnetic field has always played a fundamental protective role, by deflecting the ionized particles of the solar wind, the effects of which would be lethal to living organisms. In addition, before the magnetic field was established it is highly likely that, on the primitive Earth, the solar-wind particles would have "swept away" light elements from the atmosphere, by providing them with the energy necessary to escape from the Earth's gravitational attraction.

A Partially Molten Earth:
the Magma Ocean Assumption

The idea of a global magma ocean, that is, the idea that the outer layer of the primitive Earth could have been completely molten to a depth of several hundred kilometers (or even more), is a legacy of the first explorations of the Moon (■ Fig. 2.8).

In 1969, the *Apollo 11* mission returned to Earth a large number of lunar rocks. Among them were samples from the maria, the dark areas on the surface of our satellite that may be seen by the naked eye from Earth. Their analysis showed that these were basalts generated from a differentiated mantle, that is to say that have been previously depleted in certain chemical elements such as strontium (Sr) and europium (Eu) (■ Fig. 2.9g). Because these elements are, on Earth, particularly abundant in plagioclase feldspar, it has been concluded that the lunar mantle underwent plagioclase extraction. This theory has been confirmed three years later, when the *Apollo 16* mission returned to Earth samples of anorthosites,

◘ Fig. 2.8 The Moon, photographed by the *Galileo* probe on 7 December 1992. This image shows the two main components of the lunar surface. The light areas, known as the highlands, are richer in craters and are thus considered to be older. These primarily consist of anorthosites and correspond to the lunar crust formed by the accumulation (by floatation) of plagioclase at the surface of a magma ocean. The dark areas are known as the maria. They are less cratered and are thus younger than the highlands. They consist of basalts that originated from the melting of a differentiated mantle.

dated at 4.456 Ga. In 2011, one sample of lunar anorthosites (FAN 60025) has been dated using multiple chronometers (^{207}Pb–^{206}Pb, ^{147}Sm–^{143}Nd and ^{146}Sm–^{142}Nd). All these isotopic systems give a crystallization age of 4.360 ± 0.003 Ga, which could mean that the moon solidified later (~100 Ma) than supposed. These rocks formed through the accumulation of plagioclase crystals, and because of this, are rich in Sr and Eu (◘ Fig. 2.9b to d). So the geochemical signatures of the anorthosites and of the basalts from the lunar maria have proved to be perfectly complementary. For example, ◘ Fig. 2.9d shows that where the anorthosites have a positive Eu anomaly, which is characteristic of the accumulation of plagioclase crystals, the maria basalts exhibit a negative anomaly, indicative of the loss of plagioclase (◘ Fig. 2.9g). This complementary nature is a solid argument in favor of the differentiation of the anorthosites and the maria basalts from one and the same magma. More recent space missions (the *Clementine* and *Galileo* probes) have established that the anorthositic signature is found almost everywhere on the Moon (particularly as regards the highlands, which are the light areas of the Moon's surface, visible from Earth), and thus bear witness to the global character of their petrogenetic mechanism. But how should these observations be interpreted?

The "Textbook Case" for a Lunar Magma Ocean

To take account of all these characteristic features, it has been proposed that just after the formation of our satellite (as a result of the collision of a planetoid called Theia with the Earth, followed by the accretion of debris; ◘ Box 1.5), a very large proportion of the lunar mantle melted, so giving rise to a magma ocean. The latter subsequently steadily cooled and progressively solidified. The minerals that crystallized are olivine, orthopyroxene, clinopyroxene and plagioclase. With the exception of plagioclase, all these minerals have a higher density than that of the magma within which they formed (◘ Table 2.2). They therefore slowly sank

2

a   Sampling

b   Rock

1 cm

c   Thin Section

d   Rare-earth content

Rock content/chondrite content

1 000

100

10

1

La Ce   Nd   Sm Eu Gd   Dy   Er   Yb

Lunar anorthosite No. 67 215

e   Rock

1 cm

f   Thin Section

g   Rare-earth content

Rock content/chondrite content

1 000

100

10

1

La Ce   Nd   Sm Eu Gd   Dy   Er   Yb

Lunar maria basalt No. 10 058

Fig. 2.9 Close-up on the lunar rocks. In **a**, the geologist Harrison Schmitt is in the process of collecting lunar rocks during the *Apollo 17* mission. Among the samples returned to Earth were anorthosites (**b**) and basalts from the lunar maria (**e**). The anorthosites are almost exclusively made up of plagioclase crystals (**c**; in this thin section, appearing as grey and white minerals, occasionally showing alternating white and black bands). The basalts also contain a small amount of plagioclase, but in addition, they include pyroxene crystals (deep orange in thin section **f**) and sometimes olivine, as well as iron and titanium oxides (black crystals in **f**). The rare earth in the anorthosites (**d**) and in the maria basalts (**g**) display complementary patterns: for instance, where the anorthosites have a positive Eu anomaly, characteristic of plagioclase accumulation, the basalts show a negative anomaly, indicative of plagioclase extraction. The complementary nature of the geochemical signatures of the anorthosites and the basalts from the lunar maria provides a strong argument in favor of their differentiation from a single magma. These characteristics may be explained if we assume that the lunar mantle melted shortly after the formation of our satellite, giving rise to a magma ocean, whose cooling would have led to the formation of an anorthositic crust.

Table 2.2 Chemical composition and density of minerals likely to have crystallized from the magma ocean.

Mineral	Chemical formula	Density
Olivine	$(Mg,Fe)_2SiO_4$	3.32
Orthopyroxene	$(Mg,Fe)_2Si_2O_6$	3.55
Clinopyroxene	$Ca(Mg,Fe)Si_2O_6$	3.4
Garnet majorite	$Mg_3(Fe^{3+},Al)_2(SiO_4)_3$	3.9
Plagioclase	$CaAl_2Si_2O_8$	2.65
Talc	$Mg_3Si_4O_{10}(OH)_2$	2.7
Serpentine	$Mg_6Si_4O_{10}(OH)_8$	2.54
Basic magma = basalt		2.85
Ultrabasic magma = komatiite		2.95

and accumulated at the base of the magma ocean, giving rise to what would become the lunar mantle. The plagioclase, being less dense than the magma, floated to the surface of the magma ocean, leading to the formation of an anorthositic crust. This mechanism would have had the effect of separating the plagioclase from the other minerals (olivine + orthopyroxene + clinopyroxene). As plagioclase is, unlike olivine and the pyroxenes, a Sr and Eu-rich mineral , the anorthositic crust became enriched in Sr and Eu relative to the mantle, which itself became impoverished in these same elements (■ Fig. 2.9 d and g).

The formation of this crust and the chemical differentiation between the basalts and anorthosites that followed became possible because the mass of the Moon is one 80th of that of the Earth, which means that when the surface gravity on the Earth is 9.78 m.s^{-2}, it is just 1.62 m.s^{-2} on the Moon. As a consequence, and when compared with the Earth, the pressure inside the Moon rises with depth far more slowly and whereas plagioclase is stable on Earth only down to depths of 30 km, it remains stable down to depths of 180 km on our satellite. The crystallization of the lunar magma ocean is therefore able to generate huge volumes of plagioclase that accumulated through flotation, thus giving rise to a thick anorthositic crust (■ Fig. 2.10). The latter formed an insulating layer, through which heat could escape only by way of conduction, a mechanism that is of low efficiency (i.e., slow); as a result it was undoubtedly the presence of this anorthositic crust that slowed down and delayed cooling – and thus the crystallization and solidification – of the Moon's magma ocean. As far as the basalts of the lunar maria are concerned, they formed later, with their ages reaching 3.2 Ga. They arose through a new episode of melting in the lunar mantle as a result of the release of radioactive heat or through the heat created by meteoritic impacts.

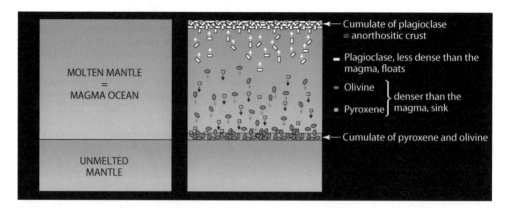

Fig. 2.10 Differentiation in the lunar magma ocean. The left side illustrates the initial state, where the upper portion of the lunar mantle is molten, forming what is known as a magma ocean, whereas the lower portion, unmelted, remains in a solid state. The right side illustrates the crystallization of the magma ocean. The crystals of pyroxene and olivine, denser than the magma, accumulate at the base of the magma ocean, whilst the crystals of plagioclase, being lighter, float and accumulate at the surface, forming the lunar anorthositic crust.

And on Earth?

The generalized melting of the lunar mantle was, basically, the result of the gravitational energy liberated through the accretion of this planetary body. But what was it like on the Earth? Given that our planet is far more massive than the Moon, the gravitational energy released during its accretion was far greater. Assuming that the release was instantaneous (which was obviously not the case), the increase in temperature could have been around 38 000 K (as against 1600 K for the Moon under similar conditions). To this source of heat, we may add the equivalent of 1500 K resulting from the core/mantle differentiation (*see above*). Finally, the decay of radioactive elements also significantly contributed to the heating of the young planet Earth. This was particularly the case with certain short-period radionuclides (i.e., through extinct radioactivity) such as ^{26}Al, which decays into ^{26}Mg with a half-life of 0.75 Ma. Taking account of the mass of ^{26}Al available at the time of the Earth's formation, it is possible to calculate that the thermal contribution given by its decay is 9500 K. As for the decay of ^{60}Fe into ^{60}Ni (half-life 1.5 Ma), it would have contributed to the extent of 6000 K. So the amount of energy available at the end of the Earth's accretion was considerable, and even if a significant amount was dissipated by radiation to space, it would have been more than sufficient to melt completely a large part of the mantle. The formation of a magma ocean was therefore potentially possible on Earth, but, once again, before concluding that it existed, we need to try to find witnesses and proofs of its existence.

As we have seen earlier, unlike the Moon, to this day the Earth has not yielded any rock that is older than 4.0 Ga. Consequently, there is no existing direct witness of the existence of a possible terrestrial magma ocean. On Earth, nothing indicates that the most ancient rocks known, might have been formed from a mantle source that had been previously depleted in Sr and Eu by the loss of plagioclase.

However, if the matter is considered more closely, this last observation is not very surprising. Indeed, as we have already mentioned, the lunar gravity corresponds to only 17 per cent of the Earth's gravity, which means that on our planet plagioclase is stable only down to a depth of 30 km. In other words, a very simple calculation shows that a hypothetical terrestrial anorthositic crust would have been about six times less thick than on the Moon. In addition, contrarily to the Moon where, under the low-pressure conditions that reign in the lunar mantle, and consequently where plagioclase was the first phase to crystallize after

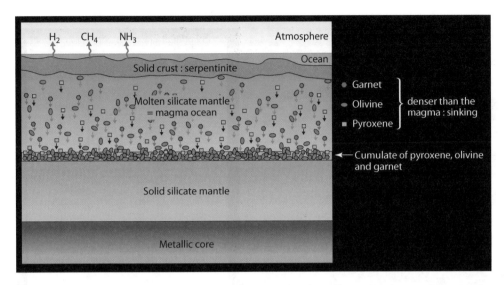

⬛ Fig. 2.11 Schematic section of the terrestrial magma ocean. The release of gravitational energy during accretion and the decay of radioactive elements was sufficient to lead to the complete melting of the outer region of the terrestrial mantle, thus giving rise to a magma ocean which must have covered the whole surface of the planet. As it cooled, olivine, pyroxenes and garnet crystallized. Because of their high density, these crystals sank to the bottom of the magma ocean, where they accumulated. The surface of the magma ocean cooled and solidified first. When in contact with the hydrosphere (with the water either in liquid or vapor state), anhydrous minerals such as olivine and pyroxenes were altered into hydrated minerals (talc, serpentine). They thus formed a superficial crust (less dense than the underlying magma). The serpentinization reactions liberated molecular hydrogen (H_2) which would have been able to combine with the carbon dioxide (CO_2) and molecular nitrogen (N_2) in the primitive atmosphere to form methane, (CH_4), and ammonia, (NH_3).

olivine (and before clinopyroxene), on Earth it was clinopyroxene that formed after olivine (and before plagioclase). So even if a small amount of plagioclase did crystallize in the terrestrial magma ocean, it was never in sufficient quantity to generate an anorthositic crust. The conclusion is that, even if it did really take place, the differentiation of the terrestrial mantle from a primitive magma ocean is not the result of plagioclase segregation. Other possible geochemical markers have thus been investigated.

Towards the end of the 1980s, researchers turned their attention to Archaean komatiites, ultrabasic volcanic rocks, typical and characteristic of the Archaean aeon that ranges from 4.0 to 2.5 Ga (the Archaean), and which we shall speak about in detail in Chap. 6. Study of them has shown that certain characteristics of the lavas that produced them (the content of rare-earths, for example) testify of a global differentiation that would have affected the whole terrestrial mantle before 3.4 Ga. Geochemists have therefore concluded that this was the result of the crystallization and segregation of garnet within the magma ocean. Unlike plagioclase, garnet has a density considerably higher than that of the magma (⬛ Table 2.2) and, like olivine and the pyroxenes, it sank and accumulated at the bottom of the magma ocean (⬛ Fig. 2.11).

More recently, in 2003, other geochemists have analyzed the isotope 142 of the element called neodymium (^{142}Nd) in basalts aged 3.872 Ga, from Isua in Greenland (*see* ▶ Chap. 6). This isotope is the product of the decay of ^{146}Sm (samarium, ⬛ Box 2.2). During their crystallization, silicates such as the pyroxenes and garnet generally incorporate more Sm than Nd, such that their crystallization modifies (decreases) the Sm/Nd ratio in the magma. Obviously, this fractionation will affect all the isotopes of Nd and Sm, including ^{146}Sm. In other words, the early crystallization of garnet and pyroxene will first result in a decrease in the content of ^{146}Sm (the parent isotope) in the magma ocean. As a result when all or part of the ^{146}Sm had decayed into ^{142}Nd, the relative abundance of ^{142}Nd in the magma would be less than in the minerals that had accumulated at the bottom of the magma ocean (the cumulate). As the life-time of the isotope ^{146}Sm is short (its half-life is 103 Ma), the ^{142}Nd can be a marker and

2

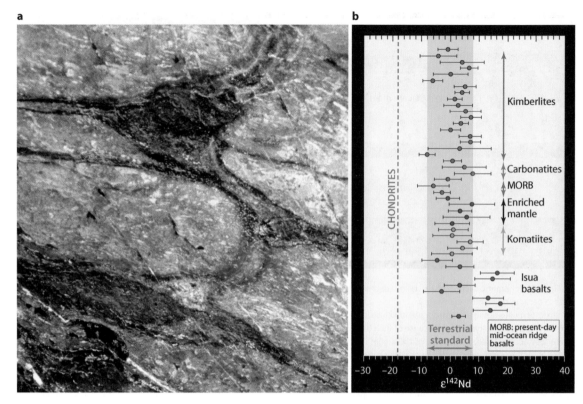

a

b

□ **Fig. 2.12 The traces of a magma ocean in the pillow basalts of Isua.** These 3.872 Ga old rocks, which outcrop in the Isua region of Greenland (**a**), are among the oldest known on Earth. The diagram (**b**) shows the ^{142}Nd content of these basalts, together with that of other terrestrial rocks generated through mantle melting. The quantity marked "$\varepsilon^{142}Nd$" is a measure of the ^{142}Nd content relative to a standard value (here the average of terrestrial values). It clearly appears that some Isua basalts have values of $\varepsilon^{142}Nd$ lying between +12 and +20, values that are significantly greater than the range of values from younger samples ($-8 < \varepsilon^{142}Nd < +8$). This excess in ^{142}Nd leads to the conclusion that the Archaean mantle from which they originated recorded and preserved the traces of its primordial chemical differentiation within a magma ocean. (Photo: M. Boyet; diagram: after Boyet *et al.*, 2003.)

a chronometer for such silicate fractionation only if the latter took place during the first hundreds of millions of years of our planet's history. Now the Isua basalts, generated by partial melting of the terrestrial mantle around 3.87 Ga ago, have an excess of ^{142}Nd relative to all other, more recent, terrestrial samples (□ Fig. 2.12). Calculations show that this anomaly may be explained by fractionation of silicates within a magma ocean during the first 100 Ma of the Earth's history.

Contrary to what happened on the Moon, the crystallization of the magma ocean did not lead to plagioclase accumulation through flotation but, on the contrary, it differentiated by precipitation of olivine, pyroxenes, and garnet at the base of the magma ocean (□ Fig. 2.11). This mechanism led to an early stratification of the Earth's mantle, with a deep layer corresponding to the cumulate made up of garnet, olivine, and pyroxenes, and a shallower layer corresponding to the residual magmatic liquid, which, after cooling, crystallized as a peridotite that was poorer in olivine and garnet. It appears that this early stratification of the mantle disappeared after 3.4 Ga, probably because of efficient mixing caused by mantle convection.

On the Moon the anorthositic crust played the part – to a certain extent – of a lid that slowed down the loss of heat and thus the cooling and then solidification of the magma ocean. On Earth, the accumulation of minerals took place through sedimentation at the bottom of the magma ocean: so there was no such anorthositic lid. Obviously, the upper layer of the magma ocean, in contact with the external, colder medium, cooled and solidified, in a manner very similar to what is observed today, on a much smaller scale, in lava lakes. This solid crust, set on the surface of the magma ocean, had a density greater than 3, whereas that of the subjacent magma, with the same composition, was about 2.9 (◘ Table 2.2). So it is perfectly logical that under the influence of its own weight, this solid crust rapidly sank into the magma ocean, leaving free space at the surface for fresh magma to rise, and which created a rapid cooling (in a few tens of thousands of years) of the magma ocean.

This scenario of very rapid cooling of the magma ocean, as much seductive it may seem, is probably not the correct one. The time required for an effective separation of the cumulate and thus for a stratified structure for the mantle is, in fact, about several tens of millions of years. In addition, as we shall see later, models for the formation of the Earth lead everyone to believe that water was available, either in liquid or gaseous form, very shortly after the formation of our planet. It could therefore immediately react with the cooled surface of the magma ocean, whose dense, anhydrous minerals (olivine, pyroxenes) it would convert into hydrated and less dense minerals such as talc or serpentine (density < 2.8; ◘ Table 2.2). When thus altered, the surface crust would have become less dense than the subjacent magma, such that it would have floated on its surface, creating an insulating layer that would have slowed down the cooling and crystallization of the magma ocean (◘ Fig. 2.11). Whatever the case may be, the Jack Hill zircons show that this ocean had completely cooled at 4.4 Ga (*see* ▶ Chap. 3).

This theory of alteration of the surface crust, proposed in 2006–2007 by the geologist Francis Albarède, is also capable of explaining the reducing nature of the Earth's atmosphere between 4.4 and 4.0 Ga (*see* ▶ Chap. 3). For example, the alteration of the ferrous component of olivine (fayalite) takes place following the reaction:

$$3\,Fe_2SiO_4 \;+\; 2\,H_2O \;\leftrightarrow\; 3\,SiO_2 \;+\; 2\,Fe_3O_4 \;+\; 2\,H_2$$

fayalite — water — silica — magnetite — molecular hydrogen

The silica reacts with forsterite, which is the magnesian component of olivine:

$$3\,Mg_2SiO_4 \;+SiO_2 \;+4\,(H_2O) \;\leftrightarrow\; Mg_6Si_4O_{10}(OH)_8$$

forsterite — silica — water — serpentine

As for the hydrogen that is liberated, it rapidly combines with the carbon dioxide, CO_2, and the molecular nitrogen, N_2, in the primitive atmosphere to form methane, CH_4, and ammonia, NH_3. These reduced molecules could combine with H_2 and H_2O to form amino acids, as was shown in the famous experiment carried out by Stanley Miller et Harold Clayton Urey in 1953 (*see* ▶ Chap. 3).

The Birth of the Outer Shells: The Atmosphere and the Hydrosphere

Nowadays, the atmosphere and the hydrosphere only represent 0.000088 per cent (5.29×10^{18} kg) and 0.022 per cent (1.35×10^{21} kg) respectively of the mass of our planet (5.98×10^{24} kg). They are essentially located at the surface, that is to say at the interface between the solid Earth and interplanetary space. It is also on that surface that life has arisen and, as a consequence, these two envelopes will play a determining role in the emergence and spread of life on Earth. The atmosphere and the hydrosphere very probably had a common origin. Indeed, at the temperatures that reigned at the period at which the Earth accreted, water was in the state of vapor in the atmosphere, and it was only after the formation of the oceans (fol-

2

lowing the condensation of the atmospheric water, *see* ► Chap. 3) that these two outer, fluid envelopes evolved separately.

The Early Degassing of the Primitive Atmosphere

It is traditionally accepted that the very first terrestrial atmosphere had the composition of the solar nebula (the protoplanetary nebula): it primarily consisted of hydrogen and helium. But, due to the turbulent history of the accretion of the Earth, there remains almost nothing of that primordial atmosphere. Indeed, the ratios between the various atmospheric components and their isotopes are very different from those measured in the Sun, which, in essence, retains the elemental and isotopic signature of the nebula that gave it birth. By contrast, the most massive planets and the most distant from the Sun better preserved their primitive atmospheres that arose from the protoplanetary nebula. By way of example, Jupiter's atmosphere consists of 81 per cent hydrogen and 18 per cent helium.

The disappearance (the erosion) of the Earth's primordial atmosphere is the result of several phenomena. First, during its accretion, Earth has been impacted by planetesimals or large meteorites (just like Theia, ◘ Box 1.5), some of the energy released by the shock was transferred to gas atoms, allowing them to exceed the escape velocity and thus break away from Earth's gravitational attraction. However, as we shall soon see, these "impactors" could equally well have brought volatile elements to our planet. As a result, we have absolutely no idea of the net balance between these collisions and the erosion of the planetary primitive atmosphere. And then, before the terrestrial magnetic field was established, the surface of the Earth was subjected to the solar wind, that is to a flux of very energetic particles that certainly "swept away" the lightest elements (H, He), enabling them to gain the Earth's escape velocity and to escape into interplanetary space. Finally, these same elements could have left the planet through simple gravitational escape. For instance, ultraviolet radiation induces the photodissociation of water and produces molecular hydrogen, H_2, and this latter dissociates into atomic hydrogen, H, which easily escapes from the Earth's gravity.

Relics of the Primitive Atmosphere in the Terrestrial Mantle!

Although most of the Earth's primordial atmosphere has disappeared, an infinitesimal fraction has, despite everything, been preserved by our planet and, as astonishing as it may seem,

---

◘ Box 2.3   How the Isotopes of Xenon Contribute to the Study of Degassing in the Primitive Earth

The study of the isotopes of a rare gas, xenon (Xe), has enabled the early nature of the degassing of the primitive atmosphere to be established. Isotope 129 of xenon (^{129}Xe) results from the radioactive decay of isotope 129 of iodine (^{129}I), the latter originating in the supernova, the explosion of which contributed to the formation of the Solar System (◘ Box 1.3). The half-life of ^{129}I is 15.7 Ma, so this element had completely decayed into ^{129}Xe in less than 150 Ma. The xenon emitted today from the mid-oceanic ridges contains an excess of ^{129}Xe relative to the atmosphere. Under mid-ocean ridge systems, the Earth's mantle ascents through convective motion; it undergoes adiabatic decompression – that is, without any exchange of heat – which causes its partial melting and the emplacement of magmas which, when solidified, will form the oceanic crust. During the course of their crystallization these magmas release the dissolved gases that come from the terrestrial mantle, thus providing information about the volatile content of the Earth's interior. If degassing of the Earth had taken place more than 150 Ma after the start of the Earth's accretion that is to say after all the ^{129}I had been transformed into ^{129}Xe, then the isotopic composition of atmospheric xenon should be identical to that of the mantle. But this is not the case, and the excess in ^{129}Xe in the Earth's mantle indicates that almost all of the degassing had finished while ^{129}I still existed, that is, less than 150 Ma after the start of Earth's accretion.

geochemists are nowadays able to find traces of it. Our atmosphere today contains about 5 ppm of helium. The isotope ^{4}He forms most of it, with the isotope ^{3}He amounting to only about 7×10^{-6} ppm (^{3}He/^{4}He = 1.4×10^{-6}). Although ^{4}He primarily results from the decay of heavy radioactive elements such as uranium and thorium, ^{3}He is considered as resulting mainly of the primordial episode of nucleosynthesis; it was thus present in the protosolar nebula. The residence time of He in the terrestrial atmosphere is about one million years, which means that the current atmospheric helium cannot by any means be a "residue" of the primitive atmosphere. However, at the bottom of the oceans, along the mid-ocean ridges (on the mid-ocean spreading centers), the ^{3}He/^{4}He ratio is, on average, 8 times higher than that in the atmosphere (in Iceland, a region that corresponds to an emergent portion of the North Atlantic Ridge, it may even be as much as 37 times greater). The only way to explain the excess of ^{3}He measured directly below the ridges is to assume that this gas arises from degassing of the magmas that are being erupted there. In other words, in the absence of any other mantle source, researchers conclude that the ^{3}He being degassed is of primary origin, that is the relict of a primitive atmosphere, trapped within the Earth's mantle. The study of another rare atmospheric gas, neon, leads to an identical conclusion: in these same mid-ocean ridges, the measured ^{20}Ne/^{22}Ne ratio is closer to that of the Sun than to that of the current atmosphere. This local excess of ^{20}Ne is similarly interpreted as the result of degassing of the mantle, through volcanism. As for the He/Ne ratio within the mantle (measured at the ridges) it is also close to that of the Sun.

All these isotopic data prove that the Earth's mantle has preserved a component of the solar type, and the composition of which is inherited from the protoplanetary nebula. Some authors, proposed that the He and Ne originally from the primitive atmosphere being found today in the mantle, were incorporated, either by being trapped in pores between the minerals, or by being dissolved in the magma ocean.

So it appears that the primordial degassing of the planet (*see later*), although it may have been significant, was not total. Even catastrophic events such as the collision with the "Theia" planetoid about 4.528 Ga ago, did not allow degassing of the deepest parts of the Earth's mantle.

An Extraterrestrial Origin for the Atmosphere of the Very Young Earth?

The smaller and consequently, the less massive planets are, the more difficult it is for them to retain light elements, because they have a lower gravitational attraction. So, as we have seen, the Earth (together with Mercury, Venus, and Mars) completely lost its primordial atmosphere, in contrast to Jupiter, Saturn, Uranus, and Neptune. In addition, the surface temperature of the planets decreases with distance from the Sun. The higher the temperature, the more significant the thermal agitation (the average velocity of the atoms and molecules) becomes, which allows more efficient gravitational escape. The Earth, both small and close to the Sun, thus fulfilled the conditions that favored rapid erosion of the primitive atmosphere.

Nevertheless, our planet still possesses an atmosphere and a hydrosphere. This implies that the light elements of which these two outer envelopes are made up must have been subsequently provided to the surface. The question that then arises is the source of this "late" addition. There are two, competing, but not mutually exclusive, theories.

The Theory of Primordial Degassing

The first theory envisages primordial degassing of the planet. It is based both on the observation of the fact that volcanoes currently release large quantities of volatile elements (mainly H_2O, but also CO_2, CO, H_2S, H_2, SO_2, N_2, Cl, rare gases, etc.) into the atmosphere, and also on the discharge of ^{3}He and ^{20}Ne by oceanic ridge systems, proving that even today, degassing of volatile elements originating in the deeper portions of the planet is still active (*see earlier*). By analogy with ^{3}He and ^{20}Ne, it has been proposed that atmospheric gases such as N_2, CO_2 or H_2O also result from significant degassing of the Earth's deep interior. The isotopic analysis of a rare gas, xenon (◘ Box 2.3) shows that such an event occurred during the first 150 million years of our planet's existence; the degassing observed today along the mid-oceanic ridges being only a pale relic of this event.

During the Earth's accretion, the hydrated mineral phases could have been destabilized by the shocks occurring between planetesimals, and the volatile elements that they contained could have been released, leading to an intensive and efficient degassing of the planet. Is it possible to suppose that the whole of the terrestrial atmosphere and hydrosphere were generated through such a primordial degassing? The answer to this question very strongly depends on the model considered for planetary formation. If, as in the conventional view, the Earth is considered to consist of 15 per cent of carbonaceous chondrites (containing 5 to 10 per cent of water), and 85 per cent of ordinary chondrites, then the amount of water contributed by these meteorites corresponds to 50–70 times the quantity of water currently present on Earth. On the other hand, if it is believed that 99.5 per cent of our planet was formed from enstatite chondrites, whose water content is only 0.05–0.1 per cent, then that source of water would have contributed only 50 to 100 per cent of the volume of present-day oceans. That is totally insufficient, because, on Earth, water is found not only in the ocean (1.35×10^{21} kg), but also in the mantle (between 5 and 50×10^{20} kg) and, in addition, the loss of volatile elements during the accretion (erosion of the atmosphere) is generally estimated to be at least 90 per cent of the Earth's initial water budget. Under these conditions, terrestrial water (as well as other volatile elements) could not have exclusively come from the primordial degassing of the planet. So it is necessary to call on extraterrestrial reservoirs.

The Theory of Cometary or Meteoritic Contributions

The planetary bodies that formed in the outer zones of the Solar System are rich in volatile elements, because these latter could condense at the low temperatures that prevailed in those regions, very distant from the Sun. Among these objects, there were, of course, the comets, whose mass consists of at least 50 per cent of water and that originate in the Kuiper belt or in the Oort Cloud (beyond the orbit of Neptune, *see* ▶ Chap. 5, ◘ Figs. 5.7 and 5.8). They are therefore potentially capable of having contributed a significant portion of terrestrial water. Some researchers consider that this cometary contribution took place very early in the planet's history, while others have suggested a continuous supply, due to a permanent flux of micrometeorites. This latter theory assumes, however, that the volume of the oceans has increased throughout the life of the Earth, which is contradicted by geological data.

An alternative theory envisages degassing of volatile elements contained within micrometeorites when they enter the Earth's atmosphere. Every year, 40 000 tonnes of such interplanetary dust fall onto the surface of the planet; these micrometeorites have an average size of about 100 μm, and 50 to 100 per cent of them consists of hydrated minerals,. Assuming that during the first 100 Ma of the Earth's history, the micrometeorite flux was 10^6 times as great as it is today (i.e., the maximum value envisaged by current models), this dust would have contributed, over that period, between 0.5 to 1.2×10^{21} kg of water to the Earth.

How to Decide?

To decide between these two theories, geochemists have once again had recourse to isotopic analysis. Indeed, they focused their research on the isotopes of hydrogen and, in particular, on the deuterium (^{2}H = D)/hydrogen (^{1}H) ratio. In the present-day oceanic water, this D/H ratio is 155.7×10^{-6}, a value that is perfectly consistent with the isotopic composition of carbonaceous chondrites and micrometeorites (◘ Fig. 2.13). These bodies could, therefore, have

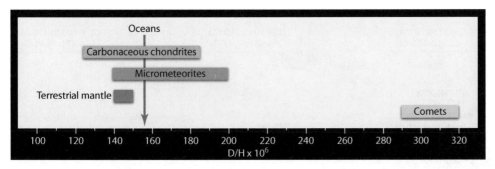

Fig. 2.13 Tracking extraterrestrial water with hydrogen isotopes: comparison of the deuterium(^{2}H = D)/ hydrogen(^{1}H) (D/H) ratio in the Earth's ocean and in different reservoirs: comets, carbonaceous chondrites, meteorites, and terrestrial mantle. Clearly, that with values of D/H > 290, comets could not have contributed significantly to the supply of volatile elements to the primitive Earth, unlike the case with the carbonaceous chondrites and micrometeorites.

supplied a substantial part of Earth's water. Yet, with a D/H ratio ranging between 290×10^{-6} and 320×10^{-6}, cometary water could not, by any means, have been the main source of terrestrial water: at the most, the cometary contribution to the Earth's oceans could not have exceeded 10 per cent.

When could extraterrestrial water have been supplied to the surface of the Earth? Here, this is the abundance of siderophile elements and in particular platinum-group elements (platinoids = Pt, Pd, Rh, Ru, Ir and Os) in the Earth's mantle that can help to answer this question. All the siderophile elements that were present in our planet when the core and mantle had segregated, migrated into the metallic core. However, the content of platinum-group metals in the present-day terrestrial mantle is by no means insignificant, which cannot be explained unless these platinoids were added to the mantle after the core/mantle separation. This is what is known as the "late veneer" theory. In addition, and consistently with the previous conclusion, the relative abundance of siderophile elements in the mantle is the same as in primitive meteorites, i.e., those meteorites, such as the carbonaceous chondrites, that have not undergone any differentiation into a metallic and a silicate phase.

The excess of platinum-group metals in the mantle enables the meteoritic flux to be assessed as 0.45 per cent of the Earth's mass. By assuming that these meteorites were chondritic, i.e., that they contained between 6 and 22 weight per cent of water, it is possible to assess their contribution as ranging between 1.6 and 6.0×10^{21} kg of water, which is more than the current mass of the oceans (1.35×10^{21} kg). Models predict that this late contribution of water of extraterrestrial origin took place between $t_0 + 10$ Ma (the minimum age for the mantle differentiation) and $t_0 + 70$ Ma. However, it is not possible to decide between the theory of a late veneer brought by a few large asteroids and that of a regular sprinkling by micrometeorites. This last theory is only realistic if the flux of micrometeorites onto the early Earth had really been 10^6 greater than its current value.

To Summarize: the Atmosphere Between 4.56 and 4.4 Ga

Primordial degassing of our planet on its own or with contributions of extraterrestrial volatile elements enabled the formation of an atmosphere that replaced the primitive atmosphere inherited from the solar nebula. Initially, the surface temperature was such that water, and also, obviously, other volatile elements (including CO_2), were in the state of vapor (■ Fig. 2.14a).

If all the mass of the current ocean water was in the form of vapor, that would produce a partial atmospheric water pressure of 270 bars (■ Fig. 2.14b). Similarly, as far as CO_2 is con-

2

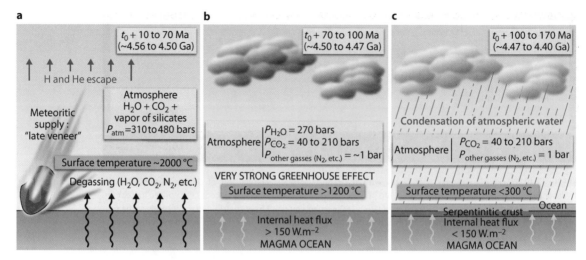

a t_0 + 10 to 70 Ma (~4.56 to 4.50 Ga)

H and He escape

Meteoritic supply: "late veneer"

Atmosphere H_2O + CO_2 + vapor of silicates P_{atm} = 310 to 480 bars

Surface temperature ~2000 °C

Degassing (H_2O, CO_2, N_2, etc.)

b t_0 + 70 to 100 Ma (~4.50 to 4.47 Ga)

Atmosphere P_{H_2O} = 270 bars, P_{CO_2} = 40 to 210 bars, $P_{other gasses (N_2, etc.)}$ = ~1 bar

VERY STRONG GREENHOUSE EFFECT

Surface temperature >1200 °C

Internal heat flux > 150 W.m^{-2} MAGMA OCEAN

c t_0 + 100 to 170 Ma (~4.47 to 4.40 Ga)

Condensation of atmospheric water

Atmosphere P_{CO_2} = 40 to 210 bars, $P_{other gasses (N_2, etc.)}$ = 1 bar

Surface temperature <300 °C

Ocean
Serpentinitic crust
Internal heat flux < 150 W.m^{-2} MAGMA OCEAN

□ Fig. 2.14 Schematic representation of the various episodes in the evolution of the atmosphere and the hydrosphere before 4.4 Ga. a. Between 10 and 70 Ma (after the beginning of the accretion), the metallic core and the silicate mantle became differentiated. The latter was largely molten and the surface of the Earth was covered with an ocean of molten silicates (magma ocean). The primordial atmosphere, inherited from the protosolar nebula (H + He) was rapidly eroded, whereas degassing of the planet and the meteoritic contribution ("late veneer") brought the elements of a new atmosphere to Earth. **b.** Between 70 and 100 Ma, the internal heat flux remained very high (> 150 W.m^{-2}) such that even the very outermost portion of the magma ocean could not crystallize. The massive atmosphere of water vapor and carbon dioxide created an intense greenhouse effect and the temperature "at the ground" exceeded 1200 °C. **c.** Between 100 and 170 Ma, the heat flux had decreased such that the formation of a solid crust on the surface of the magma ocean became possible. The Earth's outer envelopes then cooled much faster and the atmospheric water was able to condense: one or several oceans covered the Earth's surface. The concentration of atmospheric CO_2 remained very high and the temperature at the ground became less than 300 °C. (After Pinti, 2002.)

cerned, some researchers consider that all the carbon nowadays present in carbonates was in the atmosphere in the form of CO_2, from which they conclude that the partial pressure for that molecule was 40 bars. If the carbon that is now in the mantle (where it has been recycled through the subduction zones) is also taken into account, the estimated partial pressure for atmospheric CO_2 reaches 210 bars. By way of comparison, the atmospheric partial pressure of CO_2 is currently 3.5×10^{-4} bars. In addition to these gases, were both molecular nitrogen (N_2) and other rare gases, but researchers consider that their partial pressure was less than 1 bar.

Water and carbon dioxide being both efficient greenhouse gases, the radiation emitted by the Earth could not have been totally released to space. That would thus contribute to efficiently maintaining a very high surface temperature, such that, immediately after Earth accretion, the very surface layer of the magma ocean would not be able to solidify. When the flux of internal heat decreased (dropping below the threshold of 150 W.m^{-2}), a solid crust could form on the surface of the magma ocean (*see above*), resulting in an insulation of the hot magma from the outer envelopes, these latter being then able to cool more rapidly (□ Fig. 2.14c). It was this atmospheric cooling that allowed water vapor to condense and to give rise to oceans of liquid water. Is it possible to determine when this condensation took place?

All the models relating to the origin of terrestrial water converge towards the following conclusion: on the Earth, about 100 Ma after the start of accretion, the amount of water necessary for the formation of the oceans was available. In addition, as we have already indicated, the Australian Jack Hills zircons prove that liquid water was stable on the surface of the planet, 4.4 Ga ago, that is about 170 Ma after that the accretion started. Finally, the impact of the planetoid Theia, 40 Ma after the beginning of the accretion, liberated enough energy for all the liquid water present on the Earth to be completely vaporized. When taken all together these different data, lead to the conclusion that the condensation of water and the formation of a liquid ocean occurred between 4.47 and 4.40 Ga. In addition, recent models show that 400 to 700 years were sufficient for all the atmospheric water to condense and fall onto the Earth's surface. This event led to the formation of the oceans and certainly left an atmosphere that was "massive" (with an atmospheric pressure of 40 to 210 bars, depending on the model), basically consisting of CO_2, whose greenhouse effect alone would be sufficient enough for maintaining a surface temperature of around 200 to 250 °C.

The Conclusion: a Planet That Was Undoubtedly Uninhabitable

Our foray into the Earth's infancy has led us from the beginning of its accretion (4.568 Ga) to the period when geologists are certain that the oceans had condensed (4.4 Ga). Is it possible to imagine that life could have appeared on our planet during those 170 Ma? No, that seems highly improbable because the surface conditions resulted in a hostile and unfavorable environment. However, this period did see a transition between a planet that was utterly uninhabitable to a planet that was potentially habitable. Indeed, during the first hundred million years that followed the start of the Earth's accretion, the high temperatures that reigned at the surface of the Earth allowed neither the liquid water to exist, nor the surface of the magma ocean to cool and form a solid crust (and thus no stable, cool substrate). In addition, major meteoritic impacts churned up the planetary surface and vaporized, at least partially, the atmosphere and the hydrosphere. It was only subsequently, in a second stage, that the various environmental elements progressively arose that, later, would permit the start of prebiotic chemistry and the emergence of life: the birth of a magnetic field, the cooling of the magma ocean, and the formation of a solid crust, the condensation of the water and the formation of oceans, and the stabilization of the atmosphere.

2

Water, Continents, and Organic Matter...

Chapter 3

From 4.4 to 4.0 Ga: A Potentially Habitable Planet?

From 4.4 billion years BP (4.4 Ga), the Earth took a more familiar appearance: Its surface was covered with oceans from which a few continents emerged.

Its atmosphere probably contained enough greenhouse gases, such that, despite a "cold" Sun, its surface temperature remained above 0 °C.

The Earth was thus ready for life: It was potentially habitable, although it is not possible for anyone to know, yes or no, whether it was then inhabited.

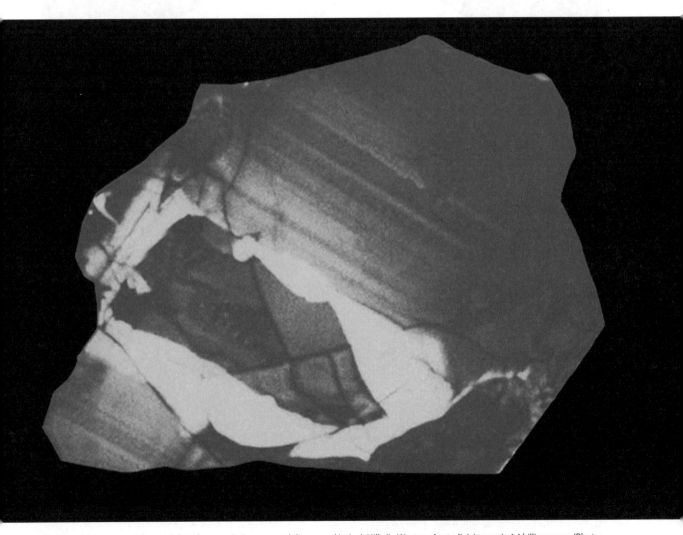

◨ **The oldest terrestrial material yet known:** A zircon crystal discovered in Jack Hills (in Western Australia). Its age is 4.4 billion years. (Photo: John Valley, University of Wisconsin, Madison, United States.)

M. Gargaud et al., *Young Sun, Early Earth and the Origins of Life*,
DOI 10.1007/978-3-642-22552-9_3, © Springer-Verlag Berlin Heidelberg 2012

3

The image of our planet such as seen by an astronaut nowadays is that of a mosaic of brown continents and blue oceans, occasionally hidden by a veil of white clouds ( Fig. 3.1). But the Earth has not always presented such an appearance – far from it! Consequently, the question that may legitimately be asked is at what period in its history it acquired these features that, since then, have been so familiar to us.

At the end of the last chapter, we have seen that 4.4 Ga ago, the Earth's surface was already covered by one or more oceans. How have geologists acquired such certainty? By conducting what amounts to a real police investigation, during which they have had to carry out enquiries and gain information from the smallest clues. These clues, as we have already said, are simple crystals of a rather common mineral, zircon. But these particular zircons are exceptional! Discovered at Jack Hills in Western Australia, they are the oldest terrestrial material

Fig. 3.1 The Earth photographed during the *Apollo 17* mission on 7 December 1972: a mosaic of brown continents and dark-blue oceans, partly hidden by a veil of white clouds. When did our planet first gain this face that is so familiar to us today? Geologists are conducting the enquiry…

3

yet identified. The oldest of them has been dated at 4.4 Ga, which means that it crystallized hardly later than 170 Ma after the beginning of our planet's formation.

In this chapter, we shall discover how the analysis of the isotopic composition of oxygen in the Jack Hills zircons (the $^{18}O/^{16}O$ ratio) has allowed us to affirm that liquid water already existed on Earth 4.4 Ga ago. We shall also see that these same crystals have allowed us to answer another fundamental question concerning the history of the Earth: when did the first continents form? We shall finish our sketch of the Hadean Earth between 4.4 and 4.0 Ga by attempting to decipher the physical and chemical conditions that prevailed at the surface: what was the composition of the atmosphere? When, where, and how did prebiotic chemistry, the set of reactions that leads to the synthesis of the more or less complex organic molecules necessary for the appearance of life, begin and then develop?

The Two Faces of the Earth's Crust

The Earth's structure is "onion-like" (◘ Fig. 3.2a): it consists of concentric layers which, going from the center outwards, successively are a metallic core (where the inner portion is solid, and the outer is liquid), then a solid, silicate mantle and, finally, a solid crust. The crust and the rigid outer portion of the mantle form the lithosphere (◘ Fig. 3.2b). The latter is divided into plates that move over the subjacent, more viscous, mantle, called the asthenosphere. The movement of plates is described and explained by a theory known as plate tectonics, whose driving force is mantle convection, a process that results from the internal production of heat (◘ Box 3.1). This internal structure is obviously not visible to an external observer who, like our astronaut, only sees oceans and continents. This is why, initially, we shall discuss the links that unite the oceans and the continents to the deep structure and the internal dynamics of the planet.

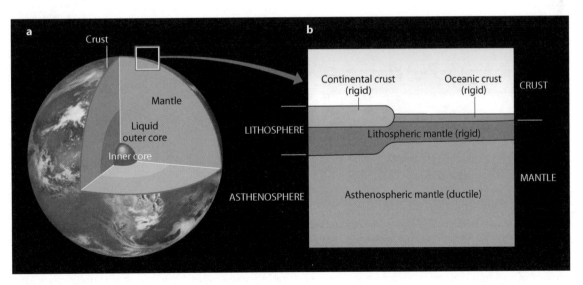

◘ **Fig. 3.2 The Earth's internal structure. a.** Schematic section of the Earth, showing its "onion-like" internal structure. Only the outer core is liquid, the inner core, the mantle, the oceanic crust, and the continental crust are solid. **b.** Schematic section of the oceanic crust, the continental crust, and the upper portion of the mantle. The crust (rigid) and the rigid portion of the mantle form the lithosphere, while the deeper portions of the mantle are ductile and form the asthenosphere.

Oceans and Oceanic Crust

An ocean consists of a layer of water, whose depth averages 5 km and that is located above an oceanic crust. With a typical thickness of 7 km, this crust mostly consists of mafic igneous rocks, basaltic in composition. It is generated at divergent lithospheric plate margins, that is in mid-ocean ridge systems (◘ Fig. 3.3). There, the basaltic lavas erupt at temperatures of about 1250 °C. After the crystallization of these lavas, the mantle convection (◘ Box 3.1), progressively moves the newly generated oceanic crust away from the ridge, where, due to its contact with the cold oceanic water, it slowly cools. As cooling progresses, the lava density (*d*) that is about 3.1 initially, slowly increases such that, inexorably, the oceanic crust becomes denser than the underlying mantle (*d* ~3.3). As a result, the dense oceanic crust sinks and returns ("subducts") into the mantle, thus giving rise to what geologists call a subduction zone.

Nowadays, this cycle takes, on average, 60 Ma, and no oceanic crust older than 180 Ma is known. Put simplistically, at present the oceanic crust's cycle does not significantly modify the chemical composition of the mantle. Indeed, most chemical elements that are extracted from the mantle to generate the oceanic crust in mid-ocean ridge systems, are reincorporated into this mantle, on average, 60 Ma later, in subduction zones.

The time when this cycle of oceanic crust started is still the subject of debate and animated controversy. It is highly probable that it could not start until the magma ocean had cooled, which happened 4.4 Ga ago (*see* ▶ Chapter 2). In addition, in 2007 scientists discovered, in Greenland, remnants of an oceanic crust dated at 3.86 Ga, preserved in a segment of continental crust, thus demonstrating that the cycle was already active at that date. Consequently, it started very early in the lifetime of our planet.

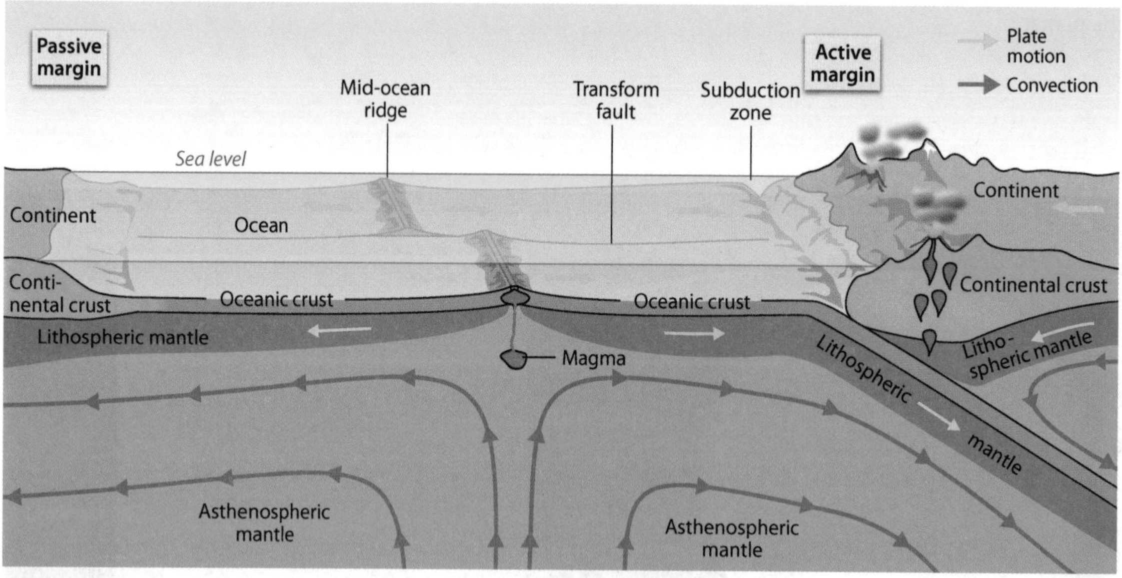

◘ **Fig. 3.3 Schematic block diagram showing the different types of relative movement between lithospheric plates.** The mid-ocean ridges are situated vertically above the ascending branches of the convection cells in the asthenospheric mantle; they correspond to divergent plate boundaries. There, the mantle peridotites undergo adiabatic decompression which causes them to melt. This process gives rise to the basalts of the mid-ocean ridges and of the oceanic crust. As it moves away from the ridge, the latter gradually cools, becoming progressively denser, such that it sinks and returns down into the mantle, forming what is known as a subduction zone (a convergent plate boundary). Such a zone corresponds to the descending branch of the convective cells in the asthenospheric mantle. In this scheme, subduction takes place beneath continental crust, but it may also occur beneath oceanic crust. In subduction zones, the magmatic activity results from dehydration of the subducted oceanic lithosphere. Finally, along transform faults, slip movements are observed. It should be noted that most terrestrial volcanic activity is localized along plate boundaries, whether they are convergent or divergent.

◘ Box 3.1   Thermal Convection

There are three ways by which thermal energy can be transported: conduction, convection, and radiation. Convection represents by far the most effective mechanism for dissipating the internal thermal energy of the Earth; it consists in a displacement of matter, which, in a way, transports its heat with it.

As we shall see, the process involved in convection is very simple. The only requirement is that the fluid under consideration has a temperature gradient: its base must be hotter than its top. Let us take the familiar example of a container filled with water and that is placed on a hotplate: it is thus heated from below. If we consider a small volume of water located at the bottom (◘ Fig. 3.4a), being heated, it will expand and will, as a result, experience a decrease in its density (mass per unit volume). Being less dense, it will then be "lighter" and will ascend towards the top of the container. As it ascends, it will exchange energy with the surrounding medium and will progressively cool. Conversely, the colder water at the top of the container will be "heavier" and will sink. This is how vertical motions are initiated (◘ Fig. 3.4b). The driving force is the Archimedes force and opposing forces are the fluid viscosity and thermal diffusivity. A dimensionless number, the thermal

Rayleigh number (Ra_T) may be used to quantify the strength of convection. It expresses the relationships between the Archimedean force (the numerator), and the viscosity and thermal diffusivity of the fluid (the denominator).

In any system, convection is initiated for $Ra_T > \sim 1700$, and will be turbulent when $Ra_T > 10^6$. The planet Earth is internally heated by radioactive sources, while in the same time; it is cooled at its outer surface; which results in a temperature gradient: the Earth's surface being colder than its interior. On geological timescales (10^7 to 10^8 years) the asthenospheric mantle rocks are plastic, they behave like fluids, and their Ra_T is about 10^8: the terrestrial mantle is thus the site of a vigorous and efficient convection. Although the coupling between the mantle's convection and the lithospheric plates is still not well-understood and is the subject of spirited debate and controversy amongst the scientific community, there is an undoubted link between the presence of the mid-ocean ridges and the rising branches of the convection cells, as well as between the subduction zones and the descending branches of the convection cells (◘ Fig. 3.3).

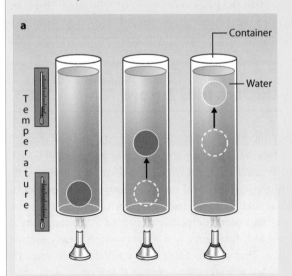

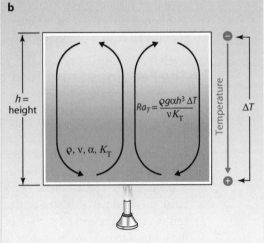

◘ **Fig. 3.4 Convection. a. Principles:** in a container, a fluid is heated from below. When heated, a small volume of this fluid, initially located at the bottom, will expand and consequently, its mass per volume unit will decrease. Being less dense, it will be "lighter" and will ascend in the container. During its rise it will exchange energy with the surrounding medium and will thus progressively cool. **b. Schematic representation of a convection** cell: the hot fluid (which is thus less dense) lying at the base of the container ascends, while the cold fluid (which is thus denser) sinks, and therefore initiates a vertical circulation of the fluids (a convection cell). ϱ = mass density, g = gravitational acceleration, α = Thermal expansion coefficient, h = height of the convection cell, ΔT = difference in temperature between the top and bottom of the convection cell, ν = dynamic viscosity, K_T = thermal diffusivity.

Continents and Continental Crust

A continent is a portion of continental crust that generally emerges above the sea level. Just like the oceanic crust, the continental crust "floats" on the asthenospheric mantle. Nowadays, this crust has an average thickness of 30 km, but it may reach 70 km beneath mountain ranges. Granitic to granodioritic in composition, the continental crust nowadays is generated at relatively low temperatures (around 800 °C) in subduction zones, as a result of melting of mantle rocks located above the subducted oceanic slab (*see* ▶ Chapter 6). Even after its complete cooling, the density of the continental crust (d = 2.76) is always less than that of the mantle. As a result, and unlike the oceanic crust, it cannot be returned and recycled into the mantle, in any significant amount. This quasi-indestructibility has two fundamental consequences. The first is that the continental crust records all the important events that have occurred on the surface of the planet, whence its essential role in recording the Earth's history. The second is that the chemical elements (Si, Al, K, Na, U, Th, etc.), that constitute the continental crust were extracted from the mantle in what may be termed a "definitive" manner, giving rise to sustainable modification in the composition of the upper regions of the latter.

This mechanism of continental crustal growth by subtracting elements that were initially in the mantle has been termed "juvenile". It results in an increase in the volume of the continental crust. The counterpart of this stability lies in the fact that the continental crust is affected by numerous mechanisms involving the surface (erosion, alteration, etc.) or deeper regions (metamorphism, anatexis = partial melting, etc.) which redistribute elements within it without any addition of new material: the volume of the crust remains unchanged. These mechanisms that operate in a closed system are generally termed "recycling". It is obvious that recycling of continental crust implies that the latter exists and that it is stable. This is the reason why, at any given location, juvenile mechanisms always precede recycling. As already mentioned earlier, one of the questions asked by geologists, is to know when the first stable continental crust appeared on Earth.

The Fabulous Story Told by the Jack Hills Zircons

Until very recently, the oldest terrestrial materials known were the Acasta gneisses, dated to 4.031 Ga, and the Nuvvuagittuq greenstone belt, with an age greater or equal to 4.0 Ga (*see* ▶ Chapter 6). In the absence of older rock materials, the most diverse theories concerning the first 500 million years in the Earth's history (the Hadean) had been proposed. In particular, it had often been stated that if rocks older than 4.031 Ga had never been found, that was because they had never existed or that they had been completely destroyed at the inhospitable and constantly reworked surface of the Earth (existence of a magma ocean, heavy meteoritic bombardment (*see* ▶ Chapter 5), etc.). That said, isotopic analysis undertaken since the 1980s on the oldest terrestrial rocks had led researchers to predict that a continental crust formed before 4.0 Ga (◘ Box 3.2). It was, nevertheless, always thought that this crust only had an ephemeral existence: it had undoubtedly been rapidly destroyed, either through return into the mantle, or by vaporization by the intense meteoritic bombardment. This view of the Earth's Hadean history was completely called into question in 2001, when zircon crystals much older than 4.0 Ga were discovered, analyzed and studied in detail.

The World's Oldest Continental Crust

Zircon is a mineral whose chemical formula is $ZrSiO_4$; it is widely used in geochronology, and this for three main reasons. First, it is abundant in felsic magmatic and metamorphic rocks.

Second, it is a very hard mineral (with a hardness of 7–7.5 on the Mohs scale, which runs from 1 to 10) and is thus very resistant to erosion and alteration. Finally, when it crystallizes, its crystal lattice incorporates uranium and thorium, whose radioactive isotopes (^{238}U, ^{235}U, and ^{232}Th) decay into lead over periods of time that are far greater than the lifetime of the Solar System. For this reason, old zircon crystals are easily datable by radiochronology using the U–Pb pair (◼ Box 2.2).

The zircon crystals that we are about to describe were discovered in Western Australia, at Jack Hills (◼ Fig. 3.5) and at Mount Narryer, in detrital sedimentary rocks (quartzites and conglomerates), deposited at about 3.0 Ga. These rocks were generated by the destruction of other, more ancient rocks, of which the zircons are the only relics that have resisted erosion and alteration.

A zircon crystal extracted from the Jack Hills quartzites has been analyzed by the U–Pb method, and the age of crystallization that has been thus calculated is 4.404 ± 0.008 Ga (◼ Fig. 3.7). This is the oldest terrestrial material known to date. This canonical age is not an exception and several other crystals, derived from the same region, have been systematically dated between 4.4 and 4.0 Ga (◼ Fig. 3.8). It is important to emphasize, however, that the age of these detrital zircon crystals is, in no case, that of the rock that contains them today. In other words, we still do not know any rock that older than 4.031 Ga.

Although the rock within which the Jack Hills zircons crystallized has completely disappeared, the latter have, at least partially, preserved its fingerprints and its geochemical signature. These are the clues that enable us to reconstruct, not just the zircon's parent rock, but also the conditions that prevailed when it has been generated.

As a general rule, zircon is one of the first minerals to crystallize, at high temperature, in acid magmas; it is, by contrast, very rare in basic magmas. Acid magmas are silicon-rich and magnesium-poor, whereas the opposite applies to basic magmas. In acid magmas, silicon

◼ **Fig. 3.5 The Jack Hills range in Western Australia.** The outcropping rocks are conglomerates and quartzites deposited about 3.0 Ga ago. They contain the oldest terrestrial materials so far reported: detrital zircon crystals whose ages go back as far as 4.4 Ga. (Photo: N. Eby, University of Massachusetts, U.S.A.)

3

▣ Box 3.2   Tracking Down the Most Ancient Continents: The Message from Neodymium

The zircons that have been recently discovered in Western Australia, at Jack Hills and Mount Narryer, and whose ages range from 4.0 to 4.4 Ga, demonstrate in an unquestionable manner the existence of a Hadean continental crust, that is to say older than 4.0 Ga. However, the presence of the latter had been predicted long before. Indeed, juvenile continental crust derives from the mantle through partial melting mechanisms. Consequently, as crust and mantle have totally different compositions, the extraction of the continental crust from the mantle has necessarily modified the composition of the latter (▣ Fig. 3.6a); the greater the volume of extracted continental crust, the more significant change in mantle composition. The theories regarding the terrestrial accretion process that have been described in Chapter 1 consider that the primitive mantle of our planet had the same composition as chondritic meteorites. That means, in particular, that the relative abundances of the isotopes of an element (here, for example, neodymium, Nd) were identical in the primitive mantle and in chondrites. Thus, it is possible to quantify the volume of the continental crust extracted from the mantle, thanks to a parameter that geochemists denote ε_{Nd}, whose value is proportional to the difference between the $^{143}Nd/^{144}Nd$ ratio measured in chondrites and the value measured from the rock being studied.

▣ Fig. 3.6b, is a diagram where ε_{Nd} is plotted *vs*. time. An $\varepsilon_{Nd} = 0$ corresponds to what is known as the CHondritic Uniform Reservoir (CHUR), in other words a mantle that has not been modified by the extraction of the continental crust.

An $\varepsilon_{Nd} > 0$ indicates that the analyzed rock has been generated from a mantle source previously depleted by the extraction of continental crust, whereas $\varepsilon_{Nd} < 0$ reflects a crustal source or an enriched mantle source. ▣ Fig. 3.6b shows that even the oldest rocks (4.0 Ga) have $\varepsilon_{Nd} > 0$; which proves that huge volumes of continental crust were extracted from the mantle before 4.0 Ga. This conclusion remains true whatever the nature of the rocks being considered: ultrabasic (komatiite), basic (basalt or gabbro), or acid (TTG or granite) (*see* the description of these rocks in ▶ Chapter 6, and the classification of rocks on p. 269).

From the early 1990s, simple calculations were able to show that about 10 per cent of the volume of the present-day continental crust must have been extracted from the mantle during the course of the Hadean (i.e., between 4.568 and 4.0 Ga). More recently, similar work, but this time based on hafnium (Hf) isotope behavior has led to perfectly identical conclusions.

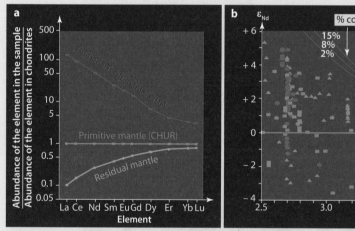

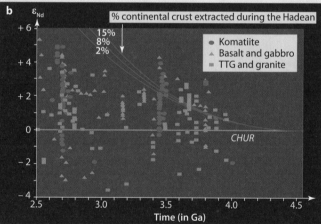

▣ **Fig. 3.6 Geochemical markers of the early production of the continental crust from the mantle. a. Rare Earth Element (REE) patterns,** showing how the extraction of a REE-rich continental crust (red curve) from a primitive mantle (green curve) impoverished the latter in this type of elements (residual mantle, blue curve). The degree of mantle impoverishment in a given REE is proportional to the content of the latter in the extracted continental crust. For instance, the residual mantle is strongly impoverished in lanthanum (La), because this element is very abundant in the continental crust. In contrast, the continental crust has relatively low lutetium (Lu) content, consequently, its differentiation had almost no effect on the Lu content of the residual mantle.. The

compositions of the continental crust and of the residual mantle are complementary. CHUR = CHondritic Uniform Reservoir, where the value 1 corresponds to the average composition of chondrites. **b.** ε_{Nd} *vs*. age diagram for Archaean terrestrial rocks. Even the oldest rocks known on Earth have $\varepsilon_{Nd} > 0$, characteristic of a mantle source whose composition has been modified by the extraction of continental crust; which proves that this extraction started before 4.0 Ga, that is, during the Hadean. The red curves represent the theoretical evolution of ε_{Nd} in the mantle for the extraction, during Hadean times, of 2, 8 and 15 volume per cent of continental crust. Komatiites and TTG are ultrabasic and acid rocks, respectively, that are typical of the Archaean Earth (*see* ▶ Chap. 6)

■ **Fig. 3.7 Age of zircon crystals from Jack Hills:** $^{206}Pb/^{238}U$ *vs.* $^{207}Pb/^{235}U$ diagram. The *Concordia* curve (■ Box 2.2) is drawn in blue and its age scale is given in Ga. Each red line corresponds to the isotopic composition of one of the analyzed zircon crystals. Among these, some are concordant, they lie on the *Concordia* curve for ages between 4.3 and 4.4 Ga. Others are discordant and all fall on the *Discordia* straight line, whose intersection with the Concordia curve also gives an age between 4.3 and 4.4 Ga. In other words, the oldest Jack Hills zircons crystallized scarcely any later than 170 million years after the start of the Earth's accretion.

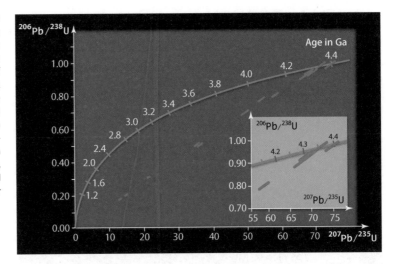

richness results in the crystallization of quartz (SiO_2). The acid magmatic rocks that crystallized at depth (i.e., plutonic rocks; *see* the classification of rocks on p. 269) are referred to as the granitoid group. This group includes the granites, the granodiorites, the tonalites, and the trondhjemites, rocks that all contain quartz, but which differ one from another in their relative abundances of plagioclase feldspar and alkali feldspar.

Taken alone, the presence of zircon crystals in the Jack Hills rocks would be enough to assume that granitoids already existed on Earth 4.4 Ga ago. This assumption is corroborated by the study of mineral inclusions within these crystals. Indeed, like all magmatic minerals, the zircons did, during the course of their growth, incorporate within their structure some of the minerals that contemporaneously crystallized in the same magma. The Jack Hills zircons contain inclusions of quartz, plagioclase feldspar, potassium feldspar, hornblende, biotite, chlorite, muscovite, rutile, apatite, pyrite, and monazite (■ Fig. 3.9). All these minerals are characteristic of granitoids. This irrefutably demonstrates that these zircons crystallized from an acid magma. This information is crucial, because granitoids are the main component of the continental crust...

The zircons from Jack Hills therefore allow us to conclude that a continental crust existed 4.4 Ga ago, less than 170 Ma after the start of the accretion of our planet. Another conclusion – a trivial one – is that this crust has not been completely destroyed during the Hadean, because relicts of it still remain today.

We shall now try to determine in more detail the characteristics of the acid magmas that gave rise to the Hadean continental crust. A magmatic liquid contains a mixture of practically all the chemical elements of the periodic table. Some are abundant (> 0.1 per cent): these are the major elements (Si, Al, Fe, Mg, Ca, Na, K, and Ti) whose polymerization gave rise to the minerals. The others are present in trace amounts (a few ppm), which often appear as trapped in mineral lattice defects. Among these, geochemists are particularly fond of the rare-earths or lanthanides. Indeed, these elements, whose atomic numbers range from 57 (lanthanum) to 71 (lutetium), are particularly sensitive to the magmatic mechanisms of crystallization and fusion (■ Box 6.1). In contrast, they are less affected by alteration and metamorphism of their host rock. They are, therefore, excellent tracers of the magma history.

So the Jack Hills zircons incorporated within their crystal lattices some of the rare earths available in the magma from which they were crystallized. Today, the analysis of their rare-earth contents shows that there are two families: those that have preserved all their magmatic characteristics (type 1) – which are those that interest us – and those that have lost them due to late hydrothermal processes (type 2). The Jack Hills zircons of type 1 have the same

3

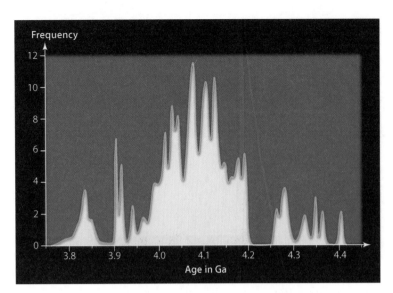

▣ Fig. 3.8 Frequency histogram for the age of 88 detrital zircon crystals from Western Australia (Jack Hills, Mount Narryer, and the Barlee Range). The zircon ages are more-or-less evenly distributed between 4.4 and 3.8 Ga. Most of them are older than 4.0 Ga, the age of the oldest known terrestrial rocks (the Acasta gneiss, *see* ▶ Chap. 6). These zircons bear witness to episodes of genesis or recycling of continental crust, which occurred regularly between 4.4 and 4.0 Ga, with perhaps a few peaks between 4.4 and 4.3 Ga, and between 4.2 and 4.0 Ga.

rare-earth contents as that of the zircons extracted from the Acasta gneisses. And we know that the latter crystallized from magmas called "TTG" magmas (tonalite, trondhjemite and granodiorite; *see* ▶ Chapter 6), typical of the Archaean continental crust, which was generated between 4.0 Ga and 2.5 Ga. In addition, based on the chemical laws that govern the distribution of rare earths between a zircon crystal and the magma within which it forms, it is possible to determine the latter's rare-earth content. The result of this calculation shows that the host magma for the Jack Hills zircons was extremely rich in light rare earths and, in contrast, strongly impoverished in heavy rare earths – a character symptomatic of TTG magmas that gave rise to the Archaean continental crust.

An exhaustive study carried out on numerous zircon crystals extracted from the detrital sediments from Jack Hills, Mount Narryer and the Barlee Range, all in Western Australia, shows that these zircon ages spread more or less regularly between 4.4 and 4.0 Ga (▣ Fig. 3.8). That means that the episode of crustal growth at 4.4 Ga was not unique but that, on the contrary, several phases of growth or recycling of continental crust took place between 4.4 and 4.0 Ga, with possibly several growth peaks at 4.4 Ga, 4.3 Ga, and between 4.2 and 4.0 Ga. This leads to one obvious conclusion: the genesis of continental crust at 4.4 Ga was not an isolated or accidental event; the crust differentiated over a long period of time between 4.4 and 4.0 Ga, even if this had occurred in an episodic manner rather than continuously. This conclusion has recently been confirmed by the fact that some samples of the Acasta gneisses in Canada that crystallized 4.03 Ga ago, contain inherited zircon cores that have been dated at 4.2 Ga, thus demonstrating that the Acasta rocks were generated by melting of older Hadean crustal materials.

Earth: A Blue Planet since its Earliest Infancy

During mineral crystallization within a magma, oxygen isotopes 16 and 18 (^{16}O and ^{18}O) segregate differently between the liquid and crystalline phases. If we know the law that governs the chemical fractionation (in other words, if we know how to determine the partition coefficient of the isotopes between two phases), it is then possible to calculate, from the $\delta^{18}O$ ratio (the $^{18}O/^{16}O$ ratio in the sample, normalized to a standard: sea water) measured in a mineral, the $\delta^{18}O$ of the parent magma. Zircon crystals from Jack Hills with ages ranging from 4.35 to 4.0 Ga have been analyzed with an ion microprobe. It turns out that the $\delta^{18}O$ ratios measured vary between 5.3‰ and 7.3‰, which enables an $\delta^{18}O$ of 6.8‰ to 9‰ to be

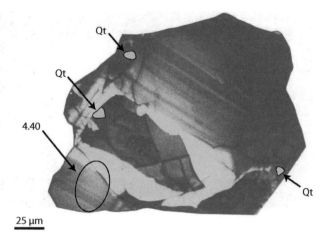

Fig. 3.9 The oldest zircon on Earth: a (false-color) cathodoluminescence image of the zircon crystal W74/2-36 hosted in the Jack Hills conglomerate (Western Australia). Ionic microprobe analyses of U and Pb isotopes were performed at the place marked with an ellipse, they enabled the computation of an age of 4.404 ± 0.008 Ga. This zircon crystal also contains small quartz (Qt) inclusions, which bears witness to its crystallization in a granitic magma and that consequently, also attests to the existence of a stable continental crust at 4.4 Ga. (Photo: John Valley, University of Wisconsin, Madison, U.S.A.).

calculated for the parent granitic magma. Typically, the average $\delta^{18}O$ of the mantle is 5.3 ± 0.3‰. Consequently, the high $\delta^{18}O$ values of the parental magmas of the Jack Hills zircons, preclude their derivation from any non-contaminated or altered mantle source. On the contrary, these values are characteristic of magmas that have reacted, either at or very close to the surface (i.e., subsurface), with liquid water. This conclusion is extremely important, because it implies that liquid water must have been stable and available in huge amounts at the surface of our planet by 4.35 Ga (and very probably by 4.4 Ga as well). In other words, about 170 Ma after the beginning of its accretion, the Earth already harbored one or more oceans of liquid water.

Several indirect arguments also indicate that a hydrosphere was present at the surface of our planet by 4.4 Ga. Indeed, zircon crystals extracted from continental rocks emplaced between 4.0 and 2.5 Ga (TTG, *see* ▶ Chapter 6) display $\delta^{18}O$ values that range between 5.0‰ and 7.5‰, values that are identical to those measured in the Jack Hills zircons. The study of the distribution of rare-earths in the zircons from Archaean TTG and the Jack Hills zircons have also led us to conclude on a probable analogy between the TTG and the parental magma of the Jack Hills zircons. TTG are generated through partial melting at high-pressure of a hydrated basalt, and, based on what we currently observe in the mid-ocean ridge systems, basalt hydration requires the presence of huge volumes of liquid water at the surface of the planet.

Another clue has allowed geologists to evaluate the abundance of water in the magma from which the Jack Hills zircons crystallized: their titanium (Ti) content. Indeed, it has been empirically shown that this latter is proportional to the zircon's crystallization temperature, which itself depends on the degree of hydration of the parental magma: the presence of water greatly decreasing the temperature of the magma's solidus (that is, the temperature at which a rock begins to melt or ceases to crystallize; ◘ Box 6.1). For example, at a depth of 15 km (~5 kbar), the solidus temperature for a granitic magma will be slightly greater than 1000 °C if it is anhydrous and just 660 °C if it is water-saturated. In 2005, a study of the Ti content of 54 zircon crystals from Jack Hills, dated between 4.35 and 4.0 Ga was carried out. It enabled determination that these minerals crystallized at 696 ± 33 °C (◘ Fig. 3.10a). Such low crystallization temperatures can only be achieved in water saturated magmas (◘ Fig. 3.10b). This conclusion is, moreover, perfectly consistent with the presence, in the same zircon crystals, of inclusions of hydrated minerals such as biotite or amphibole.

A sheaf of independent arguments thus converges on the same conclusion: large volumes of liquid water – oceans – formed very early on the surface of our planet, definitely by 4.35 Ga, but also very probably by 4.4 Ga.

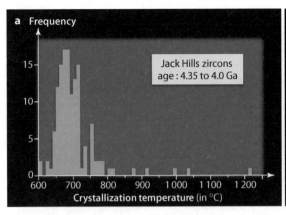

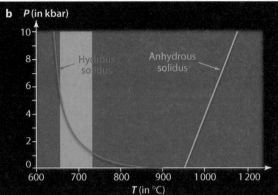

◘ **Fig. 3.10 Investigation of the crystallization temperature of the Jack Hills zircons. a.** Frequency histogram of the crystallization temperature of 54 zircon crystals from Jack Hills (the temperatures being estimated from their Ti content). **b.** A diagram of pressure *versus* temperature, showing curves for the beginning of melting (solidus) for hydrated granite (in red) and anhydrous granite (in blue). The grey zone on the diagram corresponds to the range of crystallization temperature of the Jack Hills zircons (average crystallization temperature: 696 ± 33 °C). Only magmas oversaturated in water (hydrous granites) can crystallize at such a low temperature. This oversaturation in water(in other words, the fact that the Jack Hills zircons crystallized from a magma rich in water), bears witness to the presence of large volumes of liquid water on the Hadean Earth.

A stable Continental Crust at 4.4 Ga!

The decryption of the history of the old zircons from Jack Hills is not confined solely to the detailed study of their chemical composition. It also involves a detailed analysis of the structure of these minerals.

A number of these zircons consist of an ancient core rimmed by a younger zone. For example, the core of a zircon crystal (poetically called "54–90" ...) gives an age of 4.263 ± 0.004 Ga, while its outer rim was formed 4.030 ± 0.006 Ga ago. Another crystal (54–66) has a large, rounded core, 4.195 ± 0.004 Ga old, surrounded by a zone that is dated to 4.158 ± 0.004 Ga. These figures bear witness to a complex history. They are common in more recent terranes, where they are interpreted as an evidence for at least two magmatic episodes, often linked to continental crust recycling. The ancient core is called "relict" or "inherited", because it is a remnant of an older rock that, except for the zircon crystals, has totally melted to form a new magma. And it was when the latter crystallized that the younger outer rim grew. Crustal recycling can proceed through direct melting of the ancient rock itself (direct recycling) or the latter could have been first altered and eroded, giving rise to sediments which, subsequently, almost completely melted thus generating a new magma – the crustal recycling was thus indirect, via a sedimentary stage. Whichever the recycling process, these figures imply reworking and internal rearrangement of the continental crust.

Other clues are provided by the mineral inclusions within the zircon crystals. Among these are muscovite and potassium feldspar. Muscovite is an aluminum-rich mineral ($KAl_2[Si_3AlO_{10}(OH)_2]$), which crystallizes only within peraluminous magmas. A peraluminous magma is such that its Al content is high enough for allowing muscovite growth after all the feldspars ($K[AlSi_3O_8]$) have completed their crystallization. However, magmas generated directly (by melting of mantle peridotite) or indirectly (by melting of basalt) from the terrestrial mantle are not peraluminous. These magmas (known as juvenile magmas) are metaluminous and, as a result, they cannot contain muscovite, unlike magmas generated through sediment melting, which are peraluminous. The presence of muscovite inclusions inside the Jack Hills zircons is therefore a strong evidence that during Hadean times, mechanisms of crustal recycling through a sedimentary cycle were already efficiently operating on Earth. The potassium feldspar inclusions reinforce this conclusion, indeed, juvenile magmas are rather poor in this mineral (tonalitic or granodioritic in composition), in contrast to those generated by crustal recycling (granitic in composition).

All these clues that geologists have unearthed therefore show that the Hadean continental crust had not only been relatively regularly (or even continuously) extracted from the mantle since 4.4 Ga, but also that this crust was stable. The recycling of a juvenile crust requires, in fact, that it had remained stable over a sufficiently long period thus giving time for the mechanisms of erosion and alteration to operate. The latter could then have led to the formation of detrital sediments, whose melting generated the continental crust of which the Jack Hills crystals are the ultimate remnants.

In conclusion it appears that, contrarily to what was believed, the Hadean continental crust was not ephemeral; it was not destroyed immediately after its genesis, but it was sufficiently stable to undergo these processes of maturation and recycling. Finally we may note that all these mechanisms – the genesis of juvenile hydrated magmas and their recycling – imply the availability of huge volumes of liquid water throughout the Hadean, which requires the oceans to have been permanent, and to have also remained stable throughout this whole period.

The Atmosphere Between 4.4 and 4.0 Ga: An Outline

Water, earth (the continents), fire (volcanism)... In our quest to reconstruct the face of the Earth during the Hadean, between 4.4 and 4.0 Ga, one element is missing: air, i.e., the atmosphere. Apart from the Jack Hills zircons, there are, as we already wrote earlier, no vestiges of the Hadean Earth. Although these precious minerals have revealed so much to geologists, they do not contain any clues about the composition of the primitive Earth's atmosphere. Despite this, might it be possible to outline some of its characteristics?

To start with, let us recall that nowadays the oceans and the atmosphere only represent a tiny fraction of the mass of our planet (0.022 and 0.0001 per cent, respectively). This means that a significant portion of the oceanic and, above all, atmospheric components is trapped into both the crust and the mantle – by way of example, today, all the ocean water represents only two thirds of the Earth's water. Determining the composition of the Earth's atmosphere at any given period in its history therefore assumes being able to draw up an overall balance-sheet, taking account, not only of the chemical species present at the surface, but also those that are buried in the interior. It must also be remembered that the current composition of the atmosphere is the result of intense interaction with the biosphere, that is with living beings that populate the planet for several billion years. Could the composition of the Earth's primitive atmosphere be deduced from that of the present-day atmosphere, after removing the contribution from living beings? No, because in attempting to do so, one would neglect the influence of processes that are currently masked by lifeforms, and particularly the role played by prebiotic chemistry (*see later*). Moreover, no planet that is currently accessible for analysis provides us with an analogous model for the primitive Earth. In addition, any reconstruction from details currently observed elsewhere in the universe on a body provided with an atmosphere – for example, Titan, one of Saturn's satellites – would appear to be a difficult task. A model based on a certain number of assumptions thus remains, for the time being, the only method that can be envisaged.

We have seen in the preceding chapter that the degassing of the magma ocean and the chemical species contributed by external bodies (comets, asteroids, and meteorites) enabled the formation of an initial atmosphere lying above a superheated Earth, and consisting of all of the volatile chemical species. Immediately after the condensation of the oceans (around 4.4 Ga if we believe the Jack Hills zircons), the models predict that the terrestrial atmosphere mainly consisted of carbon dioxide (CO_2), together with a still significant amount of water vapor (H_2O), as well as with other components in lesser abundance – but not of lesser impor-

3

tance – such as molecular hydrogen (H_2), molecular nitrogen (N_2), and methane (CH_4). How did this atmospheric composition evolve between 4.4 and 4.0 Ga?

Carbon Dioxide Takes Control of the Climate

After the condensation of the oceans, CO_2 became the principal component of the atmosphere, with a partial pressure that, depending on the estimations, ranges between 40 and 210 bars. This produced a significant greenhouse effect which resulted in a surface temperature of 200 to 250 °C. Under these conditions, siliceous rocks (basalts of the oceanic crust or granitoids of the continental crust) underwent leaching (at the bottom of the oceans and, where appropriate, through erosion of any emergent surface), resulting in the release of silica and bicarbonate in a reaction as follows:

$$MgSiO_3 \; + \; 2(CO_2) \; + \; H_2O \; \rightarrow \; Mg^{2+} \; + \; 2(HCO_3^{-}) \; + \; SiO_2$$

magnesium   liquid water   bicarbonate   silica
silicate

The bicarbonate HCO_3^- subsequently precipitated in the form of insoluble calcium carbonate $CaCO_3$:

$$2(HCO_3^{-}) \; + \; Ca^{2+} \; \rightarrow \; CaCO_3 \; + \; CO_2 \; + \; H_2O$$

bicarbonate   calcium carbonate

The overall outcome of these two reactions is that a molecule of atmospheric CO_2 (or of CO_2 dissolved in the ocean) was trapped in a mineral (carbonate). To prime this CO_2 "pump", all that was required was that liquid water should be available at the planet's surface. However, a very significant reduction of the atmospheric partial pressure of CO_2, required a sufficient cooling of the oceanic crust such that subduction was able to start, allowing the recycling of significant volumes of carbonate sedimentary rocks into the mantle, and consequently the long-term sequestration of the carbon that they contained. The Jack Hills zircons show us that fragments of continental crust, undoubtedly generated in a subduction geodynamic environment, existed since 4.4 Ga, and in huge amounts from 4.2–4.3 Ga onwards. From then on, the atmospheric CO_2 partial pressure decreased, leading to a diminution of the greenhouse effect, such that the temperature at the surface of the Earth became compatible with life. Depending on the scenario adopted, this process took between 10 and more than 100 Ma. However, once the subduction process had become active, trapping of CO_2 in carbonate sediments could have become so efficient that the average surface temperature of the Earth was capable of dropping below the fateful threshold of 0 °C, resulting in a global glaciation (the "Snowball Earth"). Given the low luminosity of the young Sun (about 75 per cent of its current value, ■ Box 6.2), only an efficient greenhouse effect could then have prevented such a global glaciation. The most recent atmospheric models predict that the CO_2 partial pressure, below which such an event could have occurred on Earth, would be between 0.2 and 1 bar. It must be noted that at present the total atmospheric pressure is 1 bar, of which only 3.5 $\times 10^{-4}$ bar is CO_2. Nothing can rule out the fact that such glacial episodes may indeed have taken place.

It should, however, be noted that on a totally frozen Earth, trapping of CO_2 by alteration of silicates becomes ineffective, because this gas continues to be emitted continuously through volcanic activity. This then induces an increase in CO_2 atmospheric content and a correlated increase of the greenhouse effect, which enables the Earth to rapidly escape – on a geological time-scale – from the global glaciation. The CO_2 pump responsible for the alteration of silicates could then restart, eventually leading to a new "Snowball Earth" episode, or in a less extreme fashion, to variation around a state where the two constraints remained in balance (◘ Fig. 3.11).

We can therefore see how climate regulation through carbon dioxide and episodes of glaciation – partial or total – could come to arise on the Hadean Earth. Under such conditions, it is hard to see how temperate climatic conditions could have been maintained on the Earth's

◘ **Fig. 3.11 Regulation of the surface temperature of the Earth through weathering of silicate rocks.** From 4.4 Ga, as soon as liquid water was available at the surface of our planet, leaching of silicate rocks absorbed atmospheric CO_2 (**a**). The decrease in the partial pressure of atmospheric CO_2 caused a reduction in the greenhouse effect and consequently, a reduction in the surface temperature, which could have resulted in a global glaciation. However, on a completely frozen Earth, trapping of CO_2 would be ineffective, while the gas was continuously released by volcanoes (**b**). Then the CO_2 partial pressure would have risen, together with the greenhouse effect and the surface temperature (**c**). Models based on this system of exchange between silicates and carbonates predict that, at the end of the Hadean, about 3.8 Ga ago, the Earth was extremely cold worldwide... which contradicts the analysis of oxygen and silicon isotopes in rocks of that period, which points to an ocean temperature of about 70 °C.

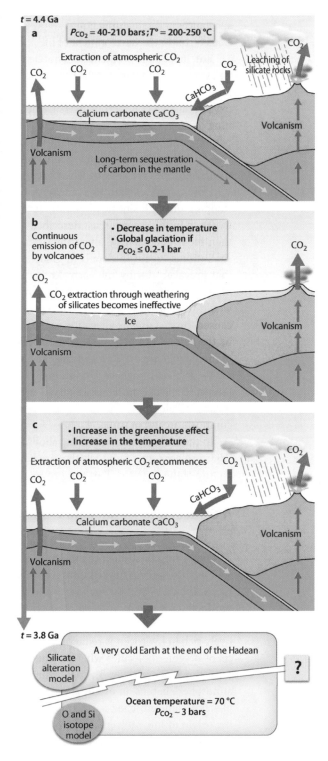

surface. In this scenario, the temperatures hospitable to life could only have arisen very temporarily and episodically, either during the initial cooling corresponding to the period when the CO_2 started to be trapped though silicate alteration, or in the wake of reheating possibly induced by large-size meteoritic impacts. However, a frozen Earth is not an insurmountable obstacle to the emergence of life. Indeed, liquid water may circulate beneath an ice-cap, and this latter is even liable to melt at the surface when close to regions with a high geothermal flux. In addition, the possible presence, within the ice, of pockets of water laden with organic substances and remaining liquid at low temperature could equally have become a very propitious factor in the development of life.

But, in fact, the problem does not lie there. Rather, it arises from the fact that this vision of an extremely cold Earth at the end of the Hadean is in complete contradiction with the values found for palaeotemperatures of surface ocean water. These latter have been determined from oxygen- and silicon-isotope measurements in sedimentary rocks dated from the very beginning of the Archaean, i.e., 3.8 Ga: these temperatures are, in fact, of about 70 °C, which, depending of the models used, corresponds to a greenhouse effect resulting of an atmospheric CO_2 partial pressure of about 3 bars. We are forced, therefore, to recognize that several elements and clues are missing when studying such a distant period in the Earth's history...

How can we explain this hiatus? A first hypothesis is that we can imagine that the CO_2 trapping through silicate alteration at the bottom of the oceans could have been less effective than in the scenario just described above. A strong cooling leading to glaciation would then require large-scale alteration of emerged continents. A second hypothesis, which we shall discuss again in details in Chap. 6, consists in assuming that the high temperatures that prevailed during the early Archaean resulted from the metabolism of methanogenic organisms capable of enriching the atmosphere in powerful greenhouse gas. As such, life would have already been present by 3.8 Ga and its action on the environment would have caused the first drastic alteration in the atmosphere composition. Another hypothesis might be the abiotic production of atmospheric methane. Nevertheless, both the early existence of methanogenic *Archaea* and the effects of high atmospheric methane content (■ Box 6.2) are the subject of debate. Undoubtedly, years of research will be required before being able to decide, all the more, as we shall see in Chap. 6, the situation in the Archaean is by no means simpler, because once again we do not have any direct record of the atmospheric composition at that time.

In the end, all that we can say about the composition of the atmosphere between 4.4 and 2.5 Ga solely rests on models that are liable to significantly evolve and change as our overall picture of the primitive Earth improves.

A Critical and Debatable Question: The Hydrogen Content

Although from a strictly quantitative aspect, hydrogen is a minor component of the atmosphere, this element nevertheless remains an essential player in the evolution of the Hadean Earth. Indeed – and we shall return to this point in some detail later – the efficiency in the production of organic substances in the atmosphere is strongly dependent on its oxidizing or reducing characteristics or, in other words, its redox potential. An atmosphere that is relatively rich in hydrogen is reducing and forms an environment that is favorable to the synthesis of organic molecules whereas, conversely, a hydrogen-poor atmosphere is neutral from the redox point of view. The determination of the hydrogen content of the primitive Earth's atmosphere is thus an essential parameter for anyone who is trying to understand how our planet became enriched in organic molecules: a question that is not unconnected with that of the origin of life. Molecular hydrogen (H_2) is released into the atmosphere through the serpentinization of the basalts of the oceanic crust, that is, through the alteration of their miner-

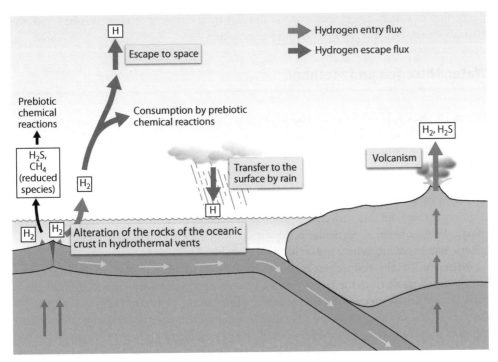

◘ **Fig. 3.12 Hydrogen flux in the atmosphere of the Hadean Earth.** The composition of the atmosphere evolved until a stable state, where the hydrogen flux (and that of other reducing gases) entering and leaving the atmosphere were in balance. The significance of hydrogen escape to space has recently been revised downwards. As a result the atmosphere of the Hadean Earth was perhaps richer in hydrogen than thought, thus being more reducing.

als by high-temperature water in submarine hydrothermal systems, The reaction is described on p. 53 and involves the transformation of an iron silicate, fayalite ($Fe_2[SiO_4]$), into silica (SiO_2), magnetite (Fe_3O_4) and molecular hydrogen (H_2). These contributions are difficult to quantify, but they were undoubtedly significant during the Hadean, in an environment where there was intense volcanic activity and thus efficient hydrothermal processes on the ocean floor (◘ Fig. 3.12).

Even before its escape, hydrogen could have contributed to atmospheric chemistry by participating to the synthesis of reduced species (H_2S, CH_4, etc.) which would have been transported to the outermost layers of the planet. The photolysis of H_2 molecules, converted them into atomic hydrogen that would reach interplanetary space through simple gravitational escape. For a long time, it was assumed that this mechanism had played a major role in the evolution of the Earth's Hadean atmosphere. Nowadays, this idea is challenged.

Indeed, escape is more efficient the higher the temperature. For instance, on Earth today, because of the presence of molecular oxygen (O_2), which absorbs solar ultraviolet radiation, the temperature of the atmosphere's outer layer (exosphere) is very high (about 1000 K) and the atoms of hydrogen that form there by photolysis of water primarily escape into interplanetary space. But if we consider that during the Hadean aeon, the upper atmosphere had a low oxygen content (limiting the absorption of ultraviolet radiation) and a high CO_2 content (able to disperse the radiative energy received from the Sun), this would imply that the exosphere was probably far less hot than nowadays. The escape mechanism would then have been very much less efficient. Under such conditions, the atmospheric hydrogen content would have evolved towards a stable state where the entering and outgoing energy fluxes were balanced. In drawing up this balance sheet, it is worth noting that prebiotic chemistry initiated in the atmosphere (*see later*) could have consumed a significant amount of hydrogen, as well as of other reducing chemical species required for the formation of organic molecules.

Be that as it may, we cannot exclude the fact that, contrary to what was believed, the Hadean atmosphere was not neutral, but was reducing.

Water, Nitrogen and Methane

Water. The atmosphere's water content depends on the temperature that prevails at the Earth's surface. After condensation of the oceans, the huge amounts of atmospheric CO_2 resulted in an intense greenhouse effect: the atmosphere was hot. As a result the partial pressure of water was raised (up to 40 bars), which reinforced the greenhouse effect. Subsequently, until the end of the Hadean, the global cooling caused by the alteration of silicates and the trapping of CO_2 in carbonates, was accompanied by a decrease in the atmospheric water content. In addition, throughout this period, the presence of water in the atmosphere resulted in meteorological phenomena, in particular precipitation, which then, as today, led to variable atmospheric water content, depending on location and altitude.

Nitrogen. Usually, researchers consider that the partial pressure of N_2 in the atmosphere has remained stable from the Hadean to the present day, where it amounts to 0.8 bar, due to early mantle degassing and other early contributions to the primitive Earth (*see* ▶ Chap. 2). But currently a significant amount of nitrogen remains trapped in the continental crust after burial of organic material originating from living beings. Nothing therefore excludes the fact that the initial N_2 content of the atmosphere may have been twice or three times as high or, on the other hand, that the atmosphere may have been impoverished by the absence of biochemical processes converting organic nitrogen into N_2.

Methane. Like other hydrocarbons and even some lipids, methane could have been produced through the alteration of magmatic rocks in the Hadean oceanic crust by CO_2-rich

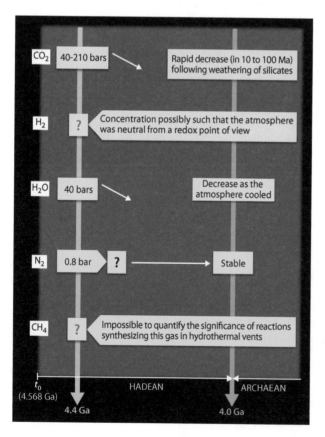

◘ Fig. 3.13 The evolution of Earth's atmosphere between 4.4 and 4.0 Ga. All our understanding of the composition of the Hadean atmosphere rests solely on models that are susceptible to change substantially as our knowledge of the primitive Earth improves. The question marks on this diagram should therefore come as no surprise.

hydrothermal fluids (via Fischer-Tropsch reaction, *see later*). However, it is almost impossible to determine to what extent these reactions have been actually active and efficient between 4.4 and 4.0 Ga. This point is, however, a key point. Indeed, on the one hand, methane is a very powerful greenhouse gas (far more powerful than CO_2) which could have contributed to prevent episodes of global glaciation, and, on the other hand, the hydrocarbons or lipids that were formed could (as we shall discuss later) have participated in the emergence or in the development of life.

3

From the Atmosphere to the Bottom of the Oceans: Was the Earth Rich in Organic Matter?

Oceans, continents, atmosphere, and climate: we now have a reasonably complete picture – even if there are still numerous dark areas – of the Hadean Earth. But there remains one essential critical question: was our planet already capable of sheltering life? Before trying to reply to this question, let us ask whether, in the scenario that we have just described, the chemistry of life and, in particular, the reactions that allow the synthesis of organic molecules that form the basis of living beings, could have taken place and, if so, where, and for the production of what kinds of molecules.

Abiotic Chemistry or Prebiotic Chemistry?

During the period that preceded the onset of biological evolution, simple organic molecules as well as more elaborate structures arose through a group of processes that are known as *prebiotic chemistry*. Behind this label are hidden the chemical pathways that could have led to the origin of life on Earth ... or elsewhere in the Universe. Organic matter and life ... to many of us, the link between the two seems to imply that they are inseparable. However, the formation of organic molecules is not necessarily synonymous with the presence of life. Organic matter is permanently synthesized in the interstellar medium, which is difficult to imagine as favorable to the emergence of life. Nevertheless, and perhaps because of the survival of vitalist ideas (■ Box 3.3), many people still consider the mere presence of organic matter as a clue of the presence of life. Imagine, for example, the consequences that would arise from the discovery of amino acids during a mission to a planetary body of the Solar-System. It is a good bet that this would rapidly be interpreted as a clue for either current or past life, whereas we already know that these compounds have been present for more than four billion years in some meteorites, objects, which as far as we know, never hosted life. This is why we make a distinction between prebiotic chemistry and *abiotic chemistry*, the latter covering the synthesis of organic matter that was produced in the past, and which is still produced today, under conditions that are not specially propitious for the emergence of life.

However, the matter is not that simple. First, because frequently the formation of organic matter on some bodies of the Solar-System or in the interstellar medium, is referred as "prebiotic chemistry", while these places are most unlikely to host life. Second, because even the definition of "prebiotic" characteristics remains very subjective, because it is linked to the conditions under which the only example of life that we know – those on planet Earth – have developed.

The organic chemistry that will be described here is prebiotic, in the sense that it takes place in an environment where, at an indeterminate time and through mechanisms that are just as uncertain (*see* ▶ Chap. 4) some of the reactions that we are about to describe could not have occurred, in a context that was not linked to the development of living systems. In other

□ Box 3.3   A Glimpse of the Evolution of Ideas About the Formation of Organic Matter

From antiquity until the end of the 18th century, vitalist ideas dominated: organic and mineral matter were fundamentally different, the former being always associated with a "vital force". Things were to change in the 19th century. In 1828, the German chemist Friedrich Wöhler showed that urea (organic molecule) can be easily synthesized from ammonium cyanate (mineral). Subsequently, during the course of the 19th and then the 20th century, thanks to the development of organic chemistry, it was realized that the most complex biomolecules that are found in living beings were not inaccessible to laboratory synthesis techniques. In the last decades of the 19th century, the vitalist concepts were thus severely discredited. Nevertheless, it took nearly a century for the abundance of abiotic organic matter in the universe to be recognized and accepted. The presence of organic matter in certain meteorites, an assumption held by some 19th-century chemists (including Friedrich Wöhler himself, and also many others, such as the Swede Jöns Jacob Berzelius and the Frenchman Marcellin Berthelot, who examined the meteorites that fell in 1806 near Alès and in 1864 at Orgueil), was still seriously doubted in the first half of the 20th century. In particular, it took the analyses carried out after the fall of the Murchison meteorite in 1969, for there to be no possible argument about the matter.

Even today, the mineral world and the organic world appear to be utterly distinct – as witness the fact that organic chemistry and inorganic chemistry are still disciplines that are taught within separate, watertight compartments. This is undoubtedly because, in our terrestrial environment, the omnipresence of lifeforms makes them appear as the almost unique source of organic molecules (whether past or present). However, the exploration of several objects in the Solar System (such as comets and Titan), as well as the progress achieved in identifying organic molecules within the interstellar medium, lead to the conclusion that the abundance of abiotic organic matter in the universe is a certainty. On the present-day Earth, the processes that produce such abiotic organic matter are difficult to detect, because of the overwhelming influence of lifeforms and because of the presence of oxygen (which, it is worth remembering, is not an element that favors the synthesis and stability of organic matter). However, it is possible that such processes do play a minor role in the synthesis of hydrocarbons through abiotic geochemical pathways.

Let us return to the end of the 19th century. The exclusion of "spontaneous generation" by Louis Pasteur's experiments accompanied the progressive disbelief in theories that assigned a "special character" to organic molecules. At the same time, these experiments ran counter to the idea that organic molecules formed in an abiotic manner could be the basis for the birth of life, because such a scenario undoubtedly formed *an exception to the impossibility of spontaneous generation*. These changes in scientific thought did, however, occur at the same time as the development of Darwinian theory which predicted the existence of a common ancestor to all living beings and thus raised the question of the origin of life. All these contradictions resulted at the beginning of the 20th century, that this question of the origin of life was, so to speak, taboo. At the beginning

words, certain reactions that we will discuss here also belong to the domain of abiotic chemistry, just as much in the interstellar medium as on Earth. So, this is almost a methodological or conventional necessity: because life arose on Earth, we shall term prebiotic chemistry, the abiotic organic chemistry that took place on Earth before the emergence of life.

Organic Matter Likes Reducing Environments

Not all environments are favorable for the abiotic formation of organic molecules. And if there is one that is not favorable, this is undoubtedly, and paradoxically, that of the present-day Earth. As inhabitants of this planet, we empirically know that, on the present day Earth, organic molecules are unstable: a number of them oxidize, degrade, and may even burn in the presence of air. In fact, the formation of organic matter is strongly disfavored in an oxidizing atmosphere, oxygen-rich like that of the Earth nowadays. So, in such an environment, these are mainly the living systems that produce organic matter, because they have acquired, through biological evolution and at the price of a significant expenditure of energy, the capacity to carry out those reactions that are disadvantaged from a thermodynamical point of view. This is probably our experience of terrestrial life that leads us, more or less consciously,

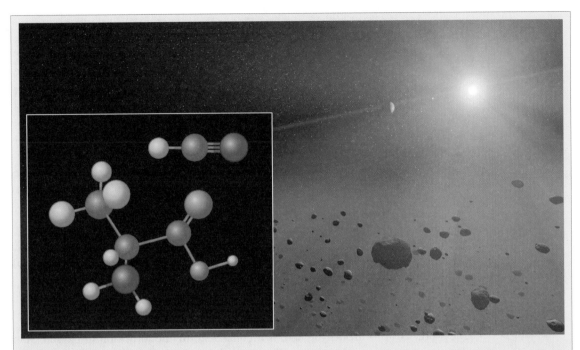

◘ Fig. 3.14 Organic molecules in space. Two organic molecules that are common in meteorites and in the interstellar medium, hydrogen cyanide, HCN, and an amino acid, alanine, are shown superimposed on an artist's view of planetary objects, accreting within the protoplanetary disk around a young star (*see* ▸ Chap. 1). Indeed, contrary to a vitalist concept that is often unconsciously present in our minds, organic matter is not inseparable from life, and the Universe is rich with organic matter, formed under abiotic conditions.

of the 1930s, Alexander I. Oparin and J. B. S. Haldane opened new avenues for research by formulating the theory – which, as will be understood, was daring at the time – that, in a reducing atmosphere, biomolecules formed by abiotic means could have supplied the first living organisms with organic matter and energy. This intuition has been confirmed only in 1953, after Miller's experiment revealed the formation of amino acids from the action of electrical discharges in a reducing atmosphere (◘ Fig. 3.17).

to associate life with organic matter. But the latter may be formed in a manner that is perfectly independent of living systems in reducing environments, where oxygen is absent.

Let us examine in greater detail the link between the formation of organic molecules and the reducing character of the environment. So doing, we will then come to understand why the question of the redox characteristics of the Earth's atmosphere during the Hadean is so significant for prebiotic chemistry.

The organic molecules correspond to a combination of atoms of carbon, hydrogen, nitrogen and oxygen [C,H,N,O], in a systematically reduced form (richer in hydrogen) relative to a mixture of CO_2, N_2 and H_2O. Using these three gases as the ultimate sources of carbon, nitrogen, hydrogen and oxygen for the formation of organic molecules requires not only a lot of energy but also a reducing agent, in accordance with the general reduction reaction (a):

3

$$CO_2 \quad + \quad N_2 \quad + \quad H_2O \quad \xrightarrow{\text{[Red]}}_{\text{[Ox]}} \quad [C, H, N, O] \qquad \text{(a)}$$

This reduction is most frequently energetically disfavored. This is quantified by using the standard Gibbs free energy of reaction, $\Delta_r G°$, which expresses the possibility that a reaction may occur spontaneously: the more $\Delta_r G^0$ is negative, the greater the likelihood of the reaction occurring; the more $\Delta_r G^0$ is positive, the more the reaction is unfavored. For example, the formation, in a gaseous phase, of formaldehyde CH_2O (the simplest of carbohydrates: it is related to sugars) from H_2O and CO_2 (b) is highly unlikely:

$$CO_2 \text{ (g)} \quad \xrightarrow[\text{1/2 } O_2 \text{ (g)}]{H_2O \text{ (g)}} \quad \underset{\text{formaldehyde}}{CH_2O \text{ (g)}} \qquad \Delta_r G^0 \text{ (b)} = 521 \text{ kJ/mol} \qquad \text{(b)}$$

The synthesis of formaldehyde here is accompanied by the formation of oxygen from the decomposition of water. Replacing water by hydrogen as the hydrogen source – in other words, carrying out this synthesis in a more reducing environment – facilitates the reaction. This latter, although it remains unfavorable, benefits from the energy liberated by the formation of water, the amount of which, being very high, contributes to the reaction (c). On the contrary, in reaction (b), the conversion of the atom of oxygen from the water into dioxygen (O_2) acts against the synthesis of formaldehyde due to its energetic cost.

$$CO_2 \text{ (g)} \quad \xrightarrow[\text{} H_2O \text{ (g)}]{2 H_2 \text{ (g)}} \quad \underset{\text{formaldehyde}}{CH_2O \text{ (g)}} \qquad \Delta_r G^0 \text{ (c)} = 63 \text{ kJ/mol} \qquad \text{(c)}$$

It is the same for the synthesis of glucose, which is energetically disfavored when it requires the conversion of water into oxygen (d), but the formation of which is, in theory, near thermodynamical equilibrium in the presence of a high hydrogen concentration (e). (In practice, this reaction is not, however, necessarily attainable through simple biochemical pathways.)

$$CO_2 \text{ (g)} \quad \xrightarrow[\text{} O_2 \text{ (g)}]{H_2O \text{ (l)}} \quad \text{1/6 glucose (aq)} \qquad \Delta_r G^0 \text{ (d)} = 478 \text{ kJ/mol} \qquad \text{(d)}$$

$$CO_2 \text{ (g)} \quad \xrightarrow[\text{} H_2O \text{ (l)}]{2 H_2 \text{ (g)}} \quad \text{1/6 glucose (aq)} \qquad \Delta_r G^0 \text{ (e)} = 4 \text{ kJ/mol} \qquad \text{(e)}$$

Of course, nowadays, the synthesis of glucose from CO_2 and H_2O (reaction d) is one of the most commonplace. But it is the prerogative of the living beings that possess a sophisticated photosynthesis equipment, which allows them to use the energy provided by light as the driving force and water as the source of hydrogen and consequently as a reducing agent. This optimum system is the result of a long biological evolution. Prebiotic chemistry was utterly incapable of attaining such a feat in the environment of the Hadean Earth. On the other hand, our present view of the Hadean environment is that of a reducing medium, which could have been far more efficient in synthesizing organic molecules.

Heat or Radiation: Activators for the Synthesis of Organic Matter

Quite apart from the fact that it is dependent on the reducing nature of the environment (the thermodynamic factor), the synthesis of organic molecules from sources of carbon such as CO_2 or CH_4; of hydrogen such as H_2 or CH_4; or of nitrogen such as N_2 or NH_3 is not spontaneous. These precursors are themselves poorly reactive. Synthesis can, therefore, proceed only through some activation, either thermal in nature (through lightning during thunderstorms, through the impact of meteorites with the Earth, at hydrothermal vents, etc.) or photochemical in nature (provided that the photons should carry sufficient energy to make up for the bonds to be broken). Under such conditions, transient, highly reactive species are formed, and random recombinations yield small organic molecules within the activated mixture.

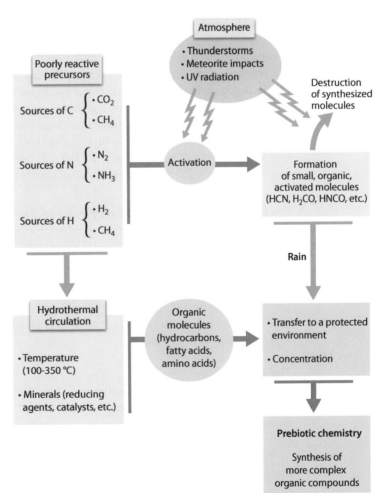

◨ **Fig. 3.15 Some of the conditions governing the prebiotic synthesis of organic matter.** The precursors of organic matter present in a reducing environment are poorly reactive, whence the necessity for their activation. This process of activation may destroy the synthesized molecules, whence the requirement to transfer the reaction products into a protected environment, where a truly prebiotic chemistry may possibly take place.

However, these organic molecules are also sensitive to activation processes: if UV radiation is able to activating mixtures of simple gases, it also has a destructive effect on organic molecules. Similarly, in hydrothermal systems, these same molecules are easily destroyed by high temperature water (often 350 °C, or even more). To be preserved, these freshly synthesized molecules must therefore be isolated from the activation system; for instance, by condensation and rain in the atmosphere, or else by circulation in hydrothermal systems. From this point of view, the different methods of activation are not equivalent (▣ Fig. 3.15).

Both the amounts and forms of energy brought into play and the time-scales of activation processes (fractions of a nanosecond for photochemistry, fractions of a second for electrical discharges, and much greater durations for hydrothermal circulations) will affect the nature of the molecules that are formed. In particular, the latter themselves will remain in an activated state if the duration of the activation is short when compared with the rate of the subsequent deactivation reactions, or of the return to an equilibrium state, as well as of the transfer rate. From this point of view, we could schematically say, that molecules formed in the atmosphere are still activated when they reach the ocean, and that they are therefore still reactive. For instance, they could directly take part in a proto-metabolism. On the other hand, molecules formed in hydrothermal systems are close to thermodynamic equilibrium and are no longer reactive, unless there is an external source of energy (for example the form that could be gained from reducing minerals originating from the mantle, *see* ▶ Chap. 6.)

In addition, any analysis of the productive or destructive nature of any activation process must take into account the efficiency of these transfers towards a protected environment, where a chemistry that could truly be described as prebiotic could develop. Such an environment would be likely to offer conditions suitable for the synthesis of the building blocks of life (biomolecules) and even for more complex assemblies. It may be noted that, in addition to this transfer, an effective prebiotic chemistry would call for a process of concentration (dilution in the ocean is of such a nature that it would make any subsequent constructive chemistry impossible, simply because any significant encounter between organic molecules would become highly unlikely, and that these latter would be fated only to degrade). It would at least require a sequestration of the molecules that had been formed within a limited space, as is the case with adsorption on the surface of a mineral (thus enabling interactions between them, and as a result, the formation of molecules of a more significant size).

Prebiotic Chemistry on the Primitive Earth: Where and to What Ends?

Let us now ask the following question: What were the environments on Earth between 4.4 and 4.0 Ga that possessed the three elements favorable for the synthesis of organic matter? Those three elements being: a reducing nature, the potential for activation of chemical precursors, and the possibility of transferring the products into a protected environment. There is no reason to believe that a unique process was involved, and, on the contrary, several paths were probably active in parallel.

The Atmosphere

The setting where prebiotic chemistry could have started is the atmosphere, where gas molecules can be activated, either through the effects of UV irradiation, or through processes that momentarily raised the temperature, such as with electrical discharges linked to thunderstorm activity, or through the impact of meteorites (▣ Fig. 3.16).

Following on from Alexander I. Oparin and J. B. S. Haldane who, in the 1920s, assumed that the atmosphere of the primitive Earth was reducing, in 1952 the chemist Harold Urey regarded it as primarily consisting of hydrogen H_2, methane CH_4, nitrogen N_2, and ammonia

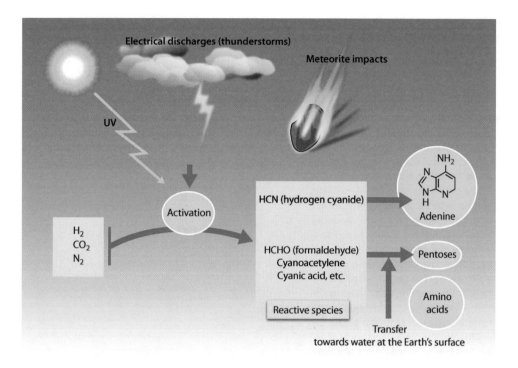

🔲 **Fig. 3.16 Prebiotic chemistry in a reducing atmosphere.** Thermal or photochemical activation of mixtures of gases produces highly reactive species capable of subsequently forming amino acids, nitrogenous bases or sugars, and of initiating the formation of far more complex systems.

NH_3. In an experiment that has become famous, in 1953 his student, Stanley Miller, showed that amino acids were indeed formed by thermal activation of a mixture of hydrogen, methane and ammonia (🔲 Box 3.4). The publication of these results led scientists to carry out similar experiments, which all showed that a wide range of biomolecules could be produced under these conditions. Among these biomolecules are some nitrogen bases found in nucleic acids. Space exploration proved the relevance of this process by demonstrating that even today, it is still at work in Titan's reducing atmosphere. All this work left its mark on the scientific community; indeed, from these researches, it became conceivable that organic molecules formed in an abiotic manner (or, more exactly, in a prebiotic fashion) could have been used as the source of both matter and energy for the first living beings (🔲 Box 3.3 and ▶ Chapter 4).

In fact, for chemists, the activation of gaseous mixtures by UV irradiation or by electrical discharges provided at least two "advantages" in terms of organic synthesis. On the one hand, chemical species that are produced under these conditions by recombination of atoms, ions or radicals (hydrogen cyanide HCN, cyanoacetylene C_3HN, isocyanic acid HNCO, formaldehyde HCHO, other aldehydes, etc.) are very reactive. Thus they could form amino acids, nitrogenous bases (such as adenine) or sugars (such as pentoses), through polymerization of hydrogen cyanide or formaldehyde, thus initiating the formation of far more complex systems, whose study is, nowadays, a major issue in prebiotic chemistry. On the other hand, the lifetimes of these species was sufficiently long for them to be transferred from the atmosphere to the ocean or the ground.

Nevertheless, the theory according to which, on the primitive Earth, the atmosphere was a favorable environment for the prebiotic synthesis of organic molecules has long been neglected by researchers. The composition of the atmosphere proposed by Urey was very quickly disputed, and a consensus arose around the idea that the atmosphere was neutral as far as redox was concerned, and that it primarily consisted of N_2, CO_2 and H_2O. Under such

▣ Box 3.4   The Miller-Urey Experiment

In 1952, the chemist Harold Clayton Urey accepted the idea, advanced at the beginning of the 1930s by Alexander Ivanovich Oparin and John Burdon Sanderson Haldane that the Earth's primitive atmosphere was reducing. He assumed that the latter consisted of hydrogen, methane, nitrogen and ammonia, and theorized that activation of such a system could lead to the synthesis of organic molecules. In 1953, his young student Stanley Miller wanted to test this hypothesis. He devised the equipment shown diagrammatically in ▣ Fig. 3.17a. He filled it with a gaseous mixture of hydrogen, methane and ammonia, then brought the aqueous phase (initially containing distilled water, which to a certain extent, mimicked the ocean) to boiling point, thus creating a steam flow. Then, the gaseous mixture was activated by electrical discharges, simulating lightning. Very soon, he identified amino acids among the reaction products. This experiment caused quite a stir because it demonstrated that the synthesis of the basic components for life from simple precursors (H_2O, H_2, NH_3, and CH_4) was possible by simulating conditions on the primitive Earth. It thus became conceivable that biomolecules formed by an abiotic method could have provided the first living organisms with organic matter and energy.

The amino acids are synthesized in aqueous solution through the Strecker reaction. In this process, α-aminonitriles (AN) are synthesized from the aldehydes (A) and hydrogen cyanide (HCN) formed in the gaseous phase by electrical discharges and ammonia. These intermediate products are subsequently hydrolyzed into α-amino acids (AA) (▣ Fig. 3.17b).

Because of progress in understanding the formation of the Earth's gaseous envelope, no one now supports,

as Urey did at the time, the idea that the primitive atmosphere consisted mainly of methane, ammonia, carbon monoxide and water. But the main conclusion of this Miller-Urey experiment remains valid for less reducing (H_2, CO_2, N_2) or completely neutral (H_2O, CO_2, N_2) mixtures : amino acids may be formed, though with varying efficiency, by abiotic pathways, as the result of the activation of the primitive atmosphere.

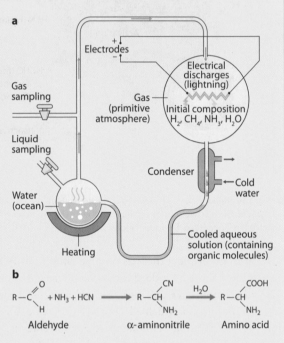

▣ **Fig. 3.17** a. The Miller-Urey experiment: a schematic representation of the equipment. b. The Strecker reaction.

conditions, it was long believed that prebiotic synthesis would have been a very inefficient process. But in the last few years, two new pieces of information have returned atmospheric prebiotic synthesis to the foreground.

The first of these is that the atmosphere remained reducing over quite a long period of time, because, as has been explained earlier, the hydrogen escape to space was overestimated. Second, was a re-examination of the production of organic matter from atmospheres that are neutral from a redox point of view. The experimental approach that concluded that this pathway was inefficient has been called into question by work published in 2008. The "productivity" of gaseous mixtures consisting of N_2, CO_2 and H_2O, has been significantly re-evaluated, even though it still remains lower than that of highly reducing systems (CH_4, NH_3, N_2, H_2O).

We may therefore conclude that the production of organic molecules did definitely take place in the terrestrial atmosphere during the whole Hadean (except, perhaps, before the condensation of the oceans, because, due to the high temperature and the water-vapor pressure, the lifetime of molecules that had been formed was probably limited), where it could have fed prebiotic chemistry. It may well have equally taken place subsequently, until the end of the

Archaean (2.5 Ga) marked by the increase in the atmosphere's oxygen content, or else, until it was overwhelmed by the extent of production by autotrophic organisms.

Hydrothermal Systems

The debate relative to the reducing nature of the primitive terrestrial atmosphere and to the efficiency of the formation of organic molecules from neutral atmospheres is not yet completed. But even if one sticks to the hypothesis that prevailed in recent decades that a rapid escape of hydrogen to space led to a non-reducing atmosphere and on the idea that the formation of organic molecules from neutral atmospheres is inefficient, other environments could have been favorable for organic synthesis. Indeed, all the assumptions presented above, only take into account the most superficial layer of the planet. As it is still the case today, the mantle of the Hadean Earth – as well as the oceanic crust derived from it – was reducing, on the whole. The synthesis of organic matter was thus possible in the close vicinity of the hydrothermal systems; indeed, these latter are located at the interface between seawater and the magmatic reducing rocks (with a high content of Fe^{2+}) of the oceanic crust (◘ Fig. 3.18).

As seen earlier, hydrothermal circulation leads, initially, to the alteration of the rocks of the oceanic crust (serpentinization) and to the release of hydrogen. In addition, in this hot and reducing microenvironment, the CO_2 dissolved in seawater may be reduced through reactions that involve liquid water (in Fischer-Tropsch-type reactions), which are likely to produce organic molecules such as hydrocarbons and fatty acids. Amino acids may also be formed in such systems.

It is often considered that the wealth of minerals and metallic ions in hydrothermal vents favors the catalysis of these reactions. According to scientists who are studying this type of organic syntheses, the surface and the pores of the minerals do thus play a prominent role. They may, simultaneously, act as catalysts and enable the adsorption of reaction products, thus preventing too great a dilution of the formed chemical species – whereas the equilibrium concentrations are very low: of about 10^{-6} mol.L^{-1} for α-amino acids – and thus favoring subsequent reactions before the molecules are dispersed in the ocean. This adsorption in a local

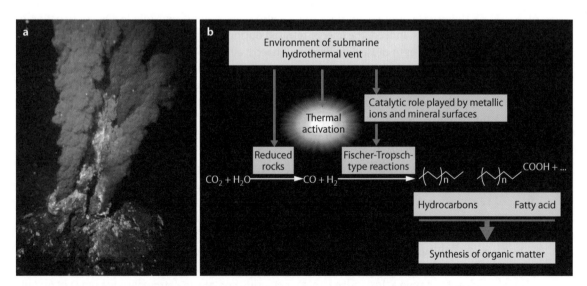

◘ **Fig. 3.18 A present-day hydrothermal vent (a) and the prebiotic reactions (Fischer-Tropsch reactions) that could have taken place in such an environment between 4.4 and 4.0 Ga (b).** Such reactions could, for example, produce hydrocarbons or fatty acids. It should be emphasized, however, that it is very difficult to determine to what extent these reactions were actually active on the Hadean Earth.

3

mineral environment would equally protect the molecules formed against the high temperatures that prevail in hydrothermal systems.

As we shall discuss in the next chapter, organic synthesis at these hydrothermal vents is considered as playing a crucial role by those chemists, biologists and geologists who subscribe to the view that life arose in such an environment, and that the first living systems were autotrophic, that is, that they synthesized their organic matter from mineral molecules. However, we have already emphasized, that it is very difficult to determine to what extent the Fischer-Tropsch reactions were actually efficient. It must also be noted that, within the context of the theory of an autotrophic origin of life, the redox gradient between rocks originating in the mantle and the atmosphere-ocean system does itself constitute a source of energy potentially capable of initiating organic chemistry processes. It could initiate a precursor of a metabolism independent of being an abiotic source of organic matter.

Meteorites

Let us leave there, the submarine depths for considering a last potential source of organic molecules that could have seeded the Hadean Earth. This source is not hiding in the water, nor in the atmosphere, but ... in space. Indeed, we have already stressed this point: the interstellar medium is a site where organic matter is synthesized and perforce by abiotic paths. Thanks to radiotelescopes and spectroscopy (that is to say, the study of the detailed characteristics of the light that is emitted or absorbed by an object), more than 150 organic species have been detected in the interstellar clouds of gas and dust, from which stars are born (*see* ▶ Chap. 1). Most of these are very simple (CO, HCN), but some may contain more than a dozen atoms (such as $H(C{\equiv}C)_5CN$). In fact, nearly the total amount (99.9 percent) of the organic matter present in the universe is concentrated within these clouds.

Irradiation is the phenomenon that underlies these organic syntheses. Indeed, even before these clouds collapse and are transformed into stellar nurseries, the interstellar medium is exposed to the intense cosmic radiation, which causes the breakdown of simple molecules present in a gaseous state, and leads to the formation of isolated atoms, free radicals, and ions. The latter are then involved in numerous chemical reactions that are studied thanks to theoretical chemistry methods, which appear as perfectly appropriate for computing the progresses of chemical reactions that take place under vacuum.

The surface of grains of dust and ice also plays a significant role in the construction of organic species. Indeed, the energy released during the synthesis of these molecules – which would be such as to cause their rupture if it were not removed – may dissipate within the solid body, which is at a very low temperature (around 10 to 50 K). It is, moreover, likely that when they attain a significant size, the molecules remain inaccessible to observation, because they are trapped within the particles (in particular within ices), whereas any possibility of detection requires their presence in a gaseous state.

The nebula within which our planet was born (the protosolar nebula) was the site of such an organic synthesis. In the outer regions of the nascent Solar System, the molecules synthesized were subsequently incorporated into comets, whereas in the inner regions, they were involved in the accretion of planetesimals, of which meteorites are the present-day representatives (*see* ▶ Chap. 1). The meteorites formed in this way and rich in organic matter are known as carbonaceous chondrites. Although accretion releases heat, the organic molecules included in objects the size of the precursors of the carbonaceous chondrites, could have been simultaneously spared from excessive heating and could have simultaneously encountered liquid water, which would have allowed them to evolve towards families of molecules such as the amino acids. Indeed, these meteorites contain an extremely wide range of organic molecules among which amino acids are well represented as well as alkyl derivatives on the α carbon, such as isovaline (Iva), whose α carbon carries a methyl group and an ethyl group (Iva is absent from the set of amino acids encoded by life.)

Both L and R optical isomers of these amino acids have been detected in carbonaceous chondrites. However, the analysis of one of them – the Murchison meteorite, which was collected immediately after its fall in Australia in 1969 – has delivered an extremely intriguing result: The L and R forms are not present in equal amounts. Should this asymmetry – the origin of which is unknown – be seen as an explanation for the fact that the proteins in the living world consist solely of amino acids of the L type (they are "homochiral")? This is what some researchers have suggested. Be that as it may, it is obvious that the organic matter that survived the fall of meteorites and comets onto the primitive Earth provided a recurring source of basic organic building-blocks, which could have been used subsequently for pre-biotic chemistry.

Was a Niche for Life Available as Early as This?

At this stage in our story, let us draw up a final summary of the main characteristics of the Earth between 4.4 and 4.0 Ga. First, we find liquid water, that is to say one of the key ingredients for the chemistry of life (■ Box 3.5). Then there are continents, which are probably sufficiently stable to have been altered by erosion. And finally, the Hadean Earth was continuously seeded by the addition of organic matter, whether endogenous – that is synthesized on Earth either from the atmosphere or from submarine hydrothermal vents (or both), - or exogenous, that is the organic materials synthesized in space and delivered to the Earth through the falls of carbonaceous chondrites.

Our ignorance of the diverse range of chemical processes that took place on Earth at that time remains considerable, but there were undoubtedly a great many of them. The presence and availability of liquid water probably allowed the development of both molecular assemblies and reaction systems that were already complex, and as such precursors of "the" transition towards life itself. However, the Earth was certainly very active geologically, which implies continuous destruction of this organic matter by heat, either by sediment burying, or by circulation of seawater in hydrothermal systems and then of the dissolved

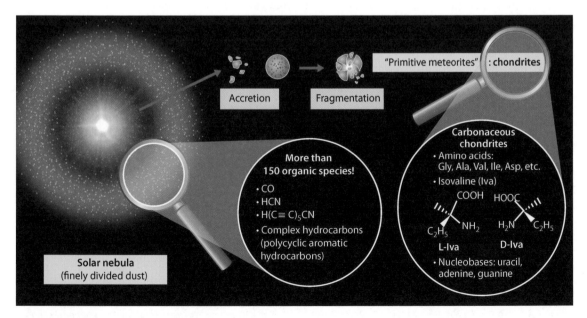

■ **Fig. 3.19 Interstellar clouds and meteorites: extraterrestrial sources of organic matter for the primitive Earth.** It is not just living beings that can synthesize organic matter. The gas and dust clouds in interstellar space are the site of intense organic synthesis. It was the same for the protosolar nebula.

◻ Box 3.5   Are Water and Carbon Necessary for Life?

Theoretically, nothing precludes the possibility that life could arise in the absence of water or of a carbon-based chemistry. However, for most researchers, the presence of these two ingredients appears to be essential for the emergence of life. There are good reasons for this.

Carbon is the chemical element that most easily forms molecular scaffolds, because it is able to link with other atoms through several covalent bonds (up to four). This partly arises through its electronic structure (four electrons in its outer shell) and from its ability to become involved in strong covalent bonds with a large number of different atoms (unlike silicon, which is also tetravalent) (◻ Fig. 3.20.). So it is not purely by chance that, among all the elements, the carbon is the one that displays the richest chemistry. In addition, this rather unique behavior is associated with a great abundance in the Universe: carbon is the fourth element by abundance in our galaxy, after hydrogen, helium and oxygen. Moreover, by far the majority of molecules detected in interstellar medium by radio astronomy are carbon based : this is not by chance!

Now water... How could one envisage a cell without liquid water? Based on this simple statement, it is often believed to be a self-evident fact that water is indispensable for life. But what is the true scientific basis for such an assertion?

The property that leads some molecules that possess one hydrophobic (non polar) end and another hydrophilic (polar) end to link together into organized, microscopic structures (such as vesicles) can only be observed in a limited number of liquids. In liquid water, this property is due to the strong interactions (known as hydrogen bonds) between the H_2O molecules (it is these bonds that allow water to remain in the liquid state up to a temperature that is much higher than that of other molecules of similar molecular mass such as ammonia or methane, both of which are gaseous at ambient temperatures). Molecules or portions of molecules that cannot fit easily into this network of hydrogen bonds are repelled by water's structure, which leads them to link together. This occurs independently of the interactions – favorable or unfavorable – that they might otherwise establish between themselves. This type of hydrophobic organization not only governs the formation of membranes, but is equally an essential element in the folding of proteins as well as for the links between molecules in which the latter are involved (the interaction between an enzyme and its substrate, for example).

From this point of view, as we have already said, substitutes for water are few and imperfect. Among the chemical species whose synthesis under prebiotic conditions is plausible, only the formamide (methanamide) is able to give rise to the aggregation of hydrophobic molecules. However the simplicity of the water molecule associated with the abundance of oxygen and hydrogen in the Universe leads to the logical conclusion that liquid water may be available in great amounts on a non-negligible number of Solar System bodies (and doubtless elsewhere); which is not the case with the other potential substitutes.

Other physical and chemical properties of water undoubtedly played a role in the emergence of life. First, this molecule may be found simultaneously in solid, liquid and gaseous form under moderate temperature conditions. A change in the water's state, either by transformation into vapor or by freezing in the form of pure ice is thus capable of very effectively concentrating solutes into the residual liquid. These concentration mechanisms, difficult to imagine in any other context than that of a change of state, will increase the probability of encounters for the dissolved molecules, thus leading to the production of larger-sized molecules. They are thus undoubtedly the determining factors in the formation of biopolymers. Moreover water possesses a high dissociative power (linked to its dielectric constant) which means that it is an excellent solvent for a great number of salts and, more generally, for chemical species carrying electrical charges. In addition, its structure based on a network of hydrogen bonds makes the proton transfer extremely rapid. This point is important because most of chemical reactions involving biomolecules imply proton migration. Biomolecules are thus more easily formed. Finally, water plays the role of chemical reagent in the numerous biochemical pathways involving hydrolysis of active species (ATP, for example). These metabolisms based on hydrolysis could also have played an important part in the emergence of life (◻ Box 4.5).

All these properties mean that, to date, apart from water, no liquid appears to be able to provide, in any credible fashion, an environment favorable for the emergence of life. It is just the same for carbon. The water molecules as well as the carbon chemistry thus appear indispensable for life, although not sufficient by themselves.

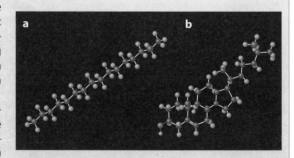

◻ **Fig. 3.20 Examples of molecules based on a carbon skeleton:** a saturated linear hydrocarbon consisting of 16 atoms of carbon (**a**); cholesterol (**b**).

3

species. Major meteorite impacts would certainly have also contributed to the destruction of organic matter (*see* ▶ Chap. 5). This means that we should think of prebiotic chemistry less as the progressive accumulation, over hundreds of millions of years, of a stock of organic matter, subsequently available "for" the appearance of life, but rather as a dynamical molecular system, completely renewed over a time scale of a few million years at the most, and within which complex processes took place. Some of those that led to the emergence of life were undoubtedly already active.

Very early in its history, perhaps from 4.4 Ga, or, more definitely, from 4.3 Ga (the time when the CO_2 pump had allowed the surface temperature to decline sufficiently), the Earth seems to have been ready to accommodate life: it was potentially habitable. However, no one knows – and doubtless no one will ever know – how much time was required for the emergence of life, nor even if it emerged "all at once" or whether it emerged repeatedly. In particular, even if the Hadean was not the hell that its name suggests (Hades was the Greek god who reigned over the underworld and the dead), and even if the Earth's face gradually evolved during this period, this does not preclude the possibility that cataclysmic events similar to the Late Heavy Bombardment (*see* ▶ Chapter 5) did take place. Assuming that such cataclysms did supervene, nothing enables us to know whether life could have survived it.

So, if we do not know when and where life appeared, can we at least answer the question "how"? Like Russian dolls, every question raises another, and we must ask an all-important question: What is life? What is a living system? And – something that is not an independent question – how could such a system have been built without an architect? These questions lie at the heart of the next chapter.

Chapter 4

Intermezzo: The Gestation of Life and its First Steps

Without our knowing exactly when it took place, a sequence of events was to have major consequences on the surface of the Earth. As a result of self-organization processes whose details have yet to be worked out, and which were as much chemical as physical in nature, the very first living beings, capable of transforming energy and matter, and of evolving, appeared. Bit by bit, they transformed their environment ...

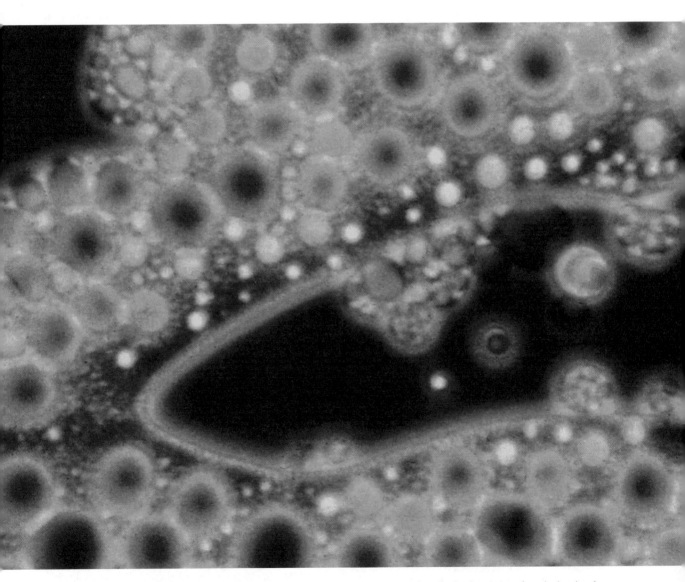

◘ **Vesicles formed by amphiphilic molecules, seen through an optical microscope.** Such molecules (consisting of a polar head and a hydrophobic tail) are capable of spontaneously assembling in a bilayer enclosing an aqueous compartment. They possibly formed the basis for the membrane of the first proto-cells. (Photo.: D. Deamer)

M. Gargaud et al., *Young Sun, Early Earth and the Origins of Life*,
DOI 10.1007/978-3-642-22552-9_4, © Springer-Verlag Berlin Heidelberg 2012

4

The transition from the non-living to the living remains one of the main enigmas faced by science. This question, which is intrinsically interdisciplinary, is one directed first at chemists and biochemists. They are confronted by a radical change over the course of an interval of time about which they know absolutely nothing. It is a bit like the interval during a play: the actors change their costumes in the wings, no one knows how, one only sees the result when they return to the stage. The result surprises us, because the scene has changed …

In just such a fashion, during the Hadean, between 4.4 and 4.0 Ga, conditions became established which seem to have been compatible with the emergence of life. When we concern ourselves with this epoch, we are therefore immediately faced with the question of the probability of this emergence. The conditions seem to have been brought together, but does that really mean that a rapid emergence of life was inescapable? Nothing can be less certain. Should we follow Jacques Monod (who won the Nobel Prize for Physiology and Medicine in 1965 with André Lwoff and François Jacob) who, in *Le hasard et la nécessité*, considered that *"the universe was not pregnant with life, nor the biosphere with man"*? In other words, was the

---

◘ Box 4.1   Self-Organization

Every living being consists of a collection of molecules that are constantly renewed and which appear to co-ordinate their evolution in space (and thus forming a system), and also in time. We are therefore dealing with organized systems, the emergence of which, perforce, implies a process of self-organization.

However, the spontaneous formation of an ordered system from disorder contradicts our everyday experience. We all know that over time, the most beautiful building is inevitably reduced to ruins. In physical or chemical terms, this tendency is expressed as a quantity, entropy, which expresses the degree of disorder in a system. The second law of thermodynamics expresses the idea that the entropy of an isolated system increases, and thus that disorder tends to increase. An isolated chemical system must therefore evolve towards an equilibrium state[1] in which the concentration of different chemical species will be determined by their individual energy levels and the laws of statistics.

So how could a system that was as disordered as that of the primitive Earth, with an incredible diversity of forms and structures, give rise to life? The answer lies in the fact that the process of self-organization, which is linked to the emergence and development of life, concerns only one part of the system. Hence, the formation of an ordered structure in a sub-system will be compensated by an increase in disorder in its environment, such that overall, the entropy does not decrease. That means then that exchanges of energy and matter are the basis of the dynamics of self-organization.

Moreover, the inescapable evolution towards disorder and the state of thermodynamic equilibrium does not predict in any way the duration of the chemical reactions involved, which may occur in a fraction of a second or, on the other hand, over a period that is reckoned in millions of years. The speed of this evolution depends on the dynamics of the reaction (the subject of chemical kinetics) and not on thermodynamics, which only predicts the sense in which it unfolds.

In chemistry, it is difficult to envisage self-organization without having recourse to the heterogeneous nature of matter on a microscopic scale, that is to the fact that matter is not indefinitely divisible. If that were the case, how could it form complex structures? It was undoubtedly this type of reasoning that led, in antiquity, certain philosophers, the best known of which remains Democritus, to postulate the existence of atoms as being the basis of matter.

The difficulty comes in passing from this microscopic heterogeneity to a single macroscopic entity that involves a coordination in the arrangement or the movement of a multiplicity of atoms or molecules (either within a three-dimensional structure or within an entire organism). The properties that molecules have of associating with one another may give rise to the formation of crystals or other macroscopic structures such as vesicles (such as those that form cellular membranes, for example, ◘ Fig. 4.23) or the micelles of surfactants. Structures that have a dynamical character may also appear through amplification mechanisms that are highly efficient, such as replication or autocatalysis (◘ Box 4.5). These mechanisms are at work in what are known as oscillating reactions, which are often considered chemical curiosities, such as the Belousov-Zhabotinsky reaction (◘ Fig. 4.1). The concentration of certain intermediates then varies until the reagents are exhausted, in a cyclic or stochastic

---

1. Note that here the term "equilibrium state" is used in its thermodynamical sense: it is the state from which an isolated system can no longer evolve. This definition is far different from the concept of dynamical stability that is found in living systems, and which is covered by the concept of a stationary state.

birth of life a highly improbable event, which was highly unlikely to occur, both on Earth and in our galactic environment? Or, on the contrary, is life, as Christian de Duve (Nobel Prize for Physiology and Medicine in 1974), believes a *"cosmic imperative"*, ready to occupy any space that is freely open to it, as soon as the conditions for its gestation are fulfilled? There is no obvious answer to these questions.

Whatever weight is given to chance, the Earth seems to have been ready, during the Hadean, to support embryonic life. But even if the transition from the non-living to the living would have actually occurred during this period, history does not provide any tangible evidence to tell us about the sequence of events. Here, as far as we are concerned, chronology comes to a stop. It may be a constant factor in major transitions that when they result from a dynamical process of self-organization, they do not comply with a strictly mechanistic description (◘ Box 4.1).

We are therefore led to abandon the description of established facts and to choose the theory, model, or reconstruction of scenarios – perforce hypothetical – starting from prob-

---

manner and in what may be very significant amounts. This variation may appear, in certain cases, as periodic changes in the color of the medium, corresponding to the coordinated evolution of a multitude of molecules. This coordination, which resembles the way in which the concentration of certain proteins evolves during the course of the cell cycle, forms an example of self-organization that reproduces at least one of the characteristics of life.

Chemical dynamics might therefore be at the basis of possible self-organization that simulate certain properties of living cells. But self-organization of life forms possesses an additional peculiarity which rests on its ability to store information that is transferred to subsequent generations during evolution. We may think that, as the Israeli chemist Addy Pross has suggested, the emergence of molecules or chemical structures endowed with this capacity and capable of self-replication (and thus simultaneously able to multiply infinitely and to perfect themselves by mutation and subsequent selection) offers a considerable kinetic advantage. They

are, in principle, capable of colonizing the environment through the action of the kinetic efficiency of reproduction (exponential growth), and to exhaust any available sources of energy and matter to their own benefit. The driving force behind the emergence of life would not then be but the result of a sort of dynamic imperative of a chemical nature. But the appearance of these structures is accompanied by a transition, whose implications are such that another scientific discipline is required to understand the phenomenon. We are moving from chemistry to biology. Henceforward, life, it goes without saying, is independent of its chemical substrate, and its evolution does not follow paths that are predictable solely based on the laws of physics.

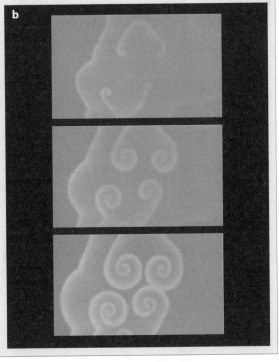

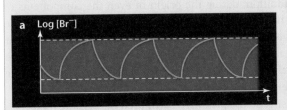

◘ **Fig. 4.1 An example of an oscillating reaction: the Belousov-Zhabotinsky reaction.** In its most common version this corresponds to the oxidation of malonic acid $CH_2(COOH)_2$ by bromate BrO_3^- in the presence of cerium(IV) sulfate. The mechanism is very complex, including an autocatalytic step, as well as numerous chemical species. In a stirred reactor (**a**), the sequence of events leads to oscillations in the concentrations of Ce^{4+} and bromide Br^-. When, however, it is carried out in a thin layer of liquid (**b**), the Belousov-Zhabotinsky reaction may produce "chemical waves" (changes of color, for example) that move through space.

4

able chemical pathways or basing our views on the analysis of characteristics that are common to life today. To resume a chronological description, incorporating biological evolution, we shall have to wait a very long time: the time until geology provides undisputable traces of past life forms. The complexity of the first confirmed traces of life (such as stromatolites, *see* ▶ Chap. 6) leaves a gaping void that it is difficult to fill: to cut it short, the oldest forms of life that we are capable of detecting may not have been the first forms of life.

So this is the situation: the essential stages of early evolution, such as the origin of life, the emergence of the principal biochemical pathways, the evolution of the last common ancestor of all modern living beings, and the first divergences from that common ancestor are impossible to date. All we are able to know is that all these events took place over a period between two particular boundaries. The first is the one when the window favorable to the origin of life opened: that corresponds to the period when favorable physicochemical and geological conditions became established, i.e., 4.4–4.3 Ga. The second is obviously the point in time when, thanks to the detection of the oldest fossil traces of life, we can be absolutely certain that life had become established, i.e., 2.7 Ga (perhaps as early as 3.5 Ga). Between these two limits, we have no idea of the time that was necessary for the establishment of the processes that led to the appearance of life, nor of that required to give rise to the major transitions of early evolution.

We have, therefore, decided to opt for tackling the process of the origin of the transition between non-life and life as soon as the first of those boundaries was reached, when the conditions on Earth became *a priori* suitable for the emergence of life and for its early evolution. This decision as to the structure of this book should not, in any way, be taken to mean that the hypothesis that "that life existed before the Late Heavy Bombardment at 3.9 Ga" is any more favored than any other. We have also decided not to dissociate this analysis from the discussion of the first stages of life, those of which we no longer have any traces. Those first life stages cannot be synchronized with the subsequent progress of evolution that led to the organisms which left the first confirmed fossil traces in our planet's "archives", namely large expanses of fossil stromatolites (over several square kilometers) dated to 2.7 Ga, and whose biological origin is not disputed, unlike more ancient traces (*see* ▶ Chap. 6).

From Chemistry to Biology

Let us linger a little on the question of the dating of the transition between the non-living and the living. It is not the habit of chemistry or biochemistry researchers to consider their work in a chronological context. In addition, dating the origin of even the components of living matter is an insoluble problem, because certain organic molecules were present in the interstellar cloud before nebulae evolved and yielded subsequently, among others, the Solar System (*see* ▶ Chaps. 1 and 3). So it is not the atoms and molecules that should be dated, but the time when they were involved in the processes leading to the emergence of life. But we have not the slightest trace of these events!

The idea that the first living organisms on the primitive Earth used the organic molecules formed by abiotic means as carbon source (*see* ▶ Chap. 3), does form the start of a chronological answer. It is reasonable to believe that the emergence of life took place on Earth[1] quite shortly after the physicochemical conditions became compatible with the survival of biopolymers such as polypeptides and nucleic acids.

---

1. The theory of panspermia will not be considered in this book, because although the possibility of the transfer of life from one planet to another is difficult to refute, it does not rest on any tangible evidence, and thus remains purely speculative in nature.

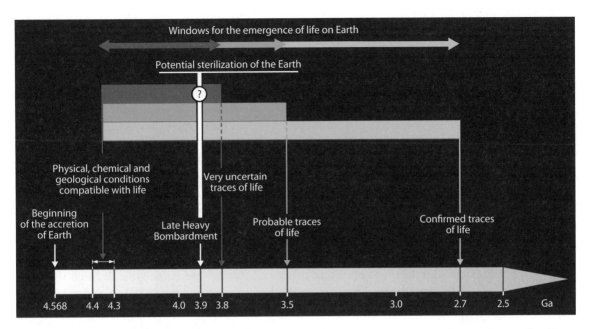

Fig. 4.2 The windows for the emergence of life on Earth are framed by two events: the time when environmental conditions became compatible with life, and the time when, thanks to the fossil record, we may be certain that our planet sheltered life. Between these two limits, we do not know – and will never know – when life appeared, nor even whether this process has been unique or whether it recurred several times.

The presence of liquid water was certainly one of these conditions, because it is involved in a number of processes of organization that are associated with life (the formation of structures and associations through hydrophobic interactions, and hydrolytic metabolisms; ▫ Box 3.5). The surface temperature must also have decreased to about 100 °C, thus becoming compatible with the most thermophilic present-day life forms, but perhaps declining to even lower levels, if one takes into account organisms that are still rudimentary and whose adaptation to very high temperatures has not yet been optimized by evolution. It was only then that the first processes of self-organization, potentially leading to life, could have started to take place. This stage, where organic prebiotic chemistry became likely to lead to the emergence of life, could have started, therefore, about 4.4 to 4.2 Ga, depending on the cooling rate of the Earth's ocean (which regulates the surface temperature).

The end of bombardment by potentially sterilizing meteorites was also necessary for life to spread. The process of evolution that led to extant life must not have been interrupted by the total destruction of the initial organisms. This implies that the period during which life emerged is marked by another unknown, with respect to the fact that 3.9 Ga ago, the Earth was subject to what might be termed a "meteoritic deluge"; the Late Heavy Bombardment (LHB, *see* ▶ Chap. 5).

If life had already bloomed within the period 4.3–3.9 Ga and the LHB was actually sterilizing, we have two possibilities: either the planet was completely sterilized and a new phase of gestation for life took place between 3.8 and 2.7 Ga, or life (or certain forms of life) managed to survive this catastrophic event. So the list of possible scenarios increases yet again ... without our having any more determining criteria to judge their validity. These possibilities will be discussed again at the end of Chap. 5.

The second limit, the most recent, that may be fixed for the non-living / living transition is, as we have said, the age of the most ancient forms of fossil life that have been identified: 2.7 Ga for the confirmed traces, 3.5 Ga for probable traces, 3.8 Ga for very uncertain traces (▫ Fig. 4.2; *see also* ▶ Chap. 6). Let us stress once more an essential distinction: the question of the end of the process of the emergence of life and that of the confirming the oldest traces of life are of completely different natures; and it is highly likely that life developed for some tens or hundreds of million years without leaving any signs that have survived until the present.

Between those two limits that we have defined, chemical evolution and the first stages of life took place, but it is futile for us to attempt to assign more specific dates to corresponding events and even to know whether they occurred just once, or if the emergence of life as we know it was preceded by other abortive attempts, followed by just as many disappearances of the resulting living beings. Even assuming that we could know the chemical stages that had to be achieved – which is far from being the case – the parameters (such as temperature and concentration of reactants) governing the rate of the chemical reactions themselves are too poorly defined for us to conclude about the duration required for each of the stages. The only reasonable prospect open to chemists and biochemists is to construct "bottom-up" scenarios based on a logical chain of stages, starting from the sources of energy and organic matter that were available on the early Earth and going as far as the accumulation of biochemical monomers and to the first cells.

And what about the evolutionary biologists? Even though the period is intimately associated with their field of research, they find themselves basically in a situation similar to that of the chemists and biochemists. The data provided by geology and paleontology should however help them by restricting the range of possibilities. However, only in exceptional

---

◘ Box 4.2   Where Was Life Born?

Although several different scenarios may be put forward for the way in which life was born, it is, on the other hand, extremely difficult, in absolute terms, to determine where it appeared. The fact is that there are multiple possibilities and the chosen environment will dictate one scenario rather than another. The two questions "where?" and "how?" are interdependent. We may, nevertheless, make a short survey of the various possibilities, which form a rather heterogeneous collection.

• **In the ocean?**
Given that water appears to be indispensable for life, the idea that the ocean could have provided an environment favorable for the emergence of life seems obvious. The vision of a primordial ocean taking the form of a prebiotic soup, rich in organic molecules, has frequently been advanced since the hypothesis for a heterotrophic origin of life was proposed by A. Oparin and J.B.S. Haldane in the years 1920–1930. It does, however, pose a problem of scale, given the volume of the ocean. Within it, organic molecules would have remained very dilute, rendering the synthesis of more complex molecules improbable. In addition, dissolved organic matter would be destroyed by the circulation of water through hydrothermal systems (at temperatures higher than 300 °C). Any accumulation could not have taken place over periods exceeding some ten million years. There remains the possibilities that poorly soluble organic material aggregated within the oceanic mass, within the coacervates that Oparin envisaged, for example (◘ Fig. 4.23), or that hydrophobic substances floating on the surface accumulated.

• **On shorelines?**
The shores of the first emerged lands were regularly swept by the tides. (After the impact that was responsible for its birth, the Moon was closer to the Earth than today, and the latter rotated more rapidly, leading to higher tidal amplitudes and a faster alternation of day and night.) These areas could have formed environments that were favorable for the concentration of organic species.

• **On dry land?**
The presence of water masses where organic matter accumulated was equally likely to have occurred on emerged continents, rather like the "warm little pond" envisaged by Charles Darwin in 1871. There, the alternation of dry and wet phases governed by the seasons and atmospheric conditions could have favored condensation through dehydration, and thus the formation of biopolymers.

• **At the bottom of the ocean or in crevasses in the oceanic crust?**
The hypothesis for the origin of life in deep-sea environments is based on the co-existence of reducing minerals and water carrying oxidized species such as CO or CO_2. Under such conditions, the synthesis of organic molecules such as hydrocarbons by the Fischer-Tropsch type of reactions (*see* ► Chap. 3), of amino acids, and even of small peptides would have been possible. The role that minerals could have played both as catalysts and in adsorbing organic molecules has often been cited as an argument in favor of such models. In addition to these properties, hydrothermal systems have also been considered as a favorable environment

cases do the traces of the most ancient life allow us to draw conclusions about the mode of life, the metabolism, or the phylogeny of the corresponding species. Biologists are therefore constrained to use indirect arguments, based on what they know of modern organisms. In particular, they compare genes and their distribution in the genomes of present-day living organisms. This could *a priori* lead to the identification of certain characteristics and metabolic pathways as being more ancient than others. Unfortunately, several types of biases (the nature of the data, evolutionary models, etc.) may affect this "top-down" approach and are likely to compromise the significance of the results. And again, even if a result – a series of events – seems compatible with observations it is impossible to draw a definitive conclusion, because evolution is not bound to follow the most parsimonious route, that is to say, the shortest route between two evolutionary stages. Nothing, for example, prevents a given characteristic, present in an ancestor, from disappearing in its descendants, and subsequently reappearing as a result of new mutations.

So, once again, we must caution the reader regarding the lack of elements that chemists and biologists have to try to date early evolutionary events that led to life as we know it, and also on the hypothetical nature of the whole set of models and scenarios that exist to explain the transition from the non-living to the living and the early evolution of life.

4

by researchers who advocate a chemolithoautotrophic origin for life, because there, the first life forms could have directly synthesized their components by means of the chemical energy drawn from the redox reactions involving reduced minerals from the mantle. The German chemist Günter Wächtershäuser has, for example, suggested that life emerged at high temperatures on surfaces within crevasses in the oceanic crust that were subject to a hydrothermal regime. The geochemist Michael J. Russell believes that life could have emerged in iron sulfide cavities in a hydrothermal chimney.

- **In clays?**

The properties of certain minerals such as clays (particularly montmorillonite) in adsorbing organic molecules at the surface of their hydrated sheets and acting as catalysts are also the origin of several hypotheses. Certain types of clay favor the formation of peptides or relatively large RNA molecules, which may argue in favor of a role for these minerals on the origin of life. The Scots chemist A. Graham Cairns-Smith has even suggested that clays were the very first substrates storing genetic information (in the form of a series of different crystalline layers). To date, this last hypothesis has been the subject of very little experimentation.

- **In suspension in the air?**

It has been suggested that droplets produced by spray from the primitive ocean could have formed chemical micro-reactors, enriched in organic substances resulting from activation of the atmosphere. These micro-reactors could have been the site of evolution towards life.

- **In ice?**

The hypothesis of a cold primitive Earth (■ Box 6.2) has led to proposals that consider that life could have developed in residual liquid inclusions trapped when ices are formed, and within which organic molecules formed in the atmosphere (or elsewhere) would have been concentrated. The presence of active species and small organic molecules at high concentrations would be favorable for the development of a proto-metabolism. Various research groups have observed that frozen solutions may give rise to the formation of nucleic bases with particularly high yields (starting with hydrogen cyanide) or the polymerization of activated nucleotides. These reactions are, because of the extremely high local concentrations of reactants, astonishingly fast, despite temperatures of some –20 °C.

- **Elsewhere?**

If the location where it was born is unknown, can we totally exclude the possibility that life might have come from somewhere else? We therefore might also consider that life might first have emerged on a body that may or may not have belonged to the Solar System, and migrated to our planet. However, apart from the problems posed by the survival of a life form during an interplanetary transfer (irradiation, impacts, etc.), this highly speculative hypothesis does not resolve any question about where life appeared, and therefore does nothing other than render the problem still more complex. In addition, how can we give it any importance when, in the Solar System, the Earth seems to provide the environments that are the most favorable for life? We may, for example, cite the presence, simultaneously, at the surface of large quantities of water in solid, liquid and gaseous states, which seems to be unique.

4

▣ Box 4.3   The Cell, the Fundamental Unit of Life

The cell is the constituent unit of living beings. Its function depends on two types of fundamental activity. On the one hand it carries information that is simultaneously expressed, such that it fabricates its own components, and also transmitted, such that it gives rise to other cells, which also carry the same information. On the other hand, the cell transforms energy and matter to meet its needs, in a manner analogous to a machine.

The information carried by the cell constitutes its genotype; this information is contained within the DNA molecules that form its genome. A cell's phenotype is the overall set of its observable characteristics (morphological, physiological, biochemical, etc.). It results from the expression of the genotype and its interaction with the environment.

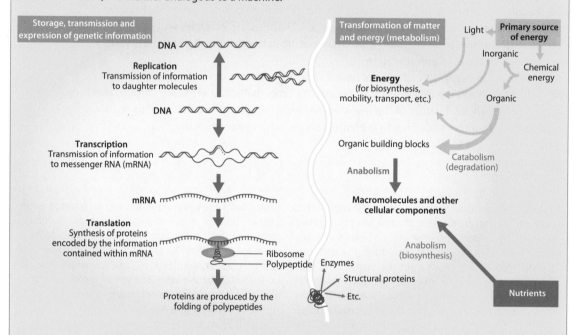

▣ **Fig. 4.3 The two aspects of how a cell functions:** storage, transmission and expression of the genetic information (*left*); the transformation of matter and energy (metabolism; *right*).

The Unavoidable Question: What is Life?

Closely associated with the chronological aspect of the transition between the non-living and the living is the question of knowing *how* the process took place. This question assumes that a precise definition of life has been chosen, and it is, moreover, uniquely within this context that the exercise becomes definitive: the choice will determine a research strategy.

What are the minimum criteria allowing to say that something is a living entity? Is it possible to define life by is major features within a unique living organism? Or should one include populations of individual entities? There are no unique responses to these questions, and they partially depend on the definition of conventions. Very diverse definitions of life have been put forward. Some are very descriptive, whereas others, on the contrary, are more abstract and attempt to determine the property or properties that are both necessary and sufficient to conceptualize life. Among the latter, there are two principal contenders: one places an emphasis on the thermodynamical aspects of life (self-organization and self-maintenance), and the other is based on the properties of replication and evolution. But too abstract an approach is found to be of little use in discussing the origin of life. From this perspective it is probably best to base one's ideas on the functioning of the cell, the fundamental unit of life (▣ Box 4.3),

Fig. 4.4 A schematic definition of life. An entity having a metabolism, a reproducible information content, and an arrangement for confining its components, may be subjected to an evolutionary process based on natural selection of variants that form its descendants. This emergence of a capacity for evolution is one of the definitions of life. One can readily imagine the diversity of scenarios that may be envisaged to describe the transition from the non-living to the living. It is possible to consider that it corresponds to the sequential emergence of three components essential for life. However, the association between two of them, or even the three, is considered by many scientists to be necessary for that transition. The scenario would thus become one of co-evolution.

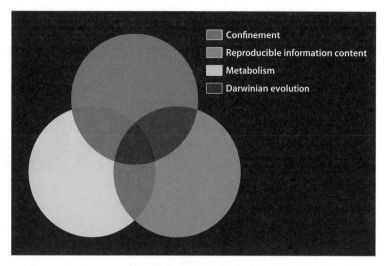

4

and to adopt a more pragmatic approach. There is agreement on the necessity for a living entity to combine three characteristics:

- a physical link between its different components, allowing a differentiation between self and non-self; this link could take the form of a confinement to a single compartment but, theoretically, other types of systems could play a part;
- the ability to maintain itself in a state that is away from thermodynamical equilibrium, within a given environmental context;
- the capacity to reproduce itself giving rise to biological (or Darwinian) evolution, that is that it should be accompanied by variations that can be sorted out by natural selection.

In the cell, these three, purely theoretical conditions are realized by the following components:

- a membrane, which confines the cellular elements
- a metabolism that allows the cell to synthesize its own components from precursors that either come from its environment, or have been synthesized by itself. In other words, the metabolism allows the making its own chemical constituents within the environment. This concerns the overall set of chemical transformations cooperating to synthesize the organic components of a living entity (anabolism) or the contrary, their breakdown and recycling (catabolism).
- a genetic material that carries replicable information encoding the cellular components. No process that is chemical or biochemical in nature can produce completely faithful replication. This process therefore introduces variations (mutations), the inevitable consequences of the imperfections of any material system.

Within populations of autonomous entities that combine the three components listed above, an evolutionary dynamic may become established based on natural selection of variants that have appeared in the descendants (◘ Fig. 4.4). The birth of this capacity for Darwinian evolution is undoubtedly the sign that marked the emergence of life on Earth.

This definition of the three essential features of life is thus an indispensable first step in tackling the question of the origin of life. If the three different features are necessary for life, the first question arises: in which order did these features make their appearance? It might be any one of them, and indeed, different scenarios suggest that either of the three were first and defined the earliest proto-organisms. However, other scenarios can be envisaged if two features, or even all three, co-evolved. We shall return to this point later.

In fact, the question that arises is rather one of knowing what would be the chances of survival for an entity that did not simultaneously possess all those elements required for Darwinian evolution. Very low, undoubtedly. We should, therefore, ask ourselves not only about the

4

◘ Fig. 4.5 Genetic system first or metabolism first? For quite a long time, this question divided the scientific community into partisans for the early appearance of the genetic system (DNA) that contains the information encoding the proteins, and in opposition, the advocates of an early appearance of peptides and proteins, responsible for the catalytic activities necessary, among others, for the synthesis of DNA.

initial steps that allowed the emergence of those three features, but also about the path that was followed and ended with an autonomous entity capable of evolution, that is, until these components were integrated into an overall system. The task is particularly hard, because each of these features has been the subject of intense debate, which may be summarized by the three questions ... literally vital ones:

- **What was the initial metabolism?** This question is essential in the context of the primitive Earth. Either one accepts, like A.I. Oparin and J.B.S. Haldane that prebiotic organic matter was abundant and led to the emergence of heterotrophic organisms (a model that was outlined in the 1920–1930s); or, on the contrary, one imagines that the first living beings were able to synthesize their components from mineral energy and carbon sources (an autotrophic origin of life).

- **Which was first: genetic systems (replicable information) or metabolism (enzymatic activities, and thus replication and expression of information)?** Modern living beings carry two different types of polymers that fulfill these different functions. Nucleic acids play the role of information carriers, coding the proteins, and thus the enzymes, but the latter are catalysts necessary for the replication of nucleic acids. The two systems as so intertwined that their coexistence appears to us to be indispensable for life (◘ Fig. 4.5). One may, however, construct models where the transition from the non-living to the living sees either the genetic system appear first, or the metabolism.

- **What individualized the first living beings?** It is essential for us to know how a link between the components of an initial organism could have arisen. The spontaneous formation of vesicles (that is to say compartments separating a volume of the medium from the exterior through the intermediary of molecules capable of joining together in bilayers, as with phospholipids) is possible. To certain authors, these vesicles would have been, initially, micro-reactors that subsequently evolved into the first cells. But other hypotheses have been advanced, beginning with the emergence of structures resulting from a simple aggregation of molecules to that of life "in two dimensions" by adsorption on the mineral surfaces.

We shall now show how the debate is fed by each of these questions. We shall first explore the different primordial systems of reactions which were likely to produce the basic organic molecules for life and to maintain the functioning of an organism. We shall then see that the discovery of RNAs having catalytic activity –ribozymes – shed new light on the question of knowing which, the genetic system or the metabolic system, appeared first. Finally, we shall ask what the first cellular membranes looked like.

The Origins of Metabolism

Present-day metabolic systems are very elaborate (◘ Box on the next page). They were probably preceded by early, simpler systems. Although their nature remains the subject of fierce debate, it is clear that they had to allow the synthesis of the various components of the first living entities: fatty acids or other chemical species that could give rise to the formation of a membrane (if it is assumed that this sort of confinement was necessary), information carriers, and agents replicating chemically those information carriers.

Replication systems could have included catalysts whose properties were linked to their structure or components, and not, unlike enzymes, being encoded. In addition, it is possible to make a distinction between catalysts that are involved in a single stage and remain outside the reaction system, and those networks whose catalytic properties (or autocatalytic, and thus likely to replicate the overall group of components that are involved) are the result of their architecture, without being bound to include (properly speaking) the catalyst (*see* the example of the formose reaction in the ◘ Box 4.5). The appearance of these networks was undoubtedly indispensable for the transition from the non-living to the living.

Which could have been the first reactions and the first molecules formed? There are three possible approaches to an answer.

If One Starts with the Organic Chemical Species Available in a Prebiotic Environment ...

The important step that was taken by the first researchers who wondered about the origin of life was to take account that the first living beings could have organic molecules synthesized abiotically at their disposal, both to produce their own components and also to utilize the chemical energy that they contained (*see* ▶ Chap. 3 and ◘ Fig. 4.8). They would then have been functioning with a heterotrophic metabolism. The capacity of living organisms to synthesize permanently their own components through their metabolism, or to use those that they found in their environment would thus have been a characteristic that was present in the very first stages of the evolution of life. Numerous workers subsequently tried to devise primitive metabolic systems, starting with organic chemical species available in a prebiotic environment (like the prebiotic soup envisaged by Oparin and Haldane – ◘ Box 4.2).

It is, however, very difficult to draw up an exhaustive list of the organic molecules formed in an abiotic manner. First, their nature depends strongly on the activation system under consideration and on the environmental conditions (*see* ▶ Chap. 3). Subsequently, if certain final products of prebiotic chemistry, such as amino acids – which may be considered final products because they are kinetically stable on a geological time scale, under the conditions for the emergence of life and in the absence of any specific process to reactivate them or destroy them – are simple, prebiotic chemistry leads also in a systematic manner to the synthesis of far more complex organic compounds often in the form of poorly defined mixtures of polymers (for example, polymers of hydrogen cyanide) resembling tars, whose detailed analysis is impossible. So, arguing solely from the analysis of the final products of prebiotic chemistry is to be confronted with a mass of information that is so enormous that it cannot be exploited.

But the most annoying aspect of this approach is not the slave labor that it involves, but rather that it denies us essential information. Indeed, if many of the final products are of limited interest for the construction of a metabolism, the same is not true of their precursors, which, thanks to their high energy content, could be usable by primitive metabolic pathways. One can thus imagine that primitive metabolisms (protometabolisms) would depend on the free energy released by the hydrolysis of activated prebiotic compounds (that is, with a high

4

◘ Box 4.4   Metabolism: Definition and Different Types

Every living being must manufacture its own components from precursors, that is to say more or less complex organic molecules that it has either found in the environment or has itself synthesized from simple inorganic species. It must also regularly recycle those components. Metabolism is the overall set of chemical transformations that cooperate in the synthesis of the organic components of a living entity (anabolism) or, doing the opposite, breaking them down (catabolism).

Another essential feature of living organisms is that they must permanently expend energy to maintain themselves in a state that is far from thermodynamic equilibrium and, in this way, avoid their components being dispersed, which would be synonymous with death. A living being thus needs free energy that may be transformed into work. Collection or production of this energy is the other component of metabolism. Present-day living organisms have evolved in such a manner that they can utilize most of the energy and carbon sources that are present on Earth.

Among the ways of obtaining energy, the oxidation-reduction reactions – or redox reactions, in which one component loses electrons (it is oxidized) which are captured by a component that accepts electrons (which is thus reduced) – that constitute respiration and fermentation have long played a dominant role. Fermentation consists of an incomplete transforma-

tion of organic molecules that takes place in the cellular cytoplasm without the addition of either oxygen nor any other external electron acceptor (ATP, the molecule that stores energy in cells is formed here by substrate-level phosphorylation, starting with highly reactive intermediates; ◘ Fig. 4.6a). Respiration involves various electron acceptors (the most efficient being oxygen, but it is not the only one) and makes use of electron transporters located in the cellular membrane (ATP is created here thanks to the action of the ATPase, a transmembrane protein which makes use of the membrane potential created by the translocation of protons across the membrane during electron transport; ◘ Fig. 4.6b). A wide variety of organic and inorganic molecules may be oxidized by respiration (◘ Table 4.1). In artificial systems, certain organisms are even able to provide electrons to an electrode acting as electron acceptor (or, on the contrary, capture them) and to link the energy produced to their metabolism.

Living organisms may be classed according to the energy and carbon sources that they use for their metabolism (◘ Table 4.2 and ◘ Fig. 4.7). Autotrophs synthesize organic matter from mineral carbon sources, whereas heterotrophs require organic matter already synthesized. If the primary energy source for metabolism is chemical (redox reactions involving either or-

◘ **Table 4.1   The wide range of electron donors and acceptors involved in the different types of respiration.** The vast majority of eukaryotes and a large portion of prokaryotes respire (carry out oxidation-reduction reactions) with oxygen (the latter is then an electron acceptor), oxidizing organic molecules. However, a number of archaea and bacteria use other electron donors and acceptors, often inorganic, during respiration.

Electron donors	Electron acceptors
Reduced organic compounds (glucides, lipids, etc.) H_2O, H_2S, $S_2O_3^{2-}$, S_0, H_2, CH_4 and derivatives of C_1, NH_4^+, NO_2^-, Mn_2^+, FeS, $FeCO_3$, HPO_3^{2-}, etc	Fumarates and other oxidized organic compounds, DMSO, $NAD(P)^+$, O_2, S_0, SO_4^{2-}, CO_2, Fe^{3+}, CrO_4^{2+}, oxidized humic acids, Mn^{4+}, UO_2^{2+}, SeO_4^{2-}, AsO_4^{3-}, triethylamine, NO_3^-, etc.

◘ **Table 4.2   The sources of carbon and energy in the living world.**

		Type of metabolism	Transformation mechanisms
Source of carbon	Organic (polypeptides, glucides, etc.)	Heterotroph	Metabolic pathways of degradation (Krebs cycle, hexose monophosphate pathway, etc.)
	Inorganic (CO_2, C_1 derivatives)	Autotroph	Metabolic pathways for the fixation of carbon (Calvin-Benson cycle, Arnon cycle, Wood-Ljundahl and hydroxipropionate pathways)
Source of energy	Light	Phototroph	Photosynthesis, other forms of phototrophy not linked to the fixation of carbon
	Chemical (oxidation-reduction reactions)	Chemotroph	Respiration, fermentation

ganic or mineral substrates), they are known as chemotrophs (this is the case with the different types of respiration), whereas if this primary source of energy is light, then they are known as phototrophs (this is the case with the organisms that carry out photosynthesis, but not only them: other organisms use light only to obtain energy and are not autotrophs). Organisms known as mixotrophs are able to combine at least

two different forms of obtaining free energy for their cell functions. Mixotrophy is also invoked sometimes when the source of energy comes from light or inorganic redox reactions and the carbon source consists of organic molecules.

a

1,3-bisphosphoglycerate

ADP
ATP

3-phosphoglycerate

Phosphoenolpyruvate

ADP
ATP

Pyruvate

b

Electron transport

Donor
Oxidized donor
Final acceptor
Reduced acceptor
H⁺ gradient
Transmembrane ATPase
ADP + P$_i$
ATP
Interior
Exterior

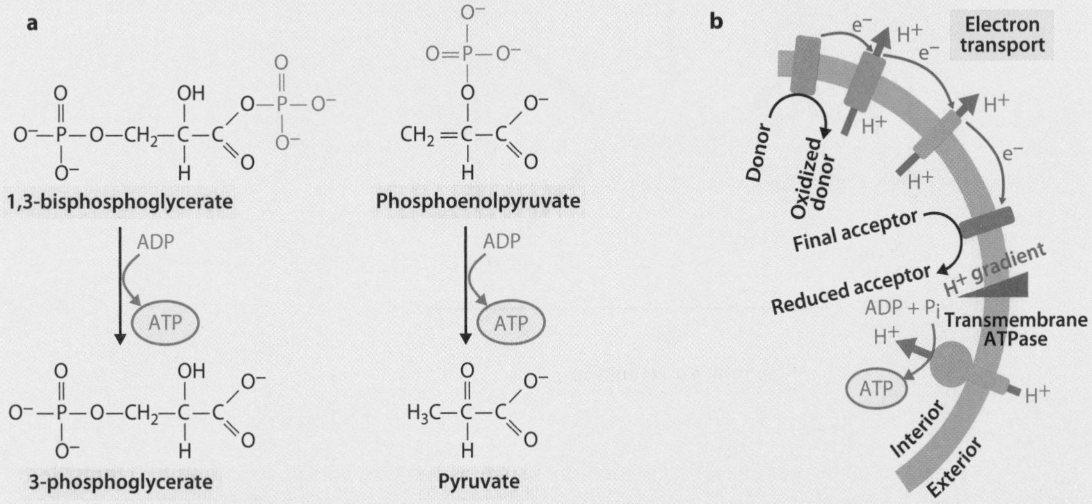

Fig. 4.6 The two major metabolic pathways for the synthesis of ATP. a. Substrate-level phosphorylation, like that involved in fermentation, is a purely chemical mechanism. It occurs twice in the course of glycolysis, during reactions involving phosphorylated species rich in energy: 1,3-bisphosphoglycerate and phosphoenolpyruvate. **b.** Oxidative phosphorylation (respiration) and photo-phosphorylation (e. g. during photosynthesis) lead to

the formation of ATP via a very elaborate physical and chemical mechanism. This includes the formation of a proton concentration gradient induced by a cascade of electron transfers, coupled with the translocation of protons out of the cell. The gradient is subsequently used in a separate process by the transmembrane ATPase (or ATP synthase) to form ATP.

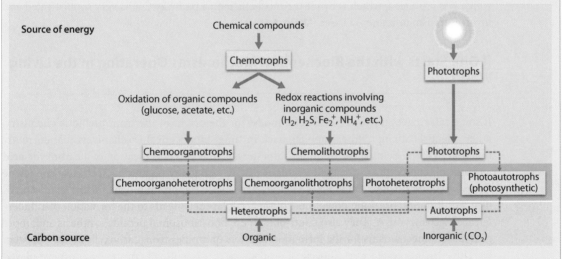

Fig. 4.7 The four major types of metabolism among living organisms. Organisms are classified according to their source of

energy, their source of carbon and the combination of the two.

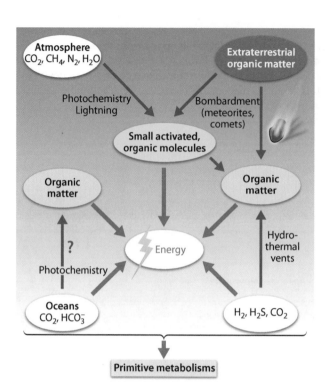

◘ **Fig. 4.8 The abiotic processes producing energy and organic materials** which could have served as a foundation for the development of the first metabolisms.

potential energy in an aqueous medium) such as the polyphosphates, aminoacetonitrile (the precursor of the simplest of the amino acids, glycine) or cyanamide (◘ Fig. 4.9).

Exploiting this process of hydrolysis, which corresponds to a loss of chemical potential energy, requires the coupling with synthetic processes that use the energy released to form something else, in a manner analogous to the process of substrate-level phosphorylation (◘ Fig. 4.6a), rather than to allow it to dissipate uselessly into the environment. It is this coupling that is likely to give rise to networks of reactions that follow a loop, and take on the character of a protometabolism (◘ Box 4.5). It is, in fact, important to bear in mind that modern energy metabolisms were not present and that each input in energy in a protometabolism posed a specific problem because of the probable lack of a universal energy currency such as ATP, and of the pathway by which it is produced, which is by far the most efficient; a pathway that is very sophisticated and involves transmembrane ATPases (◘ Fig. 4.6b).

If One Starts with the Biochemical Mechanisms Operating in the Living World …

If some researchers try to reconstruct a model for the origin of life from prebiotic chemistry, other scientists attempt "to extrapolate back in time" biochemical mechanisms that currently function in living beings. For example, thioesters, formed by esterification of a carboxylic acid and a thiol (compounds with the formula R-SH (R- being any organic radical), like coenzyme A, which is described in the ◘ Box 4.5). They play a very important biochemical role in catabolism. For instance, breaking down of the "cellular fuels" leads to the formation of a thioester, acetyl-coenzyme A. They also contribute in the non-ribosomal peptide synthesis and, more generally, in mechanisms for the formation of esters or amides from carboxylic acids. Christian De Duve has suggested that thioesters, available with prebiotic chemistry would have played a major part in the first stages of life (◘ Fig. 4.13). Other protometabolic pathways have been envisaged, starting from present-day biochemistry, such as, for example, primitive versions of

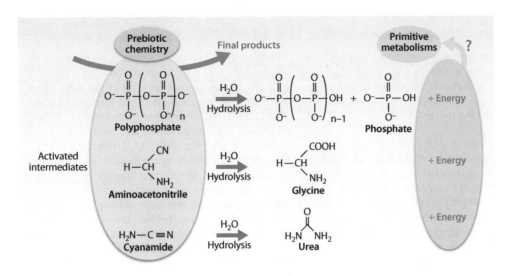

◘ **Fig. 4.9** Potential sources of energy for primitive metabolisms: the hydrolysis of prebiotic chemistry's intermediate compounds (pyrophosphates [n = 1], polyphosphates, aminoacetonitrile, cyanamide). These activated molecules (at a high energy potential) are intermediates in the reactions of prebiotic chemistry. (Their lifetime is limited and
they are thus unstable on a geological time scale.)

the tricarboxylic acid cycle, one pathway that plays an essential role in present-day cellular
metabolism.

If One Starts with the Mineral Sources of Energy Available on the Primitive Earth

There is yet one last possibility to consider in sketching a portrait of primitive metabolisms:
this relies, not on the organic species assumed to be present on the primitive Earth, but on the
mineral sources of chemical energy and, in particular, on the oxidation-reduction processes
that are capable to supplying a metabolism that produced organic molecules.

Hydrothermal systems were very active during the Hadean and the Archaean (*see*
► Chaps. 3 and 6); that is one of the features of the primitive Earth. They are rich in reduced
species (Fe^{2+}, H_2, H_2S, FeS) and, in addition, the minerals present there are likely to play
a catalytic role. Based on this assessment, a certain number of workers – the best-known
of which is the German G. Wächtershäuser (who has proposed a very detailed model of
chemical evolution) – have considered that the redox gradient between these hydrothermal
systems and an atmosphere-ocean system of a neutral redox nature could have formed a
potential source of chemical energy and that an autotrophic metabolism could have thus
emerged, starting with the reduction of CO_2 or CO into organic compounds in a context
such as:

$$CO_2 \quad + \quad N_2 \quad + \quad H_2O \quad \xrightarrow{\quad \overset{[Fe^{2+}, H_2, H_2S, FeS]}{\underset{[Fe^{3+}, H_2O, S, FeS_2]}{}}\quad} \quad [C, H, N, O]$$

Experimental analysis has, however, failed to show the capacity of these reactions to
give rise to a metabolism based on the reduction of CO_2. Nevertheless, thermodynamic

4

◘ Box 4.5   Towards Protometabolisms

The use of chemical energy by biochemical systems results from coupling a reaction that releases free energy and a reaction that uses this energy to synthesize a biomolecule or a metabolic intermediate through the exchange between the two reactions of a species that is common to both.

Let us take the formation of a thioester from carboxylic acid and coenzyme A (HS–CoA). It is a dehydration reaction (liberating water) that is linked to a hydrolysis (absorbing water) of ATP into AMP and inorganic pyrophosphate (PP_i) (◘ Fig. 4.10). In this chemical equation, water no longer appears, but the free energy content of the ATP is used to form another activated molecule (the thioester). Protometabolisms very probably included reactions of this type, the dehydrating agent could have been varied in nature, whether resembling ATP (polyphosphate) or very different.

Protometabolisms could equally have included reaction cycles, that is to say series of reactions where at least one intermediate is recycled in the same metabolic pathway. An example is the formose reaction, which allows the conversion of formaldehyde into sugars (trioses, tetroses, pentoses, etc.) by a pathway that is potentially prebiotic (◘ Fig. 4.11). This reaction offers interesting autocatalytic properties, which means that it includes several intermediate species that intervene as catalysts in a step in the reaction chain that leads to their own formation. This is something that could endow the system with remarkable kinetic performance.

Other protometabolic cycles have been suggested, for example the formation of peptides by activation of amino acids (AA) in the form of N-carboxyanhydrides, NCA (◘ Fig. 4.12).

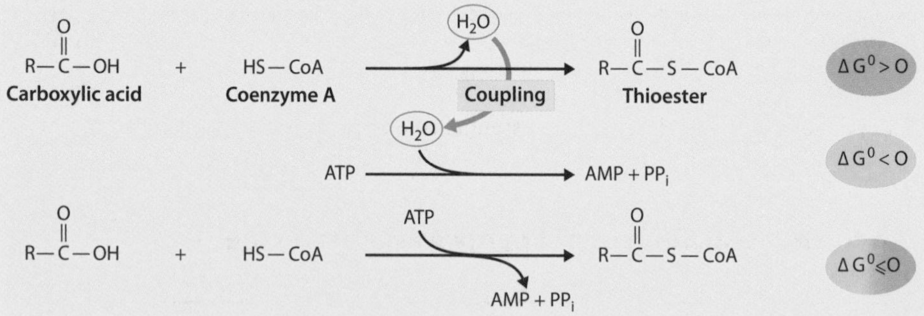

◘ **Fig. 4.10 An example of coupled reactions:** the formation of a thioester from carboxylic acid and coenzyme A. This dehydration reaction is thermodynamically unfavored ($\Delta G^0 > 0$). It is coupled with the hydrolysis of ATP, in a reaction that, this time, is favored ($\Delta G^0 < 0$) from the energy point of view. Thanks to the coupling of these two reactions through the common intermediary that is water, the reaction for the synthesis of thioester is favored overall. PP_i: inorganic pyrophosphate.

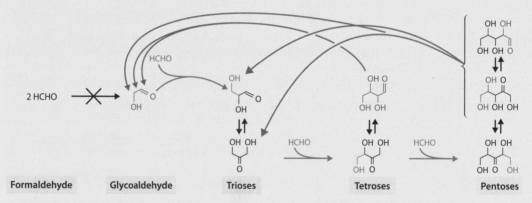

◘ **Fig. 4.11 An example of a reaction cycle: the reaction converting formaldehyde into sugars (the formose reaction).** The autocatalyic nature of this system is linked to its architecture as a network of reactions that enable the formation of sugars thanks to the incorporation of formaldehyde units by aldolization (formation of a carbon–carbon bond by the addition of a formaldehyde unit, shown in red). Autocatalysis is the result of the cleavage of sugars with 4 or 5 carbons (tetroses, pentoses) into two of their precursors by retro-aldolization (shown in blue). This cleavage permanently restores the glycoaldehyde and the trioses necessary for subsequent cycles. Whereas the direct dimerization of formaldehyde into glycoaldehyde is a reaction that is almost impossible without assistance, autocatalysis confers exceptional kinetic performance to the system.

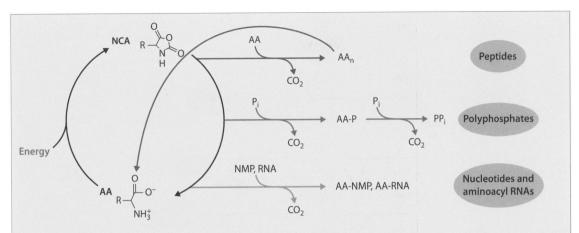

Fig. 4.12 An example of a potential protometabolism consisting of reactions involving the N-carboxyanhydrides of amino acids (NCA) based on experimental observations. Various potential prebiotic pathways turn out to be capable of leading to these forms of activated amino acids which may then react with the amino acids themselves (AA) to produce peptides (AA$_n$), thus forming a protometabolism allowing the recycling and potentially selection. They may also react with inorganic phosphates (P$_i$) to produce pyrophosphate (PP$_i$) with polyphosphates mimicking ATP and, finally, with nucleotides or RNA to produce RNA and aminoacyl nucleotides (AA-RNA and AA-NMP, respectively), which are important biochemical intermediates that we shall encounter later when analyzing the emergence of translation mechanisms.

studies have shown the feasibility of the principle of hydrothermal synthesis of organic molecules such as methane or short hydrocarbons starting with the hydrogen produced in these systems:

$$CO_2 \text{ (aq)} + 4\,H_2 \text{ (aq)} \longrightarrow CH_4 \text{ (aq)} + 2\,H_2O \quad \Delta_r G^0 \text{ (b)} = -193.73 \text{ kJ mol}^{-1}$$
$$\text{methane}$$

It is even theoretically possible to form amino acids in hydrothermal fluids. However, the low concentrations observed experimentally as well as the high temperature of these systems have led researchers to consider adsorption mechanisms. The products formed could have accumulated on the surface of minerals in an environment potentially capable of preserving them from degradation and hydrolysis at high temperatures. As it happens, these surfaces would also have been able to immobilize the various organic species present on a physical substrate, thus creating partial confinement, before the appearance of the first compartments. It is this sort of argument that led G. Wächtershäuser to propose that life was born "in two dimensions", that is on mineral surfaces at hydrothermal vents (*see later*).

An important question is that of the exploitation of the energy derived from oxidation-reduction reactions involving mineral compounds. Apart from the reduction of mineral carbon (CO, CO$_2$) into usable organic molecules, how was this energy redistributed by the primitive metabolism? Modern biochemical mechanisms coupling oxidation-reduction to the production of energy and its transport in the form of ATP are, we will recall, extremely sophisticated, because they imply the creation of a proton gradient and an ATPase (except in fermentation) (■ Fig. 4.6b). This great complexity could not have been put together by anything other than a long process of evolution, and leaves us few indications of the first pathways of the exploitation of energy and, above all, of its redistribution through reactions of the redox type. This is why many authors imagine that the organic material formed abiotically could give rise to degradation processes (fermentations) likely to lead to the production of ATP by substrate-level phosphorylation, a process that implies the formation of a high-energy phosphate bond with ADP (adenosine diphosphate), starting with an activated organic substrate, independent of membrane ATPases, (■ Fig. 4.6a).

4

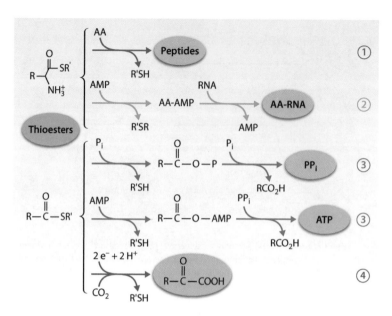

◼ **Fig. 4.13 Reconstruction of primitive metabolisms from present-day biochemical mechanisms: the thioester example.** Shown here are the essential reactions in various, hypothetical, ancestral metabolic pathways, playing a key role in several present-day metabolic pathways. These reactions, proposed by the biologist Christian De Duve, may lead to essential biochemical intermediates. The thioesters could be involved in the formation of peptides (1) and in that of aminoacyl adenylates (AA-AMP) and aminoacyl RNA (2′ or 3′-ester = AA-RNA, molecules with a key role in the translation of messenger RNAs into proteins; ◼ Box 4.8), starting with amino acid thioesters (2). They could be involved in the activation of inorganic phosphate (P_i), giving activated phosphoanhydrides like pyrophosphate (PP_i) or ATP (3), and in the formation of carbon-carbon bonds from thioesters, essential for an autotrophic metabolism (4). The first four reactions shown here are very similar to those observed experimentally for amino acid N-carboxyanhydrides (◼ box on previous page).

Apart from the redox gradient observed at hydrothermal chimneys between the reduced rocks originating in the mantle and the ocean–atmosphere system, other sources of energy were present on the primitive Earth (◼ Fig. 4.8). Accordingly, different scenarios can be envisaged for life's metabolic diversity, through listing the possible metabolisms starting with the all possible electron donors and acceptors. If one takes as a basis the variety of electron donors and acceptors functioning in present-day metabolisms (◼ Box 4.4), the range of possibilities is wide open.

The Origin of Genetic Systems

For the process of Darwinian evolution, characteristic of life, to take place, a living being must have a genetic system that, independently of its initial nature, on the one hand carries information encoding the instructions for its own production and functioning and, on the other hand, can be replicated and be modified through mutation, generating variants upon which natural selection operates.

Translated to molecular scale, that means that a minimal form of life must include a physical support for information content (genetic material), with at least one chemical activity allowing its replication. This dual nature (informative support – replicating activity) is found in modern living organisms in the form of two types of biopolymers that are specialized in each task. The role of information carrier is carried out by nucleic acids (DNA and RNA), whereas chemical

Box 4.6 The Origin of Life: Hot or Cold?

Scientists who favor a heterotrophic metabolism for the first organisms (p. 103) generally maintain the birth of life took place at low temperature, because temperatures that were too high would not have favored the emergence of organized systems, nor the synthesis and accumulation of organic molecules in a primitive soup. They would instead tend to degrade them. Nevertheless, more recent versions of these models accept that a contribution of organic molecules arising from sub-marine hydrothermal vents and synthesized by various pathways one of which is the Fischer-Tropsch reaction, could have contributed to the enrichment of a primitive soup that was not very hot. On the opposite side of the argument, the partisans of an autotrophic metabolism for the first organisms (see p. 107) support the idea that life appeared at high temperature in underwater hydrothermal systems.

activity is primarily carried out by proteins, including the enzymes responsible for nucleic acid replication. This division of labor is based on the sophisticated system that enables the translation of the message carried by the nucleic acids in proteins. The latter is itself based, on the one hand on the genetic code and, on the other, on a large number of components that cooperate within what is a whole machinery of transcription. This consists of the ribosome, transfer RNA, aminoacyl tRNA synthetases, etc. (■ Box 4.7 and ■ Box 4.8). It is clear that the genetic code and the efficient transcription machinery that we see today is the fruit of an optimization process, itself the result of a long evolution. Their degree of complexity is such that natural selection must have first been exercised on more primitive systems, where, undoubtedly, the synthesis of proteins *coded* by a genome was not yet in place. What could these first genetic systems have looked like? To reply to this question, let us try, once again, to go back in time.

The Hypothesis of the RNA World: Why and How?

The separation of the genetic information from biochemical activity is generally associated with the dual nature of DNA and proteins, as we will recall. This gives rise to the famous dilemma "which came first: the chicken or the egg?". DNA is needed to make proteins, and proteins are needed to make DNA, so was it DNA or the proteins that emerged first? Looking at the matter more closely, however, the distinction becomes less clear-cut.

First of all, a number of enzymes use nucleotides as cofactors (■ Fig. 4.15). The metabolic pathways in which they act are very varied, and it is possible to propose the hypothesis that these cofactors, far from being a simple evolutionary curiosity, were present even before the emergence of the translation machinery. They would have survived as "living molecular fossils"; a sort of relic of a stage where the biochemical reactions were not catalyzed by proteins encoded by genes.

Later, and above all since the beginning of the 1980s, it has been realized that certain RNAs possess catalytic activity. These RNA forms have been called ribozymes (in 1989, the Nobel Prize for Medicine was awarded to T. Cech and S. Altman for their work on this type of RNAs). Subsequently, it became progressively apparent that the formation of the peptide bond – which forms the skeleton of proteins by linking covalently the amino acids between them– in the ribosome was carried out by a ribosomal RNA (rRNA) and not by ribosomal proteins. In 1992, H. Noller obtained the first proof of this idea, which he had put forward in 1972. That rRNA catalyzes the peptide bond received unambiguous confirmation in 2000, with the crystallization and fine-scale resolution of the structure of the ribosome (which granted Venkatraman Ramakrishnan, Thomas A. Steitz, and Ada E. Yonnath, by their important contribution to studies on the ribosome structure, the Nobel prize in 2009), making the ribosome the most sensational ribozyme discovered to date.

◘ Box 4.7   The Genetic Code

The cell expresses the information contained in DNA (which forms its genetic material) such that it is able to construct its own proteins. This synthesis takes place in the ribosomes, where the peptide bond forms. Peptide bonds are covalent links between amino acids, giving rise to polypeptides which, in turn, form proteins. How is the information contained in a DNA sequence, that is, in a combination of nucleotides containing the four DNA bases (A, C, G, T), translated into the right combination of amino acids from among the twenty possible amino acids that can make up a protein? Thanks to the genetic code. Each of those twenty amino acids is coded by a triplet of bases (nucleotides), known as a codon. The signals for the start and end of the synthesis of a protein are equally marked by codons. The genetic code, which establishes the correspondence between a given codon and an amino acid, has three fundamental properties. It is:

– universal: every living organism uses the same code (there are some exceptions, but they correspond to minor derived changes). Certain computer simulations suggest that the genetic code corresponds to the optimum code, taking account of the physical and chemical properties of the amino acids coded by the different codons;

– redundant: several codons may correspond to a single amino acid. This is the consequence of the fact that there are sixty-four possible codons for only twenty amino acids to be coded;

– unambiguous: a single, unique amino acid corresponds to a single codon.

For the message contained within the DNA to be efficiently translated into proteins, each gene must first be transcribed into messenger RNA (mRNA). The message is then still in the form of combination of four nucleotides carrying the bases A, C, G and U (uracil, U, replaces thymine, T). It then needs an "adaptor" capable of making the link between a codon and a specific amino acid. This function is fulfilled by transfer RNAs (tRNA), specific to each amino acid. In the ribosome, positioned opposite to each codon of the mRNA comes a tRNA bearing the corresponding amino acid according to the genetic code. The tRNA recognizes its specific codon thanks to a sequence known as an anticodon (a complementary triplet sequence to the codon). In this way, a message in an alphabet with just four letters (the four bases) can be translated into an alphabet with twenty letters (the twenty amino acids).

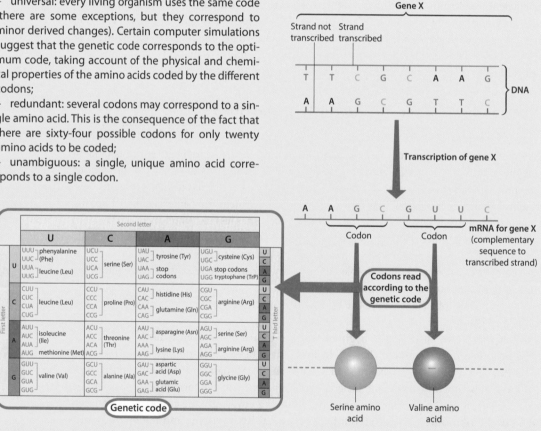

◘ **Fig. 4.14 The genetic code, or how to go from an alphabet with four letters (the four bases) to an alphabet with twenty letters (the twenty amino acids).**

This is what brought about a revolution in thinking about the emergence of life and, more generally, in biochemistry. No protein (and thus no enzyme) could form without the benefit of the catalytic activity of a nucleic acid!

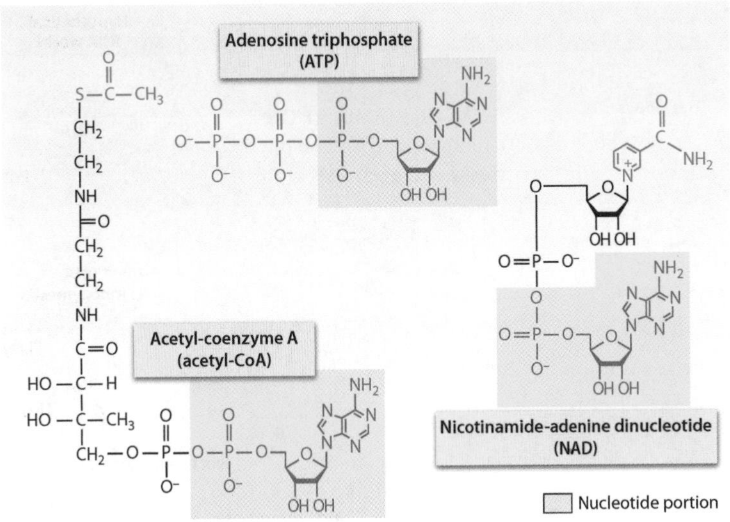

Fig. 4.15 Some cofactors that include a nucleotide portion (blue). The latter is at a greater or lesser distance from the reaction center (red). Given the multiple involvement of this type of cofactors in metabolism, it can be hypothesized that they were present even before the emergence of the translation machinery. They would then represent "living molecular fossils" from a stage in the evolution of life when biochemical reactions were not catalyzed by encoded proteins.

All these data could be associated with RNA's abilities in replicating itself in the absence of enzymes. They are, of course, limited, but they have been demonstrated experimentally on small-size model sequences. RNAs that catalyze the formation of covalent bonds between ribonucleotide sequences matched to a complementary matrix have been selected starting from assortments of random sequences produced by molecular biology methods. (It may be noted that the DNA also reveals this capacity for auto-recopying, even though DNA is less reactive than RNA.) This shows that a ribozyme can theoretically intervene in the replication of genetic information carried by RNA. We should, however, point out that a ribozyme capable of an "RNA replicase" activity – that is to say capable of forming the complementary sequence (by matching, according to Watson and Crick) starting from any strand of RNA, and then returning functional copies of the original molecule – still remains to be discovered.

This information led to the suggestion that the living world, nowadays dominated by the DNA-protein system, was preceded by a stage where RNA played a major role, both as a catalyst and as a bearer of information. This hypothesis was expressed very explicitly in 1986 by Walter Gilbert, who coined the expression "RNA world" for it, but the idea that the replication of nucleic acids preceded the formation of proteins had already been suggested at the end of the 1960s independently by three workers: Carl Woese, Francis Crick and Leslie Orgel. Ideally, one might even say that the phenotype expressed by this information consisted precisely of this capacity of direct catalysis of metabolic reactions, allowing selection of the most active sequences. The "RNA world" hypothesis (Fig. 4.16) excludes in principle any role for coded proteins, but that does not prevent most of its partisans from considering, nevertheless, that uncoded amino acids or certain peptides (formed not thanks to a translation process, but through pathways such as those discussed in the preceding section, or even – and why not? – by reactions involving RNA) were potentially present and even possessed catalytic activities.

It may be noted that, speaking evolutionarily, this idea of an RNA world is not at all revolutionary: there is no reason to think that, ever since the very early stages of life, biomolecules have played a role identical to the one that they do at present, because evolution is, by its very nature, an opportunistic process. RNA could therefore have first carried out the function of a coded catalyst, thanks to its propensity to adopt folded structures, and was thus likely to form cavities (acting as recognition sites or active sites) that could accept molecules and transition states of various reactions. Later, living systems would have evolved towards separating the information and catalytic tasks between two different types of polymers. Nucleic acids store the coded information, DNA forming the genome and RNA being a transient messenger of that information,

4

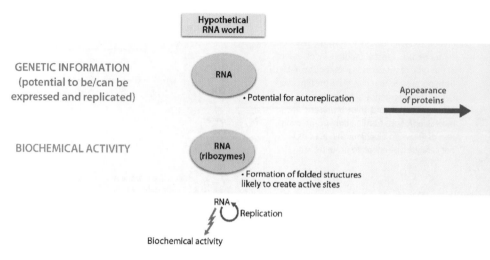

GENETIC INFORMATION
(potential to be/can be
expressed and replicated)

BIOCHEMICAL ACTIVITY

Fig. 4.16 The "RNA World" hypothesis. Proteins are required to make DNA, and DNA is required to synthesize proteins. How can this evolutionary dilemma be resolved? Based on the capacity of RNA to replicate itself in the absence of any enzymes and on the existence of RNAs that carry out catalytic activities (ribozymes), some authors suggested that the present-day world (where DNA stores genetic information and proteins carry biochemical activity)

thanks to their structure which is more or less linear and well suited for replication. The proteins carry out the catalytic function thanks to their capacity to fold and for specific recognition.

An additional argument in favor of the RNA world comes from the fact that the biosynthesis of the building blocks that form DNA, the deoxyribonucleotides, involves the reduction to deoxyribose of the ribose moiety of the corresponding ribonucleotides, and not by a direct metabolic pathway (Fig. 4.17). Similarly, thymine, that is the nitrogenous base which replaces uracil in DNA, is formed starting with uridine. UTP is reduced to dUTP, then the uracil is methylated. These biochemical indications strongly suggest that DNA is a sort of altered structure derived from RNA and that it acquired a major role in life after the establishment of a well-developed RNA metabolism.

In this context, evolution starting from the RNA world implies two major events. The first is the emergence of coded proteins based on genetic information. It involves the establishment of the genetic code and all the machinery for the synthesis of proteins (Box 4.8). The second is the change in the physical support for information from RNA to DNA. We have no precise information concerning the order in which these transitions took place, but most authors consider that the machinery for translation, involving just ribonucleotides and proteins, appeared at the stage of the RNA world, and therefore that this stage was followed by an RNA-protein stage. However, there is no certainty over this point. However that may be, the chemical structure of DNA, which provides it with better stability than RNA (Fig. 4.18), offers an explanation of the evolutionary success of the "DNA world" that we live in. DNA would have allowed an increase in the size of the genome that could be transmitted from one generation to another, and thus of the number of genes carried and the complexity of the corresponding organisms.

An RNA World? And What Came Before?

The hypothesis of the RNA world being based on an analysis of modern biochemistry, it is possible, certainly with much uncertainty, to envisage evolutionary pathways from this world to the DNA-protein stage. It is completely different with the nature of the evolutionary path that led from prebiotic chemistry to the RNA world stage. The question remains without any answer to this day, and one can only risk a few conjectures.

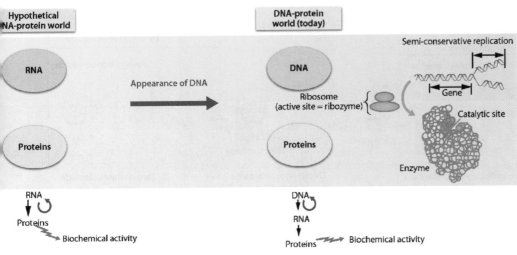

was preceded by a stage when RNA played a major role both as catalyst and as the bearer of genetic information. The RNA world hypothesis was reinforced by the discovery that, within the ribosome, it is a ribozyme that catalyzes the formation of the peptide bond. Most researchers believe that the translation machinery, involving just nucleotides and proteins, appeared at the RNA world stage, and that it was followed by RNA-protein world.

In a chronological context, it is logical to consider stages of increasing complexity. That is the classic vision of things. A first stage would correspond to the accumulation of monomers – the nucleotides – from organic sources. It would be followed by a polymerization stage, first random, then by the selection of polymers endowed with the ability to replicate themselves and evolve by mutation and natural selection. This approach seems simple, but considering the sequence of stages that it involves, there is no obvious way to find a motive force for such an evolution, because before the establishment of the RNA world (and thus of a system that was subjected to selection and capable of replicating), everything rests on a chain of very improbable events, even in the eyes of those who defend this model.

The point is all the more sensitive in that – and we shall very shortly come back to this point – a simple method of prebiotic synthesis of the building blocks of RNA (the ribonucleosides and ribonucleotides) has still not been found. The point is significant, because it then seems contradictory to assume that life developed starting with just ribonucleotides, whose presence is problematic, when we have good reason to think that amino acids, and even non-coded oligopeptides, were present on the early Earth, and undoubtedly capable of various activities including catalysis.

This lack of information is characteristic of the non-living / living transition, where researchers' only possibility is to construct scenarios whose plausibility rests on the likelihood of chemical processes and not on historical data, which are completely absent. It is then probable that the RNA world corresponded to a reality that was less strict than its name would suggest. Some workers even believe that this molecule did not form the initial bearer of genetic information.

Without going that far, the stage represented by the RNA world would have, in fact, corresponded to a subset of the molecules that were indeed present on the Earth, despite being lost in the mass, but which we would consider to be qualitatively significant among all those that led to dead-ends. Naturally, we have no information concerning possible relationships between the nucleotides and other chemical systems that have now disappeared. A close interaction between the chemistry of amino acids and that of nucleotides is also conceivable. (There are experimental arguments for this, as we shall shortly see.) Under these conditions, it becomes very difficult to discern what distinguishes a "pure" RNA world from that which is already close to the transition to an RNA-protein system. A large overlap between the different stages – corresponding to their co-evolution – would thus be possible.

4

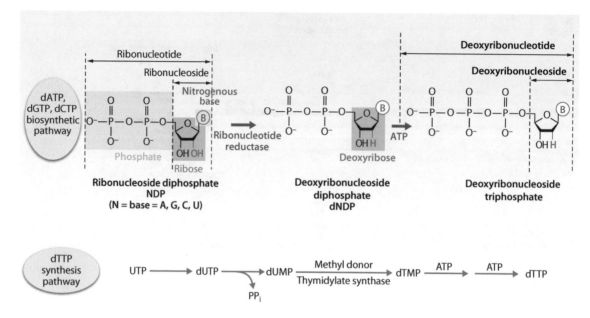

◘ **Fig. 4.17 Biosynthesis of the building blocks for DNA: the deoxyribonucleoside triphosphates.** The deoxyribonucleotides in DNA appear like sorts of "derivative products" of the RNA ribonucleotides. Indeed, the deoxyribonucleoside diphosphates dADP, dGDP and dCDP are synthesized from the corresponding ribonucleotides (ADP, GDP and CDP). They are subsequently phosphorylated to form the deoxyribonucleoside triphosphates (dATP, dGTP and dCTP) that will be incorporated in the DNA. Deoxythymidine triphosphate is synthesized through a different pathway, but still starting with a ribonucleotide (UTP). These data provide additional support for the RNA world hypothesis, because they suggest that it was from RNA that the modern world in which DNA and proteins play an essential role and RNA only – apparently – plays a subsidiary part was established.

One of the principal problems concerning the hypothesis of the RNA world is that it appears quite unlikely that a prebiotic environment could have existed containing the mixture of activated nucleotides favoring the formation and replication of ribozymes, as well as their evolution through natural selection. Even if there were several candidate reactions for an efficient prebiotic synthesis of nucleic bases, access to monomeric nucleotides by chemical pathways in fact comes up against several obstacles.

If one goes no further than mimicking the biochemical pathway, the first difficulty that occurs is that of synthesizing ribose, which is formed in just negligible quantities within the complex mixture obtained by polymerization of formaldehyde, and, what is more, has a limited lifetime. The bond between a nucleic base and ribose that produces a nucleoside is then a very difficult reaction. There still remains the matter of obtaining a nucleotide by phosphorylation, which leads to mixtures because three positions remain available on the ribose, and then there is its activation (◘ Fig. 4.20). So there are two possibilities, either to envisage an easier pathway for the prebiotic synthesis of nucleotides, or to squarely reject RNA as the initial bearer of information, in favor of an alternative bearer that has not left any evolutionary traces. These two approaches have been the subject of a certain number of studies that have had some success.

First of all, alternative synthesis pathways have been discovered, which lead not to ribose and then to its phosphorylation, but to the construction of a ribose group that is already phosphorylated by assembling elements consisting of two or three carbon atoms that could be obtained by a selective prebiotic phosphorylation pathway. This is a new example illustrating an important idea: present-day biochemical synthesis pathways are not necessarily related to those that were used in conditions where the basic building blocks were bound to be different from those derived from modern metabolic pathways.

A decisive success in this regard was obtained in 2009 by the group led by John Sutherland (a British nucleotide chemist) who showed that at least two of the set of four nucleotides can be

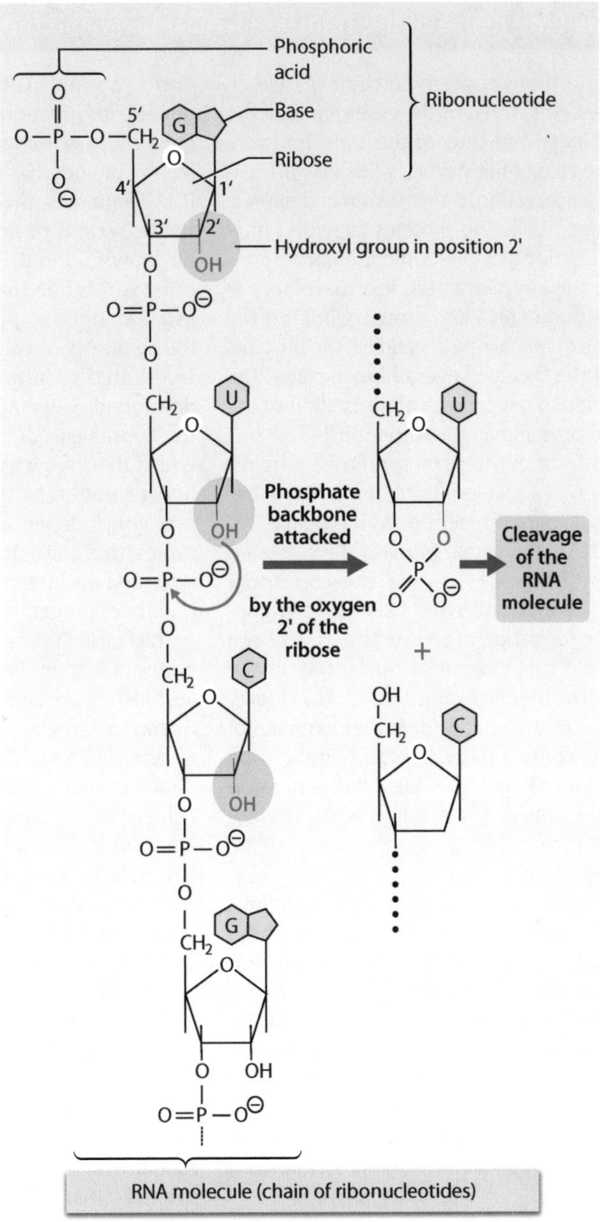

Fig. 4.18 The chemical instability of RNA. This is explained by the presence of a hydroxyl group in position 2′, which results in an easy strand cleavage through an intramolecular reaction. Such a cleavage is impossible in DNA, where the hydroxl group at 2′ is absent. The increased stability of DNA offers an explanation for the evolutionary success of this molecule to store information.

synthesized under a weakly activated form by a potentially prebiotic pathway, without involving the condensation of ribose with a free base (which by its nature is poorly reactive), and their preliminary synthesis. Indications that similar processes may yield the two other nucleotides have been reported in 2011 by this group in collaboration with that of Jack Szostak in Boston. With such results, one conclusion is obvious: we cannot definitely exclude the idea that RNA was the first bearer of genetic information. Analysis of the pathways for the synthesis of this information-carrying polymer and of its interactions with amino acids, peptides and other molecules present in the environment of the primitive Earth, must remain one of the objectives for the study of the chemistry of the origin of life in the near future.

Despite this, as we have said, many workers consider that RNA could not be the initial bearer of information, and postulate that a pre-RNA stage took place. Their approach consists of determining how the ribose-phosphate skeleton of RNA could be replaced. This way of thinking is already ancient: the ancestor of the nucleic acids may perhaps have been an-

4

◘ Box 4.8   The Emergence of the Translation Machinery and the Genetic Code

One of the key questions concerning the early evolution of genetic systems is that of the emergence of the translation machinery and thus of the transition to a system where the role of the nucleic acids was primarily linked to the conservation, transmission and expression of information, while the proteins provided multiple metabolic functions (catalysis, recognition, etc.).

The ribosome, the key element in this machinery, is a structure consisting of RNA and proteins, but it is the RNA that plays the essential role of catalyst, the proteins being responsible for "keeping everything in place". The RNA is also involved in the process of translation of the nucleic acids into proteins via tRNA and mRNA.

The mechanism of protein biosynthesis requires both the formation of a covalent bond (ester bond) between the amino acid (or the peptide as it is being synthesized) and tRNA – the whole forming the activated aminoacyl-tRNA (◘ Fig. 4.19) – and the correspondence between the anticodon and its specific aminoacyl-tRNA. It is the formation of an ester bond between an amino acid and the 3'-terminal nucleotide of the corresponding tRNA that is responsible for the fidelity of the translation. Any scenario for the emergence of the translation machinery based on the genetic code must, therefore, incorporate the molecular dimension of the interactions between the amino acids, nucleotides and RNA that are at work during the course of the translation process.

These interactions first appear in the reactions catalyzed by the aminoacyl-tRNA synthetases. These enzymes, present in a number that is at least equivalent to that of the amino acids, give rise to aminoacyl adenylate (AA-AMP), then aminoacyl-tRNA (AA-tRNA). These two species involve a covalent bond between, on the one hand, an amino acid, and on the other, a nucleotide or an RNA. Then, during translocation (on the ribosome), the tRNA is linked by a covalent bond to the whole polypeptide that is being synthesized (◘ Fig. 4.19).

Under these conditions, it is logical to consider that the translation machinery emerged starting with chemical interactions (covalent bonds) between nucleotides and amino acids. This conclusion is valid, whatever hypothesis is held about the initial basis for the genetic information. In the setting of the RNA world, in its most extreme version (an RNA-only world), molecular biology has shown that it is possible to select RNA sequences that possess the capacity to catalyze the activation of amino acids, the aminoacylation of nucleotides and the formation of a peptide bond – in other words, the three key stages in translation – thus establishing the possibility of a transition towards the synthesis of coded proteins starting from an RNA world, that is, in a context where ribozymes, already developed, were active. But if instead one holds to the hypotheses of co-evolution (see pp. 124–129) between the chemistry of amino acids and the chemistry of nucleotides, there are experimental arguments that establish, independently of any catalytic action by ribozymes, that covalent-type interactions between amino acids and nucleotides could have been established in a prebiotic context.

The feasibility of a chemical aminoacylation of nucleotides (a reaction involved in the formation of aminoacyl-RNA) has been shown in the absence of enzymes, thanks to pathways involving the N-carboxyanhydrides of amino acids (◘ Box 4.5). The thioester pathway (Christian de Duve's hypothesis, ◘ Fig. 4.13) could equally allow chemical covalent bonding between amino acids and nucleotides or nucleic acids. An evolution based on the development of interactions of this type could then have led directly to the RNA-protein stage, without have to call on the RNA-only world without proteins but with ribozymes.

There are many other questions without a definitive answer at present. How did an initial set of amino acids become established from the numerous amino acids whose abiotic synthesis is possible (about seventy have been detected in meteorites)? How did a codon come to be associated with them? Undoubtedly biochemical pathways for synthesizing what was probably a restricted initial set of amino acids developed. By subsequent evolution and diversification (possibly very late), certain codons, which at the start were redundant (that is, they coded an amino acid that was also coded by other codons), could have been assigned to additional

other organic polymer, or a mineral substrate (such as clays, as postulated by Graham Cairns-Smith), that served as a skeleton for the nucleic bases. This implies, of course, that the carrier of the genetic functions subsequently changed without leaving any trace, first, in favor of RNA, and then in favor of DNA in the DNA-protein system (a system where RNA has, however, preserved certain essential functions, which probably appeared at the RNA-world stage with, in the very forefront, its role in the translation machinery).

The nature of a possible precursor to RNA remains largely hypothetical. Logic suggests that we look for simplicity, and thus models based on structures of limited size, like non-chiral analogues of sugars or even peptides – which may seem paradoxical, because in the RNA world *sensu strictu*, the peptides do not appear until much later and only with the aid of ribozymes

amino acids (which are synthesized by living systems thanks to biochemical pathways derived from those used for the initial set), leading to the universal system that functions today.

However things may have happened, the question of the emergence of the genetic code is in itself ex-tremely difficult because the functioning of the transla-tion machinery, even simplified, requires the presence of a large part of its present-day components, among which it is difficult to identify those that could have ap-peared first. Several models have been suggested, but none has been firmly proven.

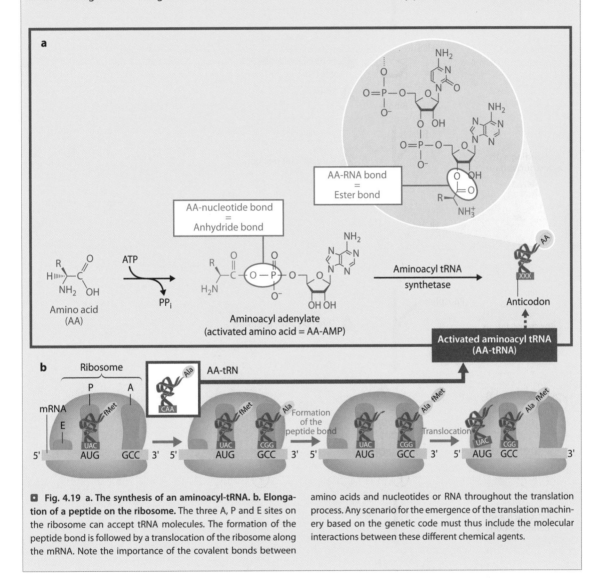

☐ **Fig. 4.19 a. The synthesis of an aminoacyl-tRNA. b. Elonga-tion of a peptide on the ribosome.** The three A, P and E sites on the ribosome can accept tRNA molecules. The formation of the peptide bond is followed by a translocation of the ribosome along the mRNA. Note the importance of the covalent bonds between amino acids and nucleotides or RNA throughout the translation process. Any scenario for the emergence of the translation machin-ery based on the genetic code must thus include the molecular interactions between these different chemical agents.

(☐ Fig. 4.21). Over the last fifteen years it has been shown that the ribose-phosphate skeleton is not indispensable in double strand formation through *Watson and Crick's base-pairing* rules. On the other hand, no functional alternative has so far been found for the set of four nucleic bases, whose properties seem to correspond to a chemical optimum in terms of the efficiency of recognition and, above all, of limiting mispairing errors. The alternative structures to RNA should equally show analogous catalytic activity. It is worth noting here that this capacity for ca-talysis is shared with DNA itself. So, from this point of view, nothing excludes DNA from being the precursor to RNA, even if available evidence make this hypothesis improbable (*see earlier*).

In summary, no definitive answer exists today as to the nature of the first genetic systems and the way in which they emerged. What Leslie Orgel (the chemist, who, like Stanley Miller, died

4

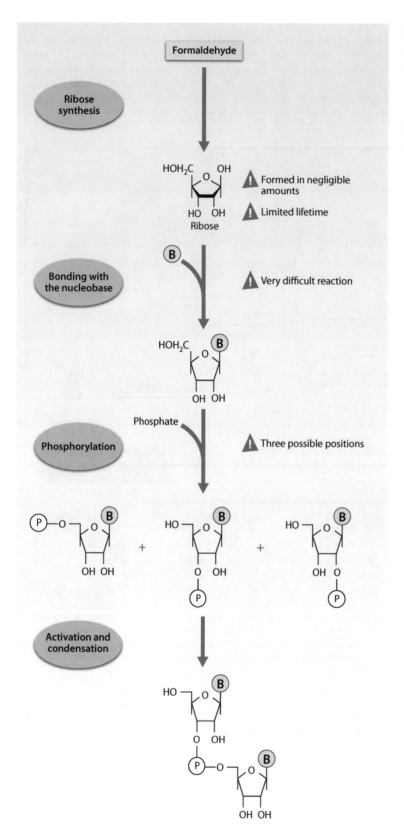

☐ **Fig. 4.20** **It's difficult to reproduce in the laboratory the biochemical pathways for the synthesis of monomer nucleotides supposedly present in the RNA world!** So it is necessary either to envisage a "simpler" pathway for the prebiotic synthesis, or else to abandon the hypothesis supporting RNA as the first bearer of genetic information. Recent research has given arguments in favor of both opinions.

in 2007, and who was among the first, in the 1960s, to put forward the role of RNA in the first stages of evolution) and his collaborator Gerald Joyce (to whom the molecular biology of ribozymes owes a lot), called the "molecular biologist's dream", namely the vision of the emergence of RNA from a mixture of monomer nucleotides, proves to be the prebiotic chemist's nightmare.

In years to come, alternative approaches must therefore be tried. We will mention two of them:

- to refine the hypothesis of the RNA world, starting either from a potential predecessor of RNA or from a re-examination of the possibilities for the prebiotic synthesis of RNA or, more precisely, of activated nucleotides. This re-examination should discard the idea that prebiotic pathways necessarily mimicked the modern biochemical pathway of synthesis;
- examining the hypothesis of co-evolutionary scenarios. We shall shortly return to this point.

However that may be, it will be noted that the hypothesis of an RNA world has the unique advantage of lending itself to experimentation with the tools of molecular biology, something that remains impossible at present with other information carriers that are usually incompatible with the enzymatic systems fashioned by the evolution of life on Earth.

The Origin of Compartments

In the process of the emergence of life, the formation of the individual, first compartments cannot be placed in a relative chronology with any certainty. For many scientists, the birth of life is linked to the *initial* appearance of compartments: this is what permitted the emergence of individual entities capable of self-maintenance (having a metabolism), of accumulating information, of reproducing and of being subjected to natural selection. For others, in contrast, populations of molecules capable of replication, not isolated from the environment, emerged first. Compartments would have appeared only at the end of the process by which life emerged, permitting the formation of sub-assemblies of molecules, that is individuals, upon which natural selection might act. Once again, we cannot decide between these two views.

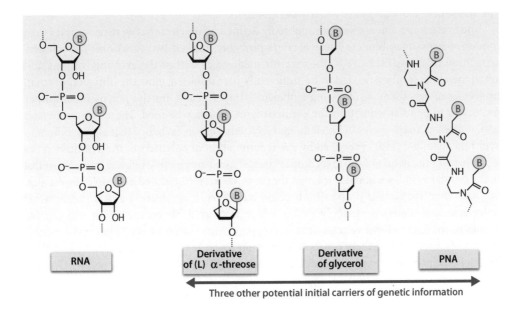

Three other potential initial carriers of genetic information

☐ **Fig. 4.21 RNA and three other molecules capable of carrying genetic information.** Derivatives of (L)-α-threose, of glycerol, and chains of β-amino acids (PNA = *Peptide Nucleic Acids*). Certain scientists believe that these molecules carried the genetic information before RNA came on the scene. The hypothetical RNA world would then have been preceded by a "pre-RNA stage".

4

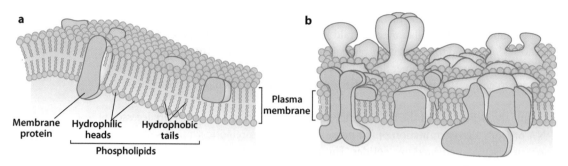

a

Membrane protein   Hydrophilic heads   Hydrophobic tails

Phospholipids

b

Plasma membrane

⬛ **Fig. 4.22 The structure of the plasma membrane of the cell. a.** the "fluid mosaic" model suggested by Singer and Nicholson in 1972; **b.** a more realistic model, currently favored, where the membrane is less homogeneous. A living system must possess an arrangement that allows the confinement of its components. In the cell, this function is served by the plasma membrane, which ensures the integrity of the cell and controls the exchanges with the surrounding medium thanks to its associated proteins.

Let us first observe present-day cells. They are surrounded by membranes that ensure their integrity, allow exchanges with the surrounding medium (with the diffusion of gases, active transport of ions and small molecules) and harbor systems for transduction (transformation) of an electrochemical energy that is not directly usable by the cell into a form of chemical energy that is. These systems create a proton gradient across the membrane thanks to primary sources of energy (from light or chemicals). Subsequently, the energy liberated through the passage of the protons along the gradient is transformed into energy that can be stored by the cell in the form of ATP, a molecule that has high-energy bonds. This universal property was advanced in 1961 by Robert Mitchell who accordingly enounced his chemiosmotic theory. For his work, Mitchell received the Nobel Prize for Medicine in 1978.

Cell membranes basically consist of phospholipids, which are amphiphile molecules, that is to say that they possess a hydrophilic head (consisting of glycerol-phosphate) and hydrophobic tail (which may be long fatty acids or isoprenoid derivatives) (⬛ Fig. 4.30). In the majority of cases these are organized in bilayers (in some organisms the two bilayers may fuse to form transmembrane bipolar lipids) within which various proteins involved in the transport and transduction of energy are located(⬛ Fig. 4.22).

The first compartments were undoubtedly defined by barriers that were far simpler than modern plasma membranes. These barriers probably fulfilled two functions: the physical separation of the self relative to the exterior medium as well as the exchange of ions and small metabolites. They needed to be sufficiently permeable to allow the diffusion of small molecules across them, while being sufficiently impermeable for the polymers and other molecules to be retained within the compartment that they defined. These two properties had, moreover, to co-evolve, with elementary systems for ion exchange that allowed them to avoid the "osmotic crisis" provoked by the accumulation of polymers in the primitive cells.

What was the nature of the first compartments? Lipid amphiphile bilayers being universal in the cellular world, we might think that the proto-cells were bounded by simpler molecules, but ones that had similar properties. It could have been long chain monocarboxylic acids (fatty acids), alcohols, or monoglycerides, which were available on the planet and capable of self-assembling to form vesicles, that is, compartments surrounded by bilayers of amphiphile molecules turning their hydrophilic portion towards the aqueous medium and their hydrophobic portion towards the inside of the bilayer (⬛ Fig. 4.23). Probably neither Oparin's coacervates (simple spherical aggregates of organic proteinaceous molecules) nor micelles (formed of an assembly of amphiphilic molecules that gather their hydrophobic portions into a core and their hydrophilic heads at the surface), participated in the emergence of cellular compartments. Both, micelles and coacervates, lack the ability to isolate any form of protocytoplasm – that is, a pocket that could contain a diversity of molecules in solution – from the exterior medium. Amphiphilic vesicles are fully capable of doing so.

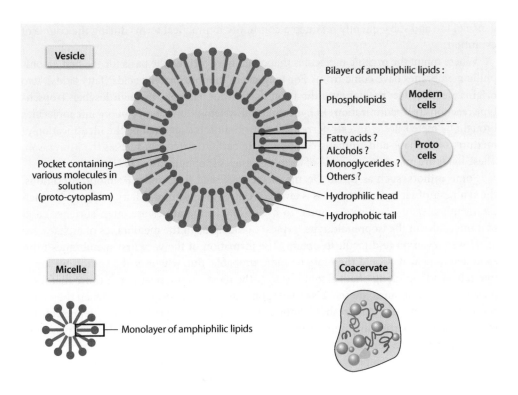

Fig. 4.23 Comparison of the structures of a coacervate, a micelle and a vesicle. Coacervates are colloidal (in suspension, unstructured) assemblages of various hydrophobic organic molecules that cohere within an aqueous medium. Micelles are ordered, often spherical, assemblies of amphiphilic molecules that form in aqueous media with the hydrophobic tails turned towards the interior, and the hydrophilic heads exposed at the surface. Above a certain concentration threshold, some amphiphilic molecules spontaneously form vesicles, that is to say bilayers of amphiphilic molecules (such as fatty acids), whose tails are oriented towards the interior of the bilayer and the polar heads towards the aqueous medium. Vesicles define an interior space that encloses a certain volume of aqueous medium, in which various molecules may accumulate. The vesicles that demarcated proto-cells doubtless consisted of amphiphilic lipids that were simpler than the phospholipids that form present-day cellular membranes.

The involvement of these vesicles in the first stages of life therefore seems highly probable. It would be compatible with the different models for the origin of life that could have preceded them, starting either from a "prebiotic soup", or from a surface metabolism like that proposed by Günter Wächtershäuser. In addition, if the synthesis and accumulation of complex organic molecules could benefit from the assistance of mineral surfaces (*see* pp. 98, 107 and 126), the role of the latter could have been far more significant: certain experiments show, in fact, that such surfaces stimulate the reproduction (division) of vesicles (*see* the experiments by Jack Szostak, mentioned *on the next page*).

There are other hypotheses, however. Within the framework of models of surface metabolism, which by definition are non-cellular, the geochemist Martin Russell and then later, the biologist William Martin have suggested that the alveoli of iron monosulfide in chimneys that form under certain conditions from alkaline hydrothermal fluids might have formed the first compartments. Such mineral membranes would, according to them, have persisted for a long time during the course of biological evolution, until the stage of the last common ancestor of present-day organisms. This hypothesis is, however, disputed, because although it seems possible that such mineral compartments could have formed initial chemical reactors, their persistence in relatively modern cells is highly improbable. In contrast, the principles of continuity and "membrane inheritance" (modern membranes do not form *de novo*, but grow and divide from pre-existing membranes) are in favor of the idea that vesicles consisting of an amphiphile bilayer were key intermediaries during the emer-

4

gence of life and subsequently retained a continuity in practical terms during the course of evolution.

Where might the organic molecules that could have formed the basis for the first amphiphilic membranes have come from? For long-chain monocarboxylic acids (fatty acids), two origins are plausible: prebiotic synthesis on Earth – for example, through Fischer-Tropsch-type reactions (serpentinization) in hydrothermal systems – or a source of organic molecules brought by meteorites. This extraterrestrial contribution is suggested by the identification of mixtures of aliphatic and aromatic compounds in certain meteorites such as the Murchison. These lipids are, moreover, capable of forming vesicles spontaneously (◘ Fig. 4.24).

Some authors, such as David Deamer, have suggested that, under prebiotic conditions, the synthesis of saturated fatty acids is favored relative to unsaturated fatty acids (with which the membranes of organisms adapted to live at low temperatures are often enriched) and that molecules of the isoprenoid type (typical of the lipids in the membranes of archaea, *see later*) were even more difficult to obtain. The formation of the very first membranes from saturated fatty acids would therefore be more probable. But, whatever the source, we know that fatty acids and long-chain alcohols (C_{16-18}, the most common length of those that form present-day membranes) do self-assemble spontaneously and form vesicles when their concentration passes a certain threshold. Shorter fatty acids (C_{14}, equally present in present-day membranes) are particularly appealing candidates, because they are less difficult to synthesize under primitive conditions and the bilayers that they form are very permeable, an advantageous property at a period when the membrane protein transport mechanisms had not yet evolved.

The team led by Jack Szostak, who has worked on this subject for a long time, showed in 2008 that fatty-acid vesicles allowed a passive and selective incorporation of ribose (◘ Box 4.9). These researchers have also discovered that the presence of impurities such as pyrroles (the first pigments, because they were available through prebiotic synthesis) in such vesicles brings about an increase in their permeability. They also established that the addition of minerals (a clay such as montmorillonite, for example) to the fatty acids catalyzed the formation of vesicles, which could then grow by the incorporation of other fatty-acid molecules, and divide. This growth could produce ionic gradients and allow the encapsulation of macromolecules (RNA and certain ribozymes, for example), which represents an experimental approximation to the synthesis of life in a test tube (◘ Box 4.9). It is thus possible to envisage, even if it remains very speculative, the existence of assemblages of various fatty-acid molecules becoming organized into vesicles capable of encapsulating, sequentially, catalytic species and a genetic system, forming in this way compartments that could undergo self-replication (◘ Fig. 4.25).

A Final Word on the Gestation of Life …

Although the appearance of molecules of RNA endowed with self-replicating capacities in a prebiotic environment, their survival, and the natural selection of the most efficient of them among their descendants constitute a plausible way to explain the emergence of a durable replicative system, the RNA world does not explain the origin of compartmentalization.

The appearance of the membrane is often seen by the partisans of this model as a belated event, which grafted itself onto the genetic system when everything was, so to speak, already in place. However, a system based on the replication of a nucleic acid on its own (or on any other chemical system relying on the self-replication of complementary sequences), represents a limit, which has been highlit by Günter von Kiedrowski, a chemist at the University of Bochum (Germany). This rests on the fact that the greater the concentration of each of the two complementary strands, the more difficult it becomes to separate them (essential

◻ **Fig. 4.24 Vesicles formed by amphiphilic lipids contained in the Murchison meteorite.** After extraction of organic molecules using organic solvents (chloroform and methanol), it was possible to separate, using chromatography, a fraction that contained amphiphilic molecules, and notably, carboxylic acids with 9 to 13 carbon atoms, as well as polycyclic hydrocarbon derivatives. It is the presence of aromatic polycyclic compounds that causes the autofluorescence in the bottom photograph. In view of the properties of the lipids contained in this meteorite as well as in others, it is possible to envisage that the molecules that allowed the formation of the membranes that defined the first proto-cells were actually of extraterrestrial origin. These molecules could, however, have also been the products of terrestrial prebiotic chemistry. (Photos: D. Deamer.)

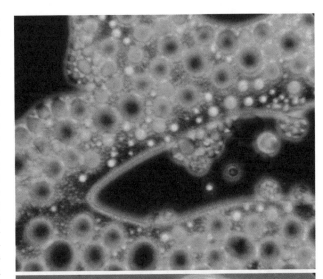

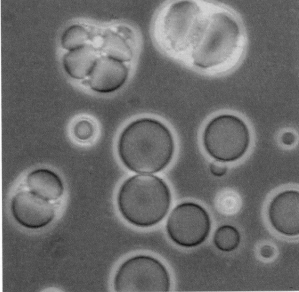

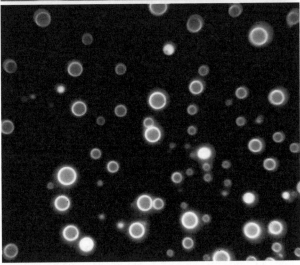

4

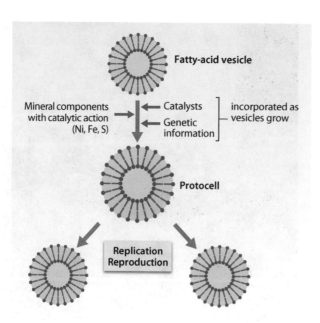

Fatty-acid vesicle

Mineral components with catalytic action (Ni, Fe, S)

Catalysts

Genetic information

incorporated as vesicles grow

Protocell

Replication Reproduction

☐ **Fig. 4.25 A possible evolutionary model of cells through incorporation into primitive vesicles of catalysts and a genetic system.** This model is based on a sheaf of experimental arguments: fatty-acid vesicles allow a passive and selective incorporation of ribose, giving a possible means of access to a hypothetical RNA world. The presence of impurities in these vesicles brings about an increase in their permeability. The addition of minerals to the fatty acids, notably iron-sulfur centers, catalyzes the formation of vesicles, which can grow and divide. These iron-sulphur centers could form a rudimentary system for electron transfer linked to the membrane.

for them to be re-copied), which inhibits their replication to a corresponding extent. That limit may be breached with the simultaneous presence of a metabolic component capable of amplifying the chemical messages delivered by the information carrier and of compartmentalization.

We can see that there are many questions that remain without answers and other hypotheses may be envisaged, and which are either alternative or complementary to that of the RNA world. We have already touched upon these. By way of concluding this first part of the chapter, let us consider them once more.

The scenarios derived from the hypothesis of a "heterotrophic" origin of life, like that of the RNA world, do not explain the emergence of energy and carbon metabolism which are regarded as being subsidiary, at least in the early stages. In addition, no present-day RNA possesses the enzymatic activity necessary for energy metabolism. Quite the opposite, these functions – notably the transport of electrons, essential for the generation of a chemiosmotic gradient across the membrane – are, above all, achieved by proteins which use cofactors involving metallic cations (primarily nickel, iron and sulfur). The hypothesis of the RNA world does not explain when, nor how these systems, crucial for life, have evolved.

Let us first recall that alongside this heterotrophic model, a model that is to a certain extent, the opposite – "autotrophic" – was proposed by G. Wächtershäuser, where the metabolism was first to appear (☐ Fig. 4.26). He advanced the hypothesis that a protometabolic network was first established on mineral surfaces (situated in crevasses in the ocean crust that were subject to a hydrothermal regime) comparable to primitive, two-dimensional organisms drawing their energy from the oxidation of hydrogen sulfide with iron monosulfide (*see* pp. 98, 107 and 128). In this model, the nucleic acids (RNA or DNA) would have first been catalysts for the formation of the peptide bond, and their function as molecules for the storage of information would have appeared secondarily. Wächtershäuser also assumed a co-evolution of replication and translation. In its turn, this genetic machinery would have co-evolved with the formation of the first phospholipid membranes. The appearance of the latter would have led to the compartmentalization of the first metabolisms and the establishment of an energy metabolism based on chemiosmosis, that is to say, involving the production of a proton gradient across the membrane and it being coupled with the synthesis of ATP. That

◘ Box 4.9   Synthetic Biology

The term "synthetic biology" covers all the research aiming to "construct life" artificially, starting with its basic components. The idea of creating life in a test tube dates from the beginning of the 20th century, but it has made a comeback nowadays with several experimental successes. The approaches being followed are of two types:

– The descending or "top-down" approach. This includes attempts to manufacture minimal cells, as far as one can imagine them, by reduction of the components of present-day cells to what is strictly indispensable to carry out their vital functions. In practice, this consists of identifying, thanks to comparative genomics, the smallest – ideally, organisms capable of autonomous growth, but in practice parasites are also used – that provides the functions that are necessary and sufficient for life. Subsequently, one could synthesize chemically a minimal genome and introduce it, for example, into a cell deprived of its own genome. A major step towards the demonstration of the feasibility of this approach was recently taken by a team at the Craig Venter Institute in the United States, which succeeded in chemically synthesizing the complete genome of the parasitic bacterium *Mycoplasma genitalum*, the sequence of which was known (572970 pairs of bases), and in inserting it into another bacterium whose genome had been removed.

– The ascending or "bottom-up" approach. This corresponds to the approaches that attempt to recreate life *de novo* starting with precursor components assumed to have been present on the primitive Earth. The life created in this way could be different from the one that we know today. However, the principles of matter self-organization observed in this instance could help us to understand how life, in its current form emerged, or at least, to test the plausibility of certain models that have been suggested to explain that emergence. Jack Szostak's team, in the United States, has been carrying out this sort of research for many years. They have analyzed the properties of vesicles of various amphiphilic molecules (phospholipids, fatty acids, alcohols of different lengths), in particular their permeability, their methods of division and their capacity to encapsulate organic molecules. Very recently, they have been able to show the replication of a genetic polymer (a single-strand DNA molecule) in a vesicle formed by simple amphiphilic molecules that was permeable to nucleotides.

4

would be equivalent to bringing about the change from a two-dimensional metabolism to a three-dimensional one. (In other words, from reactions on mineral surfaces to reactions within compartments.)

Many criticisms have been advanced against this model, in particular, its chemical feasibility is in doubt and it has difficulty explaining the transition from two-dimensional organisms to three-dimensional cells. It should, however, be noted that it offers overall coherence, and that the catalytic role that initially is played by nucleic acids is a curious convergence with the RNA world.

We have already indicated that the emergence of a translation system raises the question of an overlap between the chemistry of amino acids and that of the nucleotides. It would, in fact, be easy to imagine that RNA was not alone, but was rather a partner in a consortium of co-evolving molecules: the amino acids (which cannot be ignored, because they are molecules whose abiotic synthesis is very easy), peptides and nucleotides. This co-evolution would have led directly to a more advanced stage than the RNA world, or even directly to an RNA-protein system (◘ Fig. 4.26). The chemistry of nucleotides and the chemistry of oligopeptides would thus have been closely linked (to such a point that even the concept of the RNA world would have become ineffective). The emergence of the machinery for translation based on the genetic code does, in fact, imply that interactions (covalent bonds) can be established between amino acids, nucleotides and RNA (the same ones that are today employed in the translation process, ◘ Box 4.8). And yet both theoretical and experimental arguments show that such bonds may be established in a purely chemical context, in the absence of any catalytic action mediated by the ribosomes of the RNA world in its most advanced form. What is more, it is difficult to imagine a world of RNA on its own, without amino acids and peptides, when these molecules are easy to synthesize in an abiotic manner and, furthermore, widespread in the universe ... unlike nucleotides. Independently of this possibility of co-evolution that led more or less directly to a coded RNA-protein system, the study of the interaction of

4

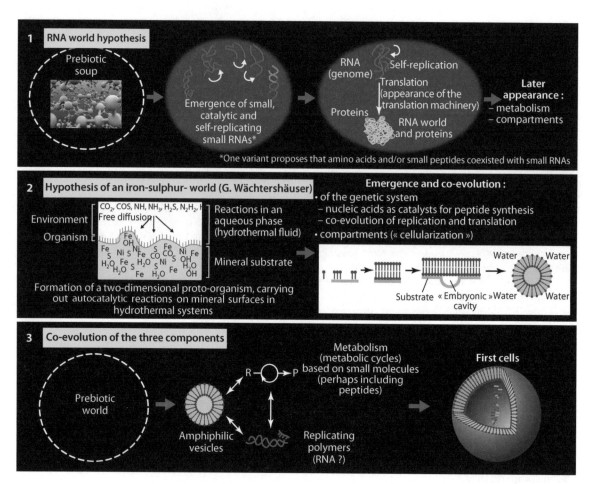

□ **Fig. 4.26 Hypotheses explaining the appearance of the three essential components of life (the genetic and metabolic systems, and compartmentalization).** The various hypotheses (RNA world, iron-sulphur surface world, and co-evolution) are discussed on pp. 127–128. The list is not an exhaustive one ...

RNAs with amino acids and peptides is, anyway, indispensable to understand the passage to an RNA-protein stage from a putative RNA-only world that excluded any initial role for peptides.

If we take into account the necessity for compartmentalization, the possibility of the co-evolution of three components – amino acids, nucleotides, vesicles – thus remains matter for research to be developed in the future, all the more in that it also offers the advantage of linking the emergence of a genetic system to that of a metabolism at the origin of life. Information acquired in the field of the prebiotic chemistry of amino acids and of peptides is extensive, and the prebiotic chemistry of nucleotides is progressing. If we add to this, the capacity for interactions that have been proved to take place between peptides and nucleotides, we come to imagine the possibility of a complex network of reactions (the precursor of a protometabolism) involving all of these elements, and having a possible role as a set of conditions that favor the emergence and replication of an information carrier. We could express this idea in a rather colorful way by considering that this future protometabolism would have formed the ecological niche that an information carrier could colonize. The birth of an initial strand capable of replicating itself in this favorable environment (which contributed help of a chemical nature) and to react to its surroundings would thus be that of life.

At the start of this chapter, we explained that a living organism must incorporate three sub-systems: genetic, metabolic and compartmental. It is tempting to conclude that it is the association of these components that is the essential factor, and thus that the emergence of life could result only from a process of co-evolution. This idea underlies a highly conceptual approach already expressed by a Hungarian scientist, Tibor Gánti who, for a long time, remained unknown outside his own country. He also insisted on the necessity for each of these sub-systems to include a measure of self-reproduction (by replication, autocatalysis or division). This hypothesis of a co-evolution of the genetic system, the metabolism and the membrane system is very appealing, because it can accommodate the idea of RNA making an essential contribution to the first stages of life, within an overall evolutionary complex with several components.

4

The Last Common Ancestor of All Existing Organisms: a Portrait

So far, we have explored one of the two possible ways of studying the transition from no life to life, which could be termed the "bottom-up" or constructive approach. It consists of drawing up hypotheses about the way in which life could have emerged, starting with chemical elements and molecules of greater and greater complexity and, of course, of testing their plausibility. The limitations of this approach are well illustrated by the multitude of solutions that may be envisaged – very frequently contradicting one another, although most are based on chemical principles that have been established experimentally – and the limitation may be summarized with predictable frustration: even if, one day, someone managed to re-create life in a test tube, that would not constitute proof that life emerged by following the scenario that would thus have been bolstered by experiment, because evolution is a historic process, and thus a contingent one. The only means of knowing with certainty how life arose would be to have a machine that could go back in time!

The alternative, and complementary way is the "top-down" or deconstructive approach. This consists of studying life forms as we know them today to uncover the basic elements that in the origin and later evolution of life played a key role and can bear witness to past events. This approach therefore allows us to work backwards in the history of life through the analysis of inherited characteristics. If we were able to follow the thread to the end, we would be able to have access to the true history of the birth of life and decide between the different hypotheses. Alas, there are limits. It is impossible to go back very far into the past starting from elements common to present-day living organisms, which are complex and optimized by billions of years of evolution. The top-down approach certainly does allow us to reconstruct the portrait of the last ancestor common to all living beings. But in essence, this can tell us nothing about the other lines that are now extinct, which probably existed and could help us to reconstruct the properties of ancestors even closer to the transition between the chemical world and the biochemical world. Nevertheless, the top-down approach does provide precious information about early evolution and this is what we will explore now.

The idea of a common ancestor for all living beings is not a modern one. In *The Origin of Species*, published in 1859, Charles Darwin stated that a logical conclusion of his theory of descent with modifications and natural selection was that "*We must likewise admit that all the organic beings which have ever lived on this earth may be descended from some one primordial form.*" During the second half of the 20th century, this intuition was reinforced, thanks to considerable progress in biochemistry and molecular biology. Briefly, it transpires that the biochemical components and the structural essentials of the cell, as well as the most fundamental metabolic reactions, are shared by organisms in the three domains of life: the archaea, the bacteria, and the eukaryotes (■ Box 4.10). This allows us to conclude that they inherited these characteristics from a common ancestor that already had all of them.

This last hypothetical ancestor common to all living beings has been given several names, of which the most common are the cenancestor (from the Greek "kainos," recent, and "koi-

4

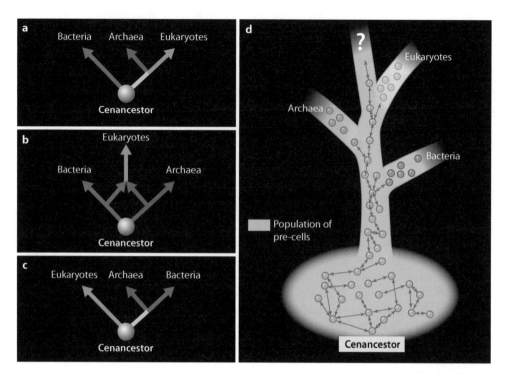

◘ **Fig. 4.27 The various models proposed to explain the evolution of the three domains of life.** Models **a** to **c** assume that bacteria, archaea and eukaryotes emerged from a last common ancestor (cenancestor) viewed as a unique cell among a community of cells that possessed most of the characteristics common to present-day organisms. Model **d** assumes that the cenancestor was a population of cells ("pre-cells") that, collectively, possessed all these characteristics. At a given time, a favorable combination of characteristics emerged from this population and gave rise to the bacterial line, then another combination gave rise to that of the archaea, and yet another to that of the eukaryotes. The question mark denotes the fact that the population of pre-cells could have continued to exist after the emergence of the three domains of life. The models **a**, **b** and **c** differ first in the degree of complexity of the cenancestor: prokaryote for models **a** and **b**, eukaryote for model **c**. They differ subsequently in the evolutionary origin of the eukaryotes. Model **a** considers that they are a sister group of the archaea, whereas model **b** considers that they are the result of a symbiosis between archaea and bacteria (*see* ► Chap. 7). Model **c** proposes that the cenancestor itself was a eukaryote. At present, models **a** and **b** are the most generally accepted among the scientific community.

nos," common), and the Last Universal Common Ancestor (or LUCA). Even though any dating is impossible, it is possible to state that the cenancestor lived a few billion years ago, before the three domains diversified. Under these conditions, reconstructing its "portrait" is a difficult and speculative task.

Most researchers agree with the idea that the cenancestor corresponded to a unique cell which, among what was probably a very varied community, lived at a given time and had most of the characteristics common to present-day organisms as well as the genes that coded for them. This cell, endowed with an advantageous combination of characteristics would have multiplied and its descendants would have eliminated the less fit lines through competition. On the other hand, others imagine a population of cells that, taken overall, was provided with all these genes, even if no cell on its own possessed every one of them (◘ Fig. 4.27). Within this population of "pre-cells," the level of genetic exchange would have remained very high until a sub-population brought together a particularly efficient combination of genes. Their isolation brought about by natural selection would have given rise to a line of organisms (still corresponding to pre-cells). In his pre-cellular theory (1994), the German microbiologist Otto Kandler, a specialist in cell walls, suggested that bacteria, archaea and eukaryotes would subsequently have emerged sequentially in this way starting from just such a population of pre-cells (◘ Fig. 4.27d).

Despite the controversy over the "unicellular" and the "population" versions of the cenancestor, there is almost a consensus for saying that it was a matter of an entity that was quite complex. That implies that before arriving at the stage of the cenancestor, life had already undergone a relatively long process of evolution since prebiotic times. In other words, the origin of life and the nature of the last common ancestor are, we repeat, two distinct evolutionary questions or are, at least, separated in time.

The level of complexity attributed to the cenancestor, even if it is relatively high, depends on the model concerned, however. For Carl R. Woese – an American researcher, who made his mark in biology in the 1970s by demonstrating, thanks to the comparison of a universal cell marker (the RNA of the small sub-unit of the ribosome), that the living world is divided into three distinct phylogenetic groups, known as domains (◼ Box 4.10) – it was a matter of an entity that was still primitive, a "progenote," that is to say "an organism that had not yet developed a link between genotype and phenotype." Others, however, believe that the progenote stage existed before the cenancestor, which was itself an almost modern cell.

Reading between the lines in these debates, there is one question: how to derive as complete a portrait as possible of the cenancestor? Several of its characteristics can be deduced from comparative studies based on biochemistry, molecular biology and, above all, on molecular genomics and phylogenetics, two disciplines that have achieved very substantial progress in recent years. These different approaches have brought sound proof regarding the very ancient nature of the machinery for the synthesis of proteins (translation), of a fairly sophisticated machinery for the synthesis of structural RNA and messenger RNA (transcription), and an energy-production process based on the generation of a proton gradient across a membrane. The features are universal: they are observed among all present-day cells and, as a result, it is fully accepted that they must have been present in the cenancestor. Other characteristics of the last common ancestor of all living beings are, on the other hand, more controversial.

Protein Synthesis: an Ancient Process

If all the genomes that have been sequenced so far – which cover organisms belonging to the three domains of life – are compared, we find that only about sixty genes are common to all. This is an extremely small number, given that the prokaryotes (the bacteria and archaea) possess between 500 and 10 000 genes and the eukaryotes have between 2000 and more than 30 000. It is important to note that the repertoire of universal genes consists, almost entirely, of genes that code ribosomal RNA (rRNA), ribosomal proteins as well as other factors involved in translation, such at the aminoacyl-tRNA synthases, those enzymes that load the specific, different amino acids onto transfer RNA (◼ Box 4.8). These genes are thus probably really ancestral. In other words, the cenancestor undoubtedly possessed ribosomes and a machinery for the formation of proteins comparable to that found in present-day organisms.

The synthesis of proteins by the ribosome is, in fact, the process that is best preserved over the course of evolution. Transcription – the transfer of the message from the genes (DNA) to mRNA – is also conserved, but to a lesser degree. Indeed, although some sub-units of RNA polymerase – responsible for the synthesis of RNA – are coded by universal genes, other RNA polymerase sub-units, as well as several transcription factors, are not. All this leads us to believe that, although transcription may be an ancestral process, the mechanisms involved in the synthesis of RNA and, more precisely, their regulation, have been optimized subsequently during the evolution of the major domains of life.

4

4

Box 4.10   The Three Domains of Life

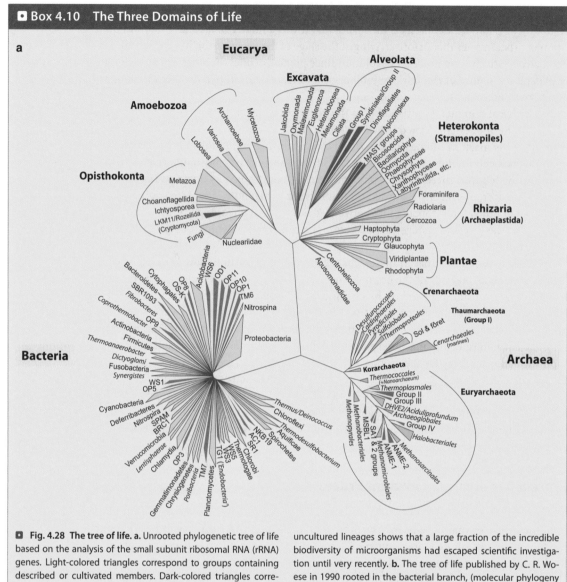

Fig. 4.28 The tree of life. a. Unrooted phylogenetic tree of life based on the analysis of the small subunit ribosomal RNA (rRNA) genes. Light-colored triangles correspond to groups containing described or cultivated members. Dark-colored triangles correspond to lineages known only from rRNA sequences directly obtained from environmental samples, and for which we lack any representatives cultured in the laboratory. The discovery of those uncultured lineages shows that a large fraction of the incredible biodiversity of microorganisms had escaped scientific investigation until very recently. **b.** The tree of life published by C. R. Woese in 1990 rooted in the bacterial branch, (molecular phylogeny based on the analysis of the rRNA genes; the position of the root was determined using sets of genes that duplicated before the diversification of the three domains of life).

The Cenancestor's Genome: DNA or RNA?

DNA is the molecule that contains the genetic information in all present-day cells. (Only a few viruses have genomes of RNA, but viruses are not part of the cellular world and, for many scientists, they are not even living organisms, but simply molecular parasites.) However, among the sixty-odd universal genes, just three code a protein that is involved in the replication or repair of DNA, namely a DNA polymerase sub-unit, anexonuclease (an enzyme that removes deoxyribonucleotides from DNA starting at one of the ends), and a topoisomerase (an enzyme that untangles the "knots" and "supercoils" produced in the DNA molecule during transcription or replication). Another remarkable fact is that the genes involved in the

Cellular organisms are divided into three distinct phylogenetic domains: Bacteria, Archaea, and Eucarya (eukaryotes) (■ Fig. 4.28a). This phylogenetic division of the living world was discovered in 1977 by Carl R. Woese, who established it in a more official manner in 1990. Based initially on the phylogenetic analysis of the RNA of the small subunit of the ribosome, this tripartite classification has been confirmed by other gene markers, by genome sequences, as well as by biochemical and structural characteristics. Bacteria and archaea are prokaryotes, unicellular microorganisms with simple structure and morphology, but possessing a great diversity of metabolisms. The eukaryotes have a more complex cellular structure, although their metabolism is much less diverse than that of prokaryotes. Most eukaryotes are unicellular, even though multicellular organisms have appeared in several groups (metazoans, land plants, and certain green, red and brown algae).

Where is the root of this universal tree of life? The tree published by Carl R. Woese in 1990 was rooted in the bacterial branch by using phylogenetic information provided by genes which duplicated before the diversification of the three domains (■ Fig. 4.28b). In this tree, the lowest branch point represents the position of the last common ancestor (the cenancestor). However, the phylogenetic relationships between the three domains of life as well as the position of the root of the tree of life are still the subject of debate (■ Fig. 4.31 and p. 140). This is why it is currently preferable to work with unrooted universal trees of life.

b

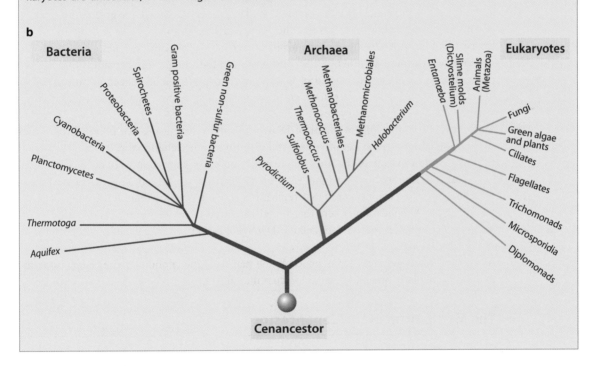

replication of DNA in the archaea (prokaryotes) greatly resemble those of eukaryotes, but are very different from those of bacteria (which are also prokaryotes). Several hypotheses have attempted to explain this situation.

Some suggest that the genome of the cenancestor was not made of DNA, but of RNA, which would explain why most DNA replication and repair proteins are not universal. DNA and its replication would therefore have appeared twice, independently: once in the bacterial line, and another in the line that led to the archaea and the eukaryotes (■ Fig. 4.27a) or in the line of archaea (■ Fig. 4.27b), according to the most widely accepted models of the diversification of the domains of life. Some authors holding the idea that the cenancestor was eukaryotic in nature (■ Fig. 4.27c), have also suggested that its genome consisted of RNA. In their eyes, the large number of small RNA molecules (sRNAs: nucleolar RNAs, interfering RNAs, etc.) in eukaryotes would attest for their ancestry. However, these sRNAs also exist in prokaryotes, notably in the archaea, but in smaller quantities.

4

The idea that the cenancestor possessed a genome of RNA is not, however, unanimously accepted. Indeed, the existence of some universal proteins and protein domains (conserved regions that are observed in several proteins), involved in the metabolism of DNA, does itself suggest that the cenancestor possessed a genome made of DNA. In addition, the synthesis of RNA is accompanied by a mutation rate that is far higher than that for DNA. The individual molecules of RNA cannot, therefore, exceed a certain size (~30–50 kb), otherwise the accumulation of errors during their replication leads to what Manfred Eigen (a German physical chemist and biochemist, awarded the Nobel Prize for Chemistry in 1967), called "a replication catastrophe." This limit of Eigen's is well illustrated by the fact that the genomes of present-day RNA viruses do not consist of more than about 30 kb, whereas those of DNA viruses (and cells) may exceed 1 Mb. Yet the cenancestor was a fairly complex organism, and it must undoubtedly have possessed about 600 to 1000 genes, according to theoretical estimates based simultaneously on the comparison of completely sequenced genomes, and on estimates of the minimal number of functions that the simplest cell that can be imagined would have required. The number of RNA chromosomes necessary to provide such a genome would therefore have posed serious problems of stability and equal distribution between daughter cells. Indeed, the greater the number of chromosomes, the less likely it is that they were equally distributed between daughter ancestral cells that had not developed mechanisms of segregating the chromosomes.

Based on all these arguments, a second series of models therefore considers that the genome of the cenancestor consisted of DNA. Several hypotheses have been advanced to explain the important differences between the replication system of bacteria and that (closely related) of archaea and eukaryotes:

- the evolution and improvement of DNA replication during the evolution of these two lines, starting with a system that was still primitive in the cenancestor;
- the retention in each of the lines of a distinct sub-set of genes that were devoted to DNA replication from two systems that would have been already present in the cenancestor;
- a far more rapid evolution of DNA replication genes in one line than in the other, starting from a DNA replication system that was already well-developed in the cenancestor;
- the substitution, in bacteria, of cellular DNA replication genes by genes of viral origin.

As you will have gathered, DNA or RNA, the nature of the genome of the cenancestor remains a particularly hot, and debatable, scientific subject.

The Metabolism of the Cenancestor

Trying to sketch an outline of the metabolism of the cenancestor is particularly difficult, because most of the genes now involved in the energy-production pathways or in the transformation of carbon (or both) are not conserved between the different groups of organisms. These genes most often form large multigenic families, whose various members have been recruited to carry out extremely varied functions (in the different groups of organisms and even within an individual line). Let us take the example of the super-family of the dehydrogenases, enzymes involved in the process of oxidizing molecules by eliminating hydrogen atoms. It includes various members which have become specialized, during evolution, in the oxidation of different substrates, going from fatty acids to sugars, via various enzymatic cofactors (coenzyme A, NADH[2], etc. This diverse range of substrate specificity is explained by what is called enzymatic recruitment: a gene duplication produces an additional gene coding a dehydrogenase; the latter is not subjected to strong selective constraints (because there is one functional

---

2. NADH = Nicotinamide adenine dinucleotide, a coenzyme found in all living cells.

Fig. 4.29 The production of ATP in the cenancestor. It is very difficult to reconstruct the metabolism of the cenancestor. The presence of a membrane ATPase that is highly conserved in all cellular organisms, however, allows us to identify one characteristic of its energy metabolism. The cenancestor was capable of converting a proton gradient across the membrane into free energy stored by the cell in the form of ATP. (The mechanism involved in present-day organisms is shown in Fig. 4.6.) The protein (or proteins) that established this gradient are not known with any certainty.

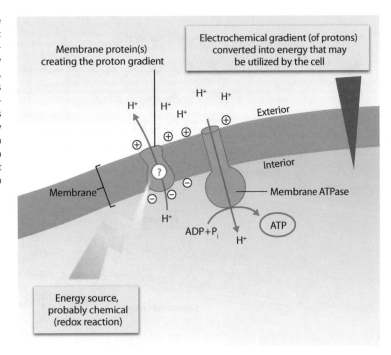

4

copy at work) and accumulates mutations. This may lead to a change in the enzyme's affinity over the course of evolution, and it then becomes capable of oxidizing a new substrate.

To this we may add the fact that the genes involved in the metabolic pathways are often affected by the phenomenon of "horizontal gene transfer" – they pass from one group of organisms to another – because they may confer an immediate selective advantage to the recipient organisms (if they allow, for example, the use of a given source of carbon, or resistance to an antibiotic). This obviously greatly complicates any attempts at the phylogenetic reconstruction of ancestral metabolic pathways.

Despite all these difficulties, we can determine one characteristic of the energy metabolism in the cenancestor. The presence of a membrane ATPase, highly conserved in all cellular organisms, indicates that the cenancestor possessed the capacity to synthesize ATP by exploiting a proton gradient across the membrane (Fig. 4.29).

The source of energy used by the cenancestor to create this gradient was probably chemical (redox reactions) because phototrophy (the exploitation of the energy of light), probably did not appear until later in the course of evolution, in the bacterial branch. What were the molecules acting as electron donors and acceptors for these redox reactions? Were they organic, inorganic, or both at the same time? The answer to this question remains uncertain, but it is possible that a variety of molecules were used.

Recent studies have examined the phylogeny of terminal cytochrome oxidases, which are universally conserved proteins. These bear witness to the existence, in the cenancestor, of an electron transport chain associated with the membrane, and which are involved in the redox reactions linked to respiration. They suggest that the cenancestor was capable of "respiring" molecular oxygen and, in all probability, nitric oxide (NO). Nitric oxide – whose structure is very similar to that of molecular oxygen – as well as its derivatives nitrite, NO_2^-, and nitrate, NO_3^-, were available from volcanic emissions on a primitive Earth whose atmosphere was deficient of any significant quantity of oxygen. It is equally probable that the cenancestor was able to ferment certain organic substrates in the cytoplasm to produce ATP by substrate-level phosphorylation (Fig. 4.6a). Finally, with regard to the metabolism of carbon, whether the cenancestor was an autotroph, a heterotroph or possessed both types of metabolism remains open to this day.

4

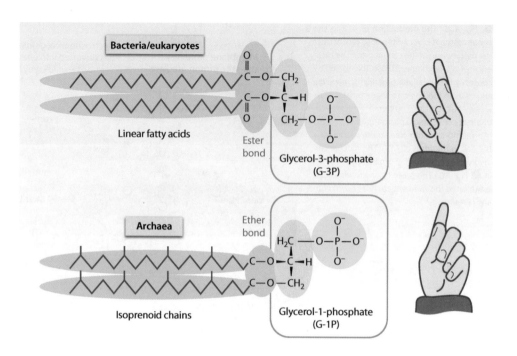

🔲 **Fig. 4.30 The structure of membrane phospholipids in the three domains of life.** The type of stereo-isomer of glycerol-phosphate distinguishes, with no known exception, between the phospholipids of bacteria and eukaryotes, and those of the archaea: glycerol-3-phosphate in bacteria and eukaryotes, glycerol-1-phosphate in archaea. Other differences are observed in the hydrophobic chains and the bond between the latter and the glycerol-phosphate, but there are exceptions. Some phospholipids in archaea have chains of fatty acids, and some phospholipids in bacteria include ether bonds.

The Cenancestor's Membranes

All cells are surrounded by a plasma membrane consisting of phospholipids, generally organized into a bilayer. One might therefore conclude that there is no doubt that the cenancestor already had a phospholipid bilayer.

But when one starts to look at the nature of the phospholipids the picture becomes more complicated. The composition of the phospholipids in the archaea is very different from that observed in the bacteria and eukaryotes. The differences are first seen in the hydrophobic chains – isoprenoids in the archaea, fatty acids in the bacteria and eukaryotes – and in the bond between the lateral chains and the glycerol-phosphate that forms the hydrophilic part of the molecule – an ether bond in the archaea, ester in the bacteria and eukaryotes. However, there are exceptions from these differences: bacterial lipids with ether bonds and archaeal lipids with fatty acids.

The fundamental distinction (with no known exceptions) between the phospholipids of the bacteria and eukaryotes, and those of the archaea rests in the type of stereo-isomer of glycerol used; glycerol-3-phosphate in the bacteria and eukaryotes, glycerol-1-phosphate in the archaea (🔲 Fig. 4.30). The pathways by which these two stereo-isomers are synthesized are so different that, for certain authors, the cenancestor did not have membranes and was an acellular organism or, according to yet others, it had mineral membranes consisting of iron monosulfide. The idea of an ancestor without lipid membranes, however, comes up against one piece of evidence: there are membrane proteins that are universally conserved, like the ATPase. A less radical hypothesis, which had recently been supported by molecular phylogeny analyses, would be to imagine a cenancestor with a heterochiral membrane, that is to say possessing a mixture of phospholipids constructed with glycerol-1-phosphate and glycerol-

3-phosphate. Additional molecular phylogenetic studies further support the idea that the cenancestor had a complete toolkit to make phospholipids of either fatty acids or isoprenoid chains. Evolution would have subsequently led to opposite stereospecificity and choice of lateral chains in the bacteria and in the archaea.

Other Open Questions About the Cenancestor

4

Membranes, metabolism, and nature of the genome. These are not the only question marks hanging over the last common ancestor of all organisms. Among the numerous other questions that are the subject of debate in the scientific community about the cenancestor, we have chosen two.

Did the Cenancestor Live at a High Temperature?

The idea that the cenancestor lived at a very high temperature was proposed in the aftermath of the discovery, in the years 1960 to 1980 of bacteria and, above all, of archaea whose optimum temperature for growth was above 80 °C. These organisms, called hyperthermophiles (❑ Box 4.11) were discovered in continental hot springs (Yellowstone in the United States and Solfatara in Italy) and at the hydrothermal chimneys associated with mid-ocean ridges. In the first phylogenetic trees of life based on small-subunit ribosomal RNAs, these hyper-

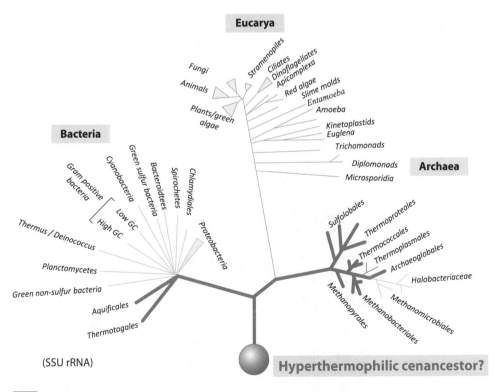

❑ **Fig. 4.31 One of the first rooted universal phylogenetic tree.** In this type of phylogenetic tree based on small subunit rRNA genes and rooted on the bacterial branch, hyperthermophilic organisms (red) branched out closest to the root (this conclusion would also be reached in models where the root is placed between archaea and bacteria, with eukaryotes resulting from a symbiosis between the two). This tree topology would support the idea that the cenancestor was also hyperthermophilic. However, this idea is highly debated today because, whereas the basal position of hyperthermophilic archaea appears robust, the placement of hyperthermophilic bacteria at the base of the bacterial branch could be the result of an artifact of phylogenetic reconstruction.

4

◘ Box 4.11   Life Under Extreme Conditions

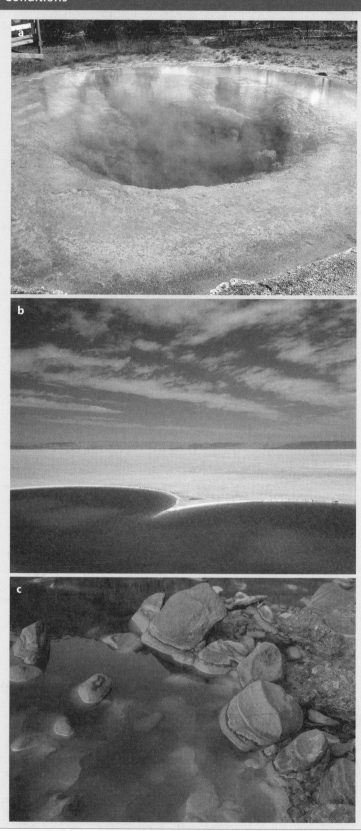

◘ **Fig. 4.32 Some environments populated by extremophiles. a.** A hot spring in Yellowstone Park in the United States (temperature: 80–90 °C). **b.** The salt desert of Chott-el-Jerid in Tunisia, where salt marshes are colonized by halophilic organisms (whence their reddish color). **c.** The Rio Tinto in Spain, an acid river (pH = 2–2.5) with high metal concentrations (iron and arsenic in particular). (Photos: P. López-García, K. Benzerara.)

Several times during the course of evolution, living organisms have adapted to environments where the temperature, pH, salinity, pressure, dryness, doses of radiation, or even metal concentrations are extreme. These organisms, which develop their life cycles close to the physical and chemical frontiers of life, beyond which life is not possible, are known as extremophiles. They populate extremely varied environments, from the depths of the oceans to searing-hot springs (◘ Fig. 4.32), including permanently frozen soils (permafrost), salt mines, deserts, industrial and even radioactive waste. ◘ Table 4.4 provides a brief identity card for the principal extreme champions found in each of the three domains of life.

◘ Table 4.4

"Extreme" physicochemical parameter	Type of organism	Definition (optimal growth conditions)	Example habitats	Distribution in the three domains of life and examples		
				Archaea	Bacteria	Eukaryotes
	Hyperthermophile	> 80 °C	Submarine and continental hydrothermal systems Geysers	Up to 113 °C *Pyrolobus, Methanopyrus,* etc.	Only up to 95 °C *Aquifex Thermotoga*	None
Temperature	Thermophile	60–80 °C	Solfataras Deep oceanic and continental sub-surface	*Thermoplasma, Sulfolobus, Archaeoglobus,* etc.	*Thermoanaerobacter, Chloroflexus,* etc.	Certain algae and fungi, up to 60-64 °C
	Psychrophile	< 5 °C	Deep ocean Polar caps and ice floes Snow and high mountains Permafrost	*Nitrosopumilus, Cenarchaeum,* several uncultured lineages from the deep ocean	*Psycrophilus, Flavobacterium,* many uncultured lineages	Several lines of protists (cilliates, algae, etc.)
pH	Acidophile	pH < 2–3	Mines Acid hot springs Solfataras	*Picrophilus, Thermoplasma, Sulfolobus,* etc.	*Acidithiobacillus, Leptospirillum,* etc.	Several protists (algae, heliozoans, etc.)
	Alkaliphile	pH > 9–10	Soda lakes Alkaline hot springs	*Natronobacterium, Natronococcus,* etc.	Many cyanobacteria	Some protists and fungi
Salinity	Halophile	Strong concentration of salt (~2–5 M NaCl)	Salt marshes Certain soda lakes (such as Lake Natron) Marine brines Evaporites, salt mines	*Halobacteirum, Haloferax, Natrialba,* etc.	*Salinibacter, Halomonas,* etc.	*Dunaliella salina, Artemia salina*
Pressure	Barophile (piezophile)	High pressure	Deep ocean Deep subsurface	Several lineages of thermophile and psychrophile archaea from the deep ocean and subsurface locations	*Shewanella, Colwelia,* several uncultured lineages from the deep ocean	Abyssal fauna, various protists

4

"Extreme" physicochemical parameter	Type of organism	Definition (optimal growth conditions)	Example habitats	Distribution in the three domains of life and examples		
				Archaea	Bacteria	Eukaryotes
Hygrometry	Xerophile	Extreme dryness	Hot and cold deserts Solar salterns	Halophile archaea	*Deinococcus, Metallogenium, Pedomicrobium,* etc.	Fungi, lichens
Radiation exposure	Radio-resistance (radiotolerance)	Withstands high levels of ionizing, or UV radiation, etc.	Radioactive waste Naturally radioactive mines Deserts, salt marshes, high mountains	*Thermococcus gammatolerans*	*Deinococcus radiodurans*	Certain fungi
Metal concentration	Metallo-tolerant	Withstands high concentrations of metals	Mines Aquifers contaminated with metals Industrial waste	*Acidianus, Thermoplasma,* etc.	*Acidithiobacillus, Leptospirillum,* etc.	Several fungi and algae

◘ Table 4.4 (continued)

thermophiles occupied the lowest branches, that is to say the closest, in evolutionary terms, to the cenancestor. If the root of the tree of life lies on the bacterial branch (and equally in the models where the root lies between the archaea and the bacteria, the eukaryotes being a chimera between the two), that would imply that the cenancestor was a hyperthermophile, because the branches that first emerged corresponded to hyperthermophilic bacteria and archaea (◘ Fig. 4.31). This interpretation seems, in addition, to go along with models suggesting that life itself arose at high temperature.

These conclusions are greatly disputed. With regard to a hot origin for life, first, RNA and other macromolecules are fragile at very high temperatures (RNA is, nevertheless, quite stable at temperatures as high as 90 °C if it is in a saline solution or bound to clays). It would therefore have been impossible for life to have been born under such conditions, especially if one is considering the RNA-world hypothesis. To get over this difficulty, some scientists maintain that life arose at lower temperatures, but that the cenancestor itself was hyperthermophile, because only hyperthermophiles, living in deep-sea vents and in the oceanic crust would have survived the rise in temperature that followed the Late Heavy Bombardment (*see* ▶ Chap. 5), and which could have caused the vaporization of a large part of the ocean.

Other critics are concerned with the phylogenetic trees on which the hypothesis of a hyperthermophile cenancestor is based. Indeed, although there appear to be sound arguments for the basal position of hyperthermophilic archaea (which supports the hypothesis that the archaeal common ancestor was hyperthermophile), this is not the case for the basal location of hyperthermophilic bacterial branches. The latter might be the result of an artifact of phylogenetic reconstruction, arising from the restricted number of sequences that were included at the time in the analysis and, above all, of the use of too-simplistic evolutionary models.

More recent analyses suggest that, although it remains probable that the ancestor of the archaea was hyperthermophile, that of the bacteria was not. Present-day hyperthermophile bacteria would have, in fact, adapted secondarily to life at very high temperatures. Under these conditions, the cenancestor could quite well have been a hyperthermophile, a meso-phile (living at moderate temperatures, between about 10 °C and 45 °C), or a moderate ther-mophile (living between 45 °C and 80 °C). This last possibility would be in good agreement with data that suggest that the average temperature of the ocean in the Archaean was about 70 °C.

In 2009, a team of researchers from Lyon took the debate a stage further. Using a strategy of reconstructing ancestral RNA ribosome sequences, they concluded that the ancestors of the archaea and bacteria were hyperthermophilic, whereas the cenancestor was mesophilic or simply thermophilic. The debate over the temperature at which the cenancestor lived is therefore far from over.

4

What Degree of Complexity Did the Cenancestor Have?

To the vast majority of researchers, the cenancestor was an organism with simple structure that resembled present-day prokaryotes, which would be in agreement with the location of the root of the tree of life being between the bacterial branch and a branch leading to the archaea and eukaryotes or between the bacteria and the archaea (depending on the model, ◘ Fig. 4.27a and b.).

Other authors, however, have suggested that the root should be placed between the eu-karyotic branch and a branch leading to the prokaryotic lines (◘ Fig. 4.27c). Such a root would still be compatible with a prokaryotic cenancestor but it also leaves open another pos-sibility: that the cenancestor was an organism that was structurally more complex, possess-ing characteristics present in modern eukaryotes, such as a nucleus with a membrane that isolates the genetic material within the cell, or having numerous small RNAs (which might possibly be vestiges of the hypothetical RNA world). In this last view of things, a line leading to the prokaryotes would have then evolved towards a reduction in complexity before split-ting into the line of bacteria and that of the archaea.

According to this hypothesis, the prokaryotes would thus form a monophyletic group. This view comes up against several objections, including the uncertainty in placing the root of the tree of life, the discovery of numerous small RNAs (sRNA) in prokaryotes – which invalidates the idea that this characteristic is a trait exclusively found in eukaryotes which would bring them closer to a presumed RNA world – and the difficulty of explaining how evolution could have led to a high level of structural complexity (of the eukaryote type) be-fore the cenancestor stage.

In theory, one could certainly imagine placing the root of the tree of life on each of the branches that lead to the three domains of life. (The root could quite well have been on the ar-chaeal branch, but few researchers have suggested this.) However, there are two types of con-straints linked to observation, which refute the hypothesis of an ancestor of eukaryote type.

The first constraint is imposed by the fossil record. (We shall speak more about this in Chap. 6.) Even if the authenticity of the most ancient traces of fossil life (with ages greater than 2.7 Ga) are in dispute, all the traces of fossil life dated more than 1.8 to 2.1 Ga are of a prokaryote nature. That includes:

– isotopic-type traces: for example, fractionation of sulfur isotope from 3.5 Ga, which bear witness to the use of this element in reactions to gain energy which are observed exclusively in prokaryotes (archaea and bacteria), or else very negative carbon isotopic deviations, which in-dicate the presence of methanogenic archaea and methanotropic prokaryotes (*see* ▶ Chap. 6);

– macrofossils and microfossils: the presence of stromatolites (the result of the activity of microbial communities mainly consisting of prokaryotes, dominated by photosynthetic bac-teria as primary producers) from 2.7 Ga and perhaps even from 3.5 Ga; in the same interval of time, the presence of morphologically probable fossils, similar to prokaryotes in size and in the absence of decoration; the presence of incontestable and varied prokaryote microfossils from 1.9 Ga (the Gunflint formation in Canada);

4

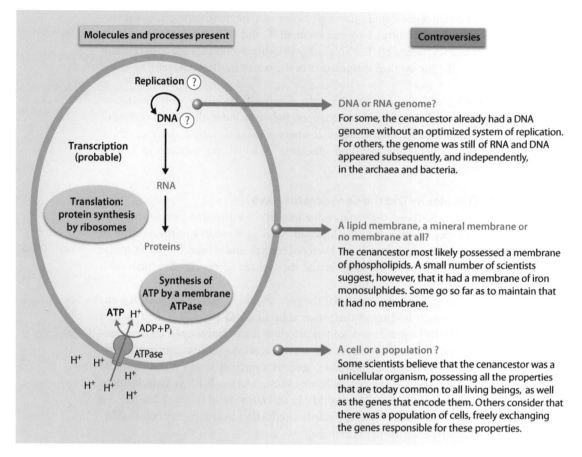

Molecules and processes present

Replication ?

DNA ?

Transcription (probable)

RNA

Translation: protein synthesis by ribosomes

Proteins

Synthesis of ATP by a membrane ATPase

ATP H^+

ADP+P_i

H^+

ATPase

H^+ H^+

H^+

H^+ H^+

H^+

Controversies

DNA or RNA genome?
For some, the cenancestor already had a DNA genome without an optimized system of replication. For others, the genome was still of RNA and DNA appeared subsequently, and independently, in the archaea and bacteria.

A lipid membrane, a mineral membrane or no membrane at all?
The cenancestor most likely possessed a membrane of phospholipids. A small number of scientists suggest, however, that it had a membrane of iron monosulphides. Some go so far as to maintain that it had no membrane.

A cell or a population ?
Some scientists believe that the cenancestor was a unicellular organism, possessing all the properties that are today common to all living beings, as well as the genes that encode them. Others consider that there was a population of cells, freely exchanging the genes responsible for these properties.

▣ **Fig. 4.33 "A" photo-fit of the cenancestor.** All the organisms in the living world share a certain number of biochemical and molecular characteristics that they have inherited from the cenancestor (last universal common ancestor or LUCA). Apart from the list of the processes and molecules present in this ancestral organism that have been established from these characteristics, several aspects of the biology of the cenancestor remain controversial.

– the oxygenation of the atmosphere from 2.45–2.32 Ga (*see* ▶ Chap. 7): as far as this may be said to be a consequence of the activity of cyanobacteria, these must therefore have appeared before 2.45–2.32 Ga;

– biomarker fossils, notably the hopanes, typically prokaryote, from 2.15 Ga.

As for the oldest trace fossils (microfossils) that are indisputably eukaryotes, they date from about 1.6 to 1.8 Ga (2.1 Ga-old fossils recently discovered in Gabon have been interpreted as potential multicellular eukaryotes, but they might be as well interpreted as bacterial colonies). The same applies to biomarkers that are probably from eukaryotes, the steranes, the oldest of those that are definitely confirmed only date from 1.7 Ga (*see* ▶ Chap. 7). All these traces are thus between one and two billion years younger than the oldest fossil traces of prokaryotic organisms!

The second constraint that argues in favor of a recent origin for the eukaryotes is biological and phylogenetic in nature. In addition to a nucleus that separates the genetic material from the rest of the cell, the eukaryotes have a second common and exclusive characteristic. This is the presence of organelles known as mitochondria (or organelles derived from mitochondria) where respiration takes place. There is compelling evidence that mitochondria derive from ancient symbiotic bacteria belonging to the alpha-proteobacteria sub-division of the Proteobacteria, which are, themselves, a group of highly derived bacteria. That means that the ancestor of present-day eukaryotes incorporated a bacterium that had already experienced a long process of evolution (in which it became an endosymbiont), before diversifying

4

and giving rise to the different modern groups of eukaryotes. (We shall discuss this point in more detail later in Chap. 7.) As a result, the eukaryotes that we know today are definitely the result of an evolutionary history that started after the emergence of bacteria.

These two types of constraints, linked to the arguments given above have led to the major part of the scientific community accepting the idea that the cenancestor was prokaryotic in nature. This is therefore the hypothesis that we will accept in the last part of this chapter. Starting from this probable prokaryotic ancestor, what are the selective forces that led to the diversification of life? Here again, we shall see that the hypotheses are very varied and sometimes contradictory.

The Earliest Diversification of Life

We do not know when the cenancestor lived, but it is from it that modern organisms have evolved. The first stage of this diversification of life was the separation of the two principal lines, leading to the bacteria and the archaea (or the bacteria and a lineage that, later would divide into the archaea and the proto-eukaryotes, ◨ Fig. 4.27), because by far the majority of scientists agree on the fact that the eukaryotes emerged and then evolved well after the diversification of the prokaryotes (*see above*). It is reasonable to think that the specialization of the pathways for the synthesis of membrane phospholipids and DNA replication pathways was linked to the separation of these two lines. The evolution of the energy and carbon metabolism was probably another key element that accompanied the major phases of diversification among organisms, starting from the initial division into two.

The Diversification of Metabolisms and the Major Prokaryote Lineages

Energy Metabolism

As far as energy metabolism is concerned, fermentation is probably very ancient. Indeed, this oxidation-reduction process, where the electron acceptors and donors are organic molecules, and where oxidation is incomplete (it does not lead to complete conversion of carbon into CO_2, but to the production of alcohols or organic acids), is universally observed in the organisms in the three domains of life. In addition, fermentation requires only enzymatic functions that were already present in the cenancestor, like those of kinases, which are responsible for the substrate-level phosphorylation leading to ATP, but also for many other transformations. In particular, fermentation does not involve any of the membrane systems required for oxidative phosphorylation. However, it is difficult to draw any definite conclusions about the cenancestor's ability to carry out fermentation, precisely because there is no family of enzymes that is specific to this process.

In all the organisms in the three domains of life, we find the same conversion of chemical (redox) or light energy into free energy that may be utilized by the cell, through the chemiosmotic coupling of a transmembrane electron-transport chain and the activity of an ATPase (a proton pump) exploiting the proton gradient created across the membrane. However, although the ATPases are strongly conserved, the electron-transport chains are extremely diverse from one organism to another, and from one type of energy metabolism to another (phototrophy if the primary source of energy is light; chemotrophy if the primary source of energy derives from redox reactions associated with respiration). If the final electron acceptor is oxygen, the respiration is called aerobic, and anaerobic when any other electron acceptor is used; ◨ Box 4.4). As a result, a large number of the components of electron chains must have evolved after the cenancestor stage.

4

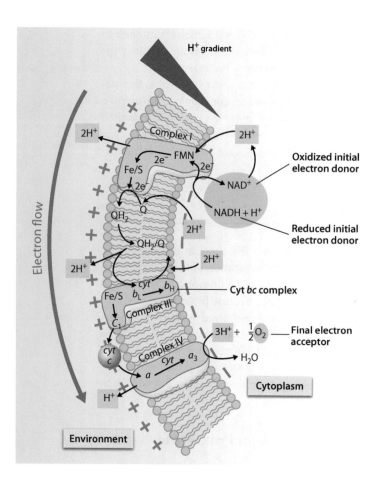

◻ **Fig. 4.34 An example of the production of a proton gradient (a chemiosmotic gradient) by an electron-transport chain during aerobic respiration (the final receptor of electrons is oxygen).** Certain elements of the complex *bc* are universal (present in the three domains of life). However, the electron-transport chains are extremely diverse among organisms and types of energy metabolism. The + and − charges on the two sides of the membrane correspond to an excess of H^+ and OH^- ions, respectively. FMN: flavine mononucleotide; cyt: cytochrome; Q: quinone; Fe/S: iron/sulfur protein.

Some enzymes and cofactors involved in these electron-transport chains are, however, quite widespread on the tree of life, which bears witness to their ancestral nature. This is the case with the complex cytochrome *bc*, common to chemotrophic bacteria (which carry out anaerobic or aerobic respiration) and to phototrophic bacteria (photosynthetic bacteria, and also other bacteria that use light as a source of energy, but which do not fix CO_2). Certain elements of this complex *bc* (◻ Fig. 4.34) are also present in the archaea, and, of course, in the mitochondria (former bacteria) of eukaryotic cells, whose membranes are the site of oxidative phosphorylation. They are thus universal. We may mention specifically cytochrome *b* and certain terminal cytochrome oxidases.

Among the cofactors, NADH (nicotinamide-adenine-dinucleotide) is universal as are cofactors based on cyclic tetrapyrrols. It is interesting to note that these tetrapyrrols are relatively easy to synthesize under abiotic conditions. Four cycles of pyrroles form a macrocycle, called a porphyrin, in the center of which is an atom, the nature of which varies. This is iron in haeme, which serves as cofactor in the cytochromes (haemoglobin also includes a haeme group), magnesium in the chlorophylls and bacteriochlorophylls, nickel in coenzyme F430 (involved in methanogenesis, *see later*) and cobalt in cobalamin (vitamin B12, involved in isomerization reactions and the transfer of methyl groups, especially in the synthesis of methionine; ◻ Fig. 4.35). Because cytochrome *b* is universal, it is possible to hypothesize that the electron-transport chains involved in the production of energy in the cenancestor contained tetrapyrrole nuclei linked to iron. During subsequent evolution, the duplication of genes involved in the synthesis of this cofactor and the replacement of the metallic atom would have led to the bacteriochlorophylls and chlorophylls in photosynthetic bacteria, and to cofactor F430 in methanogen archaea.

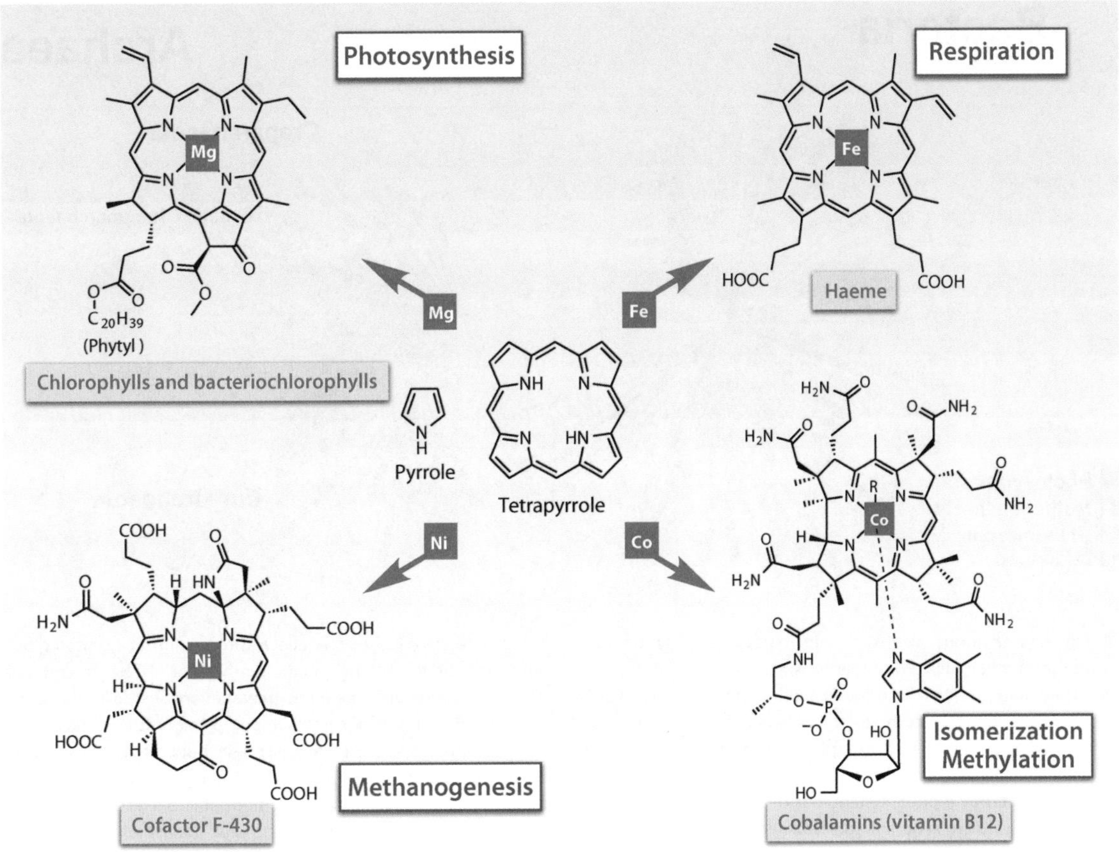

Fig. 4.35 Different porphyrin cofactors formed from a tetra-pyrrole structure. Four cycles of pyrroles form a macrocycle known as a porphyrin, which hosts a central metal atom of various types. The haemes are, among others, the cofactors of cytochrome *b*. The latter being universal, it is possible that the electron-transport chains involved in the production of energy in the cenancestor could have had iron-linked tetrapyrrole centers.

Because the cenancestor appears to have possessed an electron-transport chain leading to a cytochrome b that possessed a haeme group, it must have been capable of respiration (that is to say, carry out redox reactions within a membrane complex). However, as we have seen earlier, the type and number of electron acceptors involved remains uncertain. Although recent phylogenetic analyses have led to the suggestion that oxygen was the final receptor, others allocate this role to nitric oxide, NO (in fact, nowadays cytochrome b can also use this receptor in the absence of oxygen), and its nitrite and nitrate derivatives.

Carbon Metabolism

As regards the carbon metabolism, it is extremely difficult to reconstruct the evolution of the chemical pathways for the synthesis and breakdown of organic molecules in the three domains. There are three reasons for this. We have already mentioned the first two: the heterogeneous distribution of the various metabolic pathways in the tree of life and the extent of horizontal gene transfer. The third is that we do not know the metabolic pathways used by more than half of the major lineages of archaea and bacteria: those that we have not been able to cultivate and analyze in the laboratory!

All that we can say is that certain catabolic pathways, like glycolysis or the citric-acid cycle (the Krebs cycle), are quite widely distributed among known living organisms. We can additionally list four pathways for the fixation of mineral carbon, but none is universal. These pathways are the Calvin-Benson cycle (reductive pentose-phosphate cycle), the Arnon cycle

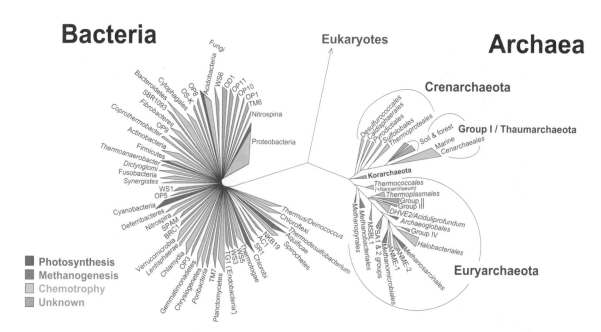

◻ **Fig. 4.36 Distribution of the major types of energy metabolism in the prokaryotes.** Photosynthesis, that is, obtaining free energy from light and its link to the fixation of CO_2 is observed only in certain lines of bacteria: the cyanobacteria, green sulfur bacteria, green non-sulfur bacteria, and purple sulfur bacteria (certain proteobacteria). In some firmicutes and acidobacteria, energy from light is captured by a photosystem, but these organisms are photoheterotrophs: there is no link to the fixation of CO_2. Other forms of phototrophy that are not coupled with the fixation of carbon do not use a photosystem, but a single protein of the rhodopsin family: proteorhodopsin or bacteriorhodopsin. The latter is a proton pump sensitive to light: it therefore allows the creation of a proton gradient across the membrane and the synthesis of ATP by a membrane ATPase. Environmental genomic (metagenomic) studies show that proteorhodopsin is widespread among planktonic bacteria and archaea. It appears to be of bacterial origin, but is easily transmissible by horizontal gene transfer. Eukaryotes inherited their energy metabolism from endosymbiotic chemotrophic bacteria with aerobic respiration: the mitochondria are the evolutionary relics of these endosymbionts. Photosynthetic eukaryotes additionally incorporated endosymbiotic cyanobacteria, which became the chloroplasts.

(reductive citric-acid cycle) the Wood-Ljundahl cycle (reductive acetyl-coenzyme A cycle), and the hydroxy-propionate pathway (and its variants). The Calvin-Benson cycle seems to have appeared relatively late in the evolution of the bacteria. (We do not, however, know whether it first appeared in photosynthetic bacteria or in the chemo-litho-autotrophs.) Each of the other three carbon fixation pathways has been suggested as the most ancestral within the framework of the various hypotheses. This merely underlines the difficulty in retracing the evolutionary history of carbon metabolic pathways.

Domain-specific Metabolisms: Photosynthesis and Methanogenesis

The only metabolisms on the tree of life that probably appeared after the archaea and bacteria diverged are photosynthesis in bacteria and methanogenesis in the archaea (◻ Fig. 4.36). As for the energy metabolisms in the eukaryotes, they vary little compared with the vast range that is observed in the prokaryotes, and they have undoubtedly been inherited from bacteria: aerobic respiration is carried out in the mitochondria, organelles that, as will be recalled, derive from endosymbiotic alpha-proteobacteria. Photosynthesis takes place in chloroplasts, which are themselves derived from endosymbiotic cyanobacteria. There is, therefore, no form of energy metabolism that is specific to eukaryotes.

Methanogenesis in Archaea

Methanogenesis is a form of anaerobic respiration in which dihydrogen, H_2, is oxidized into H_2O through a complex process in which CO_2 (or in certain cases, acetate) is reduced to methane ($\blacksquare$ Fig. 4.37). It is carried out only by a few groups of archaea belonging to the Euryarchaeota branch. We do not know if, in this branch, methanogenesis is an ancestral metabolism or not. However, we do know that certain euryarchaeota (such as *Archaeoglobus*, for example) possessed the ability to form methane but have almost completely lost it. It may also be noted that euryarchaeota of the ANME groups – which form symbiotic assemblages (or consortia) with sulfate-reducing bacteria in oceanic sediments ($\blacksquare$ Fig. 4.38) – do carry out anaerobic oxidation of methane, a type of metabolism that was discovered at the beginning of the 21st century. They probably use a reverse methanogenesis pathway to carry out this particular metabolism known as anaerobic methanotrophy.

Photosynthesis in Bacteria

Photosynthesis transforms light energy, which is captured by pigments of the chlorophyll or bacteriochlorophyll type, into chemical energy that is used for the biosynthesis of organic molecules starting with inorganic carbon ($\blacksquare$ Fig. 4.39). Photosynthetic organisms are thus able to fix carbon from CO_2 to fabricate organic matter.

It should be remembered that there are other forms of phototrophy that are not linked to the fixation of carbon, which are based on different photosensitive molecules (such as proteorhodopsin or bacteriorhodopsin). In this case, the energy from light is transformed into chemical energy in the form of ATP – which is used for different functions within the cell

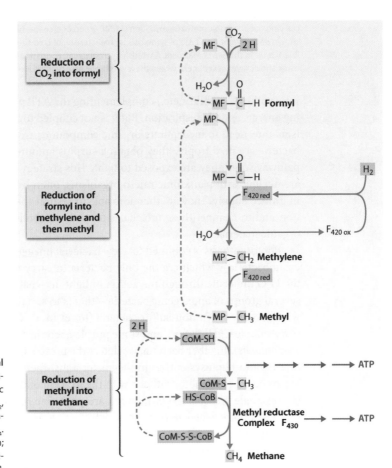

$\blacksquare$ **Fig. 4.37 An energy metabolism typical of archaea: methanogenesis.** Methanogenesis is a form of anaerobic respiration specific to archaea in which molecular hydrogen, H_2, is oxidized into H_2O through a complex pathway where CO_2 is reduced to methane CH_4. MF: Methanofurane; MP: methanopterin; CoM: coenzyme M; F_{420red}: coenzyme F_{420} reduced; F_{430}: coenzyme 430; CoB: coenzyme B.

4

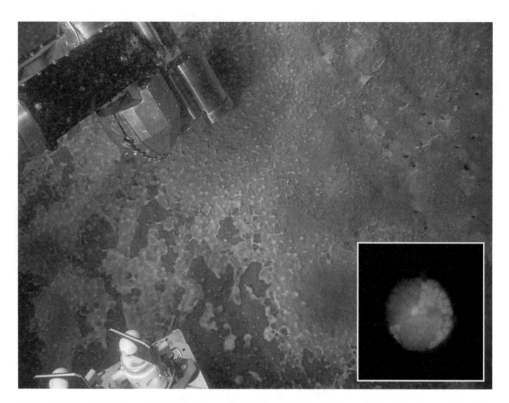

□ **Fig. 4.38 Symbiosis at great depths.** A microscopy photograph of a symbiotic consortium of archaea and bacteria carrying out anaerobic methane oxidation (ANME group of archaea, pink cells), and sulfate-reducing bacteria (green) is shown *bottom right*. This type of symbiosis is characteristic of cold seeps, like this photographed by Ifremer's submersible *Nautile* in the Sea of Marmara. A whitish microbial mat, consisting of bacteria that oxidize sulfur, is observed on the black anoxic layer of sediment where these symbiotic associations are found.

(biosynthesis, mobility, etc.) complementing the ATP produced by other pathways for obtaining energy, notably respiration. But it is not coupled to the fixation of carbon and these organisms thus need to incorporate organic compounds from their environment. In reality, these bacteria are mixotrophs: they obtain a surplus amount of energy by an efficient, alternative pathway when they are exposed to light. This strategy is all the more interesting because we are dealing with planktonic micro organisms, often living in conditions where nutrients are in limited supply. They are therefore able to restrict the use of organic molecules for respiration and to channel these molecules towards catabolism and the biosynthesis of their own organic material.

Photosynthesis is confined to a few bacterial lineages. Strictly speaking, it is present in the cyanobacteria – which are the only bacteria to carry out oxygenic photosynthesis, splitting the H_2O molecule through the action of light, in what is a controlled photolysis – as well as several groups of anoxygenic bacteria – that is to say, that do not produce dioxygen – among which are the green non-sulfur bacteria (for example *Chloroflexus*), the green sulfur bacteria (for example *Chlorobium*), and some purple bacteria or Proteobacteria (for example *Rhodopseudomonas*). In other bacteria, we detect the presence of a similar complex apparatus to capture light (a photosystem) as in the bacteria just mentioned, but it has not been possible, as yet, to detect true photosynthesis with the fixation of carbon. This is the case with Firmicutes (for example *Heliobacterium*), which are photoheterotrophs, and in the Acidobacteria (for example *Chloracidobacterium*), in which the presence of a photosystem was recently discovered, but where we do not yet know whether they are capable of fixing carbon. Sequencing the genome of these organisms may allow us to answer this question.

All these bacteria are characterized by the presence of a photosystem. This macromolecular complex, which is the basis of the photosynthetic machinery, consists of photosensitive pigments responsible for capturing light, together with numerous proteins that form an electron transport chain. Two types of photosystems are known, called I and II.

Only one or other of these photosystems is present in bacteria that carry out anoxygenic photosynthesis. Anoxygenic photosynthetic bacteria use H_2, H_2S, S^0, $S_2O_3^{2-}$ (thiosulfate), Fe^{2+}, As(III), NO_2^- (nitrite) and perhaps other electron donors (◘ Fig. 4.39). These electrons are transferred to a cyclic electron-transfer chain, which creates a proton gradient and, eventually, ATP. This is what is known as cyclic photophosphorylation, in which an electron at the reaction center of the photosystem is excited by light and goes through a series of electron transporters, to return, at a lower energy level, to the starting reaction center. In reality, high-energy electrons feed the pool of quinones or ferredoxins, intermediate electron carriers, which are able to transfer those electrons to other transport chains (*see later*). In that case, however, they are replaced by external reduced donors. There is, therefore, neither a net contribution nor a net consumption of electrons (◘ Fig. 4.40a). However, to reduce CO_2, hydrogen atoms must be available, that is to say, protons and electrons, which is called the reductive power. What do these bacteria do to obtain this reductive power? The various lineages of anoxygenic photosynthetic bacteria possess their own specific strategies involving external donors.

If we turn to oxygenic photosynthesis, the cyanobacteria (and their derivatives, which are the chloroplasts) possess the two types of photosystems (and not only one), which allows them to couple the two electron-transport chains together in a way that is not cyclic (◘ Fig. 4.40b) and to produce not only energy (in the form of ATP), but also reductive power (in the form of NADPH). Oxygenic photosynthesis in cyanobacteria, unlike anoxygenic photosynthesis, couples the two different, though interconnected, photochemical reactions: the lysis of the H_2O molecule, which is very unfavorable from a thermodynamic point of view (it takes place through the action of light: that is, photolysis), and the creation of a proton gradient through an electron flow, which produces energy and reductive power.

The evolution of photosynthesis is complex, and there are several hypotheses to explain the patchwork distribution of the photosystems among the groups of photosynthetic bacteria. Several elements, notably the phylogeny of the genes that code the enzymes for the biosynthesis of the Mg-tetrapyrrole cores that are the key components of photosynthetic pigments, suggest that anoxygenic photosynthesis is ancestral. However, to explain the distribution of the type I and II photosystems in present-day groups, it is necessary either to invoke lateral gene transfer events, or to imagine a photosynthetic ancestor that possessed the two photosystems and lost one or the other during the evolution of diverse anoxygenic

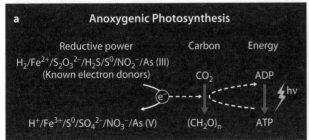
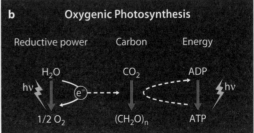

◘ **Fig. 4.39 Sources of energy and electrons (reductive power) for the synthesis of organic matter in oxygenic and anoxygenic photosynthetic organisms.** To fix mineral carbon into organic matter, photosynthetic organisms simultaneously need energy in the form of ATP and "reductive power", that is, reduced molecules capable of providing electrons, the formation of which involves an external electron donor. **a. In anoxygenic photosynthesis**, different electron donors are used depending on the organisms that carry out this type of metabolism (◘ Fig. 4.40a). **b. In oxygenic photosynthesis**, the electron donor is a water molecule (◘ Fig. 4.40b).

photosynthetic bacteria. This example of photosynthesis illustrates again how difficult it is to reconstruct the evolution of metabolisms. However, the comparison of genes and genomes from a larger and larger variety of organisms will perhaps, one day, allow us to better understand the origin and diversification of metabolisms and, as a result, the organisms that utilize them.

The Tempo and Mode of Prokaryotic Evolution

So far we have seen some of the elements that may have contributed to the diversification of the major lineages of prokaryotes since the last common ancestor. It is reasonable to think that specific characteristics present either in all the archaea, or in all the bacteria (such as the nature of the membrane phospholipids and the DNA replication mechanisms) are ancestral traits that were present in their last, respective, common ancestor.

It is equally logical to envisage that complex characteristics, such as certain energy metabolisms (photosynthesis and methanogenesis) that are found confined to certain groups within the bacterial and archaeal domains are the result of later evolution. Strictly speaking, however, one could also imagine that photosynthesis and methanogenesis existed, respectively, in the common ancestor of the bacteria and in the common ancestor of the archaea – or even in the cenancestor – and that they were then selectively lost in different lines. However, this hypothesis does not have many adherents, because it is difficult to explain how a

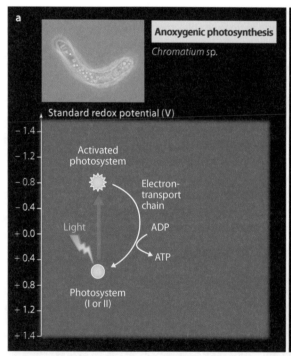
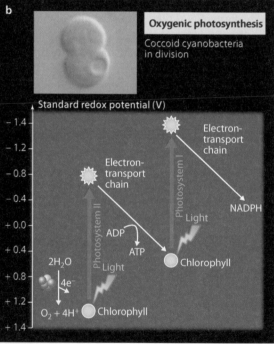

Fig. 4.40 Schematic, comparative diagrams of the photosystems in oxygenic photosynthesis and anoxygenic photosynthesis.
a. Bacteria carrying out anoxygenic photosynthesis (like the proteobacterium *Chromatium* sp., shown inset) have just one photosystem (PS I or PS II). ATP is produced through the chemiosmotic gradient created by a cyclic flux of electrons, in which there is neither a net contribution nor consumption of external electrons. The reductive power necessary to fix mineral carbon is obtained from NADH, or from ferredoxins, which are reduced through the oxidation of a wide

variety of reduced electron donors, some of which consume energy. **b.** Bacteria carrying out oxygenic photosynthesis (like the coccoidal cyanobacterium, shown dividing inset), have two coupled photosystems (PS I and PS II). Following photolysis of water, the electrons are transferred to a transport chain involving these two photosystems which, from the energy of light, are able to produce simultaneously reductive power in the form of NADPH and energy usable by the cell in the form of ATP (through chemiosmotic coupling and the membrane ATPase, ▪ Fig. 4.6b).

metabolism that is as complex and providing so many advantages as photosynthesis should have been uniformly lost in a whole series of lineages without leaving any traces. Observation of the distribution of photosynthesis in the tree of life, and even more, that of phototropy based on a proteorhodopsin, rather indicates the opposite. Once an advantageous adaptive metabolism appeared – such as obtaining free energy from light – it has a tendency to spread through horizontal gene transfer.

Horizontal gene transfer seems, however, to be more often found in certain pathways than others, and the reconstruction of the evolutionary origin of one pathway would, of course, be that much more difficult if it had been affected by this sort of transfer. So horizontal gene transfer probably was more important among the different types of respiration (genes that, for example allowed the use of various electron donors and acceptors). On the other hand, they had less effect on photosynthesis, and even less on methanogenesis, which require numerous proteins, coded by a large number of genes and which, in addition, must occur within a precise context in a membrane. It is probably for these reasons that their distribution is, despite the phenomenon of horizontal gene transfer, more restricted.

Nevertheless, in studying the topology of the present-day tree of life (◘ Fig. 4.28), the beginning of a reply may be offered as regards the relative order, and the relative speed of divergence among groups. If we look at the bacterial branch, there is a multiplicity of lines, almost all of which coming from the same point of divergence. This reflects what is called an evolutionary radiation. The major lineages of bacteria separated from one another over a short period of time, such that it is very difficult to determine if some emerged before others or whether all appeared more-or-less simultaneously. This means that when one group – that of the cyanobacteria, for example (◘ Fig. 4.41) – appeared and diversified, most of the other groups of bacteria were also present and in the process of diversifying. Similarly, we can say that at that time, bacteria were already separated from the line that would lead to the archaea (or the line that led to the archaea and proto-eukaryotes, *see* ▶ Chap. 7).

The situation is very different with the archaea. On the one hand, their branch is less extensive, which could also well be explained by a less significant diversity than in the bacteria or ... by our lack of knowledge of their real diversity. On the other hand, we do not see a radiation similar to that of the bacteria, but basically, a bifurcation between two principal lineages – the Crenarchaeota and Euryarchaeota – within which we then see a succession of lineages instead. This is particularly noticeable in the euryarchaeotal branch. At the base, it is dominated by hyperthermophilic organisms (including some methanogens), then we see a succession of thermophilic lineages and, finally, at the end of the branch, mesophilic and halophilic lineages (◘ Fig. 4.32). We may thus conclude that halophilic archaea appeared after the hyperthemophilic and moderate thermophilic archaea along this evolutionary path.

We do not know why the tempo and the mode of evolution in archaea and bacteria are so dissimilar. Some authors have suggested that these differences might be in part linked to the nature of the membranes and to the energy metabolism of the cells, but this remains highly speculative.

Finally, because most of the metabolic diversity is found in the prokaryotes, we might think that most of the ecosystems available on the early Earth (the oceans, but also continental water and soils, in short, all environments, or almost all, that allow life) were colonized once the diversification of the archaea and bacteria and the radiation of the bacteria had taken place. However, as we shall see later (Chap. 7), new ecological niches arose when multicellular eukaryotes made their entry on the evolutionary scene.

4

4

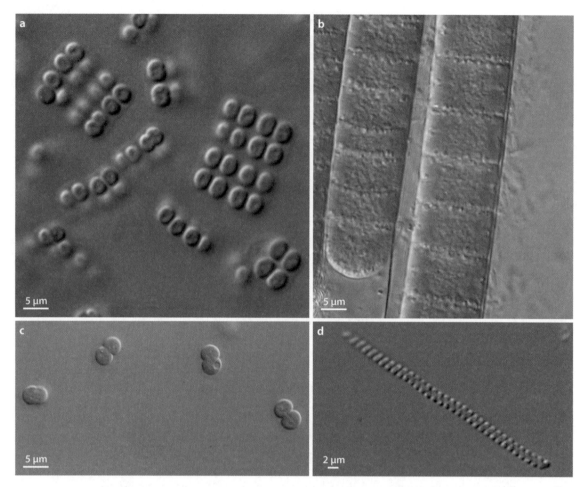

□ **Fig. 4.41 Morphological diversity among cyanobacteria: a.** colonial coccoidal cyanobacteria (*Merismopedia* sp.); **b.** filamentous cyanobacteria (*Lygnbya*-like); **c.** coccoidal cyanobacteria in division (Chroococcales-like); **d.** *Spirulina* sp. The oldest unambiguous signature of biological evolution in our planet are the 2.7 Ga-old fossil stromatolites from Fortescue in Australia, although more ancient traces might exist (stromatolites back to 3.5 Ga, *see* ▶ Chap. 6). Stromatolites are constructed of complex microbial communities that include a wide variety of photosynthetic bacteria, including cyanobacteria, as well as heterotrophic bacteria. Given the massive nature of the Fortescue stromatolites – they extend over several square kilometers – and the start of the oxygenation of the atmosphere around 2.4 Ga, one might imagine that the cyanobacteria, whose oxygenic photosynthesis is far more efficient than that of anoxygenic photosynthesizers, were already part of these structures. However, the oldest biomarkers diagnostic of cyanobacteria yet found (and undisputed) only date back to 2.15 Ga. To the extent that the early evolution of the bacteria is marked by a radiation (□ Fig. 4.28a), that means that by that date – and very probably from 2.7 Ga – the major bacterial lines had already emerged. (Photos: P. López-García.)

The Particular Case of Viruses

Viruses are "selfish" genetic elements, which exploit, like plasmids and other genetic elements (including genes), the cellular machinery to replicate but, in contrast to the later, are able to redirect the cell's metabolism to their own profit and infect new cells, thanks to an extracellular transmission form, the virion. Viruses are entities on the border of the definition of life. To certain researchers they are "living organisms" because one stage of their biological cycle is independent of the cell and because they are capable of evolving. To others, on the contrary, they are not true living beings because they are completely dependent on cells, lack any form of energy and carbon metabolism and, what is more, cannot evolve by themselves, but are, so to speak, "evolved" by cells.

There are several types of viruses, which vary in form, size, and genetic structure. They may contain genomes of RNA or DNA, single or double stranded. The virions are formed by the genome being encapsulated in a protein capsid (nucleocapsid). Certain viruses may be additionally surrounded by an envelope of lipids and glycoproteins (■ Fig. 4.42). Viruses are extremely abundant in nature. In the ocean, for example, their number is an order of magnitude greater than that of the bacteria. They are also very important from an ecological point of view, because they exert a control the population of the various microorganisms that they infect. From the evolutionary point of view, on the one hand they can transfer genes that they "pinch" from their hosts to different lineages of organisms, and on the other, they increase the evolutionary rate of these genes, because viral polymerases introduce many mutations (are error-prone) and each cycle of infection produces a very large number of virus particles.

What is the origin of viruses? Three principal hypotheses have been advanced. The first postulates that viruses are primitive entities that appeared before cells; the second that they derive from ancient cells that parasitized other cells, by the reduction and loss of ribosomes as well as of the energy and carbon metabolism; the third that they are fragments of cellular DNA or RNA that have become autonomous. These three hypotheses have been severely criticized. The first, because being obligatory parasites, viruses could not have appeared before cells. The second, because we do not know of any possible intermediaries between viruses and cellular parasites (even the most limited cellular parasites have retained the basic cellular characteristics). The third is generally the favorite hypothesis of many biologists, but it does not explain very well the mechanism by which the fragments of RNA or DNA escaped from cells and acquired a capsid. It is interesting to note, though, that vesicles budding off cells and encapsulating cellular material (including nucleic acids) have been recently discovered in different bacteria and archaea. More recently, several researchers have maintained that viruses appeared in the hypothetical RNA world. The first viruses would have been RNA viruses, and present-day RNA viruses would be distant descendants. Patrick Forterre, professor at the université de Paris-Sud and at l'Institut Pasteur, has even suggested that viruses "invented" DNA, because of the fact that several DNA viruses possess uracil (like RNA) instead of thymine, which could be seen as an intermediate stage between RNA and DNA.

Some go even further by proposing that viruses are part of the tree of life. This idea is, however, severely damaged by a whole series of scientific arguments that have accumulated over many years. We may cite, for example, the impossibility for viruses to replicate or evolve

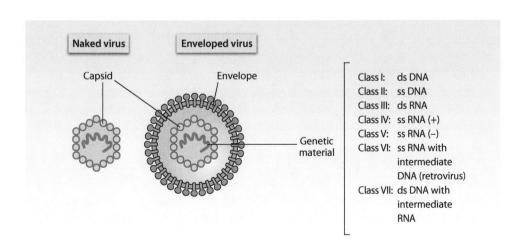

■ **Fig. 4.42 The principal types of viruses depending on the nature of their genetic material and their envelopes.** The origin of viruses remains a highly controversial subject. Are they primitive entities that appeared before cells? Remnants of ancient parasites? Fragments of DNA or RNA that have become autonomous? The debate remains open. Ss: single strand; ds: double strand; (+) or (−): sense in which the viral genome is read within the virion. The RNA of RNA(+) viruses may be directly used as mRNA by ribosomes within a cell, whereas that of RNA(−) viruses must first be transcribed into complementary RNA (+) for it to be translated into proteins.

in the absence of cells, or the fact that the only genes that they share with cells (and which could possibly allow them to be included with cellular organisms in phylogenetic trees) have been obtained, for the most part, by horizontal transfer from their respective hosts. These are therefore cellular genes and not viral ones. Whatever the situation may be, the origin of viruses is a highly controversial subject.

4

The Late Heavy Bombardment

Chapter 5

From 3.95 to 3.87 Ga: A Temporarily Uninhabitable Planet?

Between 3.95 and 3.87 billion years (Ga) ago, an extremely intense meteoritic bombardment brought an end to a fairly calm period that had lasted 400 million years.

This event – the marks of which we can still observe on the Moon's surface – must have substantially modified the surface of the Earth.

Then, the question that arises is to know whether, in the case where life had arisen before 3.95 Ga, it was able to survive this catastrophic episode.

◘ **A rain of meteorites descends on the Earth** (artist's impression).

M. Gargaud et al., *Young Sun, Early Earth and the Origins of Life*,
DOI 10.1007/978-3-642-22552-9_5, © Springer-Verlag Berlin Heidelberg 2012

The two preceding chapters have enabled us to discuss the potential habitability of the Earth between 4.4 and 4.0 Ga (Chap. 3), as well as to think about what is life and in which way it may have emerged (Chap. 4). We were led to formulate a key statement : it is absolutely impossible to determine whether life appeared and developed on Earth before 4.0 Ga. Whatever the case may be, at this period in its history, the Hadean Earth, inhabited or not, is about to undergo a major cataclysmic event: a heavy meteoritic bombardment. In this chapter, we will investigate the traces of this cataclysm, and then we will discuss its causes. We shall finally address the consequences for the primitive Earth and for the life that it may already have sheltered.

In Search of the Lost impacts

When we observe the Moon, Mercury or Mars using just an optical telescope, we are immediately struck by the impressive number of craters that pepper the surface of these Solar-System objects (■ Fig. 5.1). Obviously, the Earth does not have such a cratered surface. Nevertheless, it must not be concluded from this that, unlike the other terrestrial planets, it was spared from the meteorite falls. This aspectual difference comes from the fact that the surface of the Earth is ceaselessly modified and remodeled, in particular by plate tectonics: the Earth is a living planet! For example, we may remember that the oceanic crust, generated in the mid-oceanic ridge systems, returns into the mantle at the subduction zones, and that to this day, no oceanic crust older than 180 Ma is known. This implies that any possible marks of meteoritic impact on oceanic crust older than 180 Ma have all been irremediably erased. Contrarily to the oceanic crust, the continents are not involved in subduction processes.

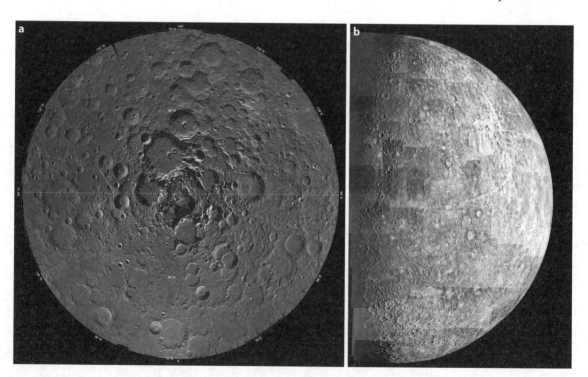

■ **Fig. 5.1 a. The North Pole of the Moon; b. a view of Mercury.** In these two photographs we can see that the surface of these two planets – where plate tectonics is absent – is riddled with extremely numerous meteoritic impact craters of all sizes, which attest to an intense meteoritic bombardment. The Earth also underwent this meteoritic bombardment, but plate tectonics and erosion have destroyed and erased all traces of it.

5

Fig. 5.2 A comparison of the stratigraphic columns for the first 1.5 billion years for the Moon and the Earth. The Moon has 1700 craters larger than 20 km in diameter. They are all dated to between 4.0 and 3.85 Ga: they were formed from the Nectarian until the Early Imbrian. These craters bear witness to an intense meteoritic bombardment about 4 billion years ago: the Late Heavy Bombardment (LHB). The arrows in continuous line correspond to isotopic ages, whereas the broken arrows are ages deduced from relative chronology. The anorthositic crust has been generated by the crystallization of pagioclase at the surface of the lunar magma ocean (*see* ▶ Chap. 2). *Serenitatis, Crisium, Nectaris, Imbrium, Orientale* and *Schrödinger* are the names of the main, large impact basins. On the Moon, the oldest traces known of melting triggered by meteoritic impacts are all more recent than 3.95 Ga. For details of the terrestrial stratigraphic column, it will be useful to refer to ▶ Fig. 2.2 in Chapter 2 and to Chapter 6. (After Ryder *et al.*, 2000.)

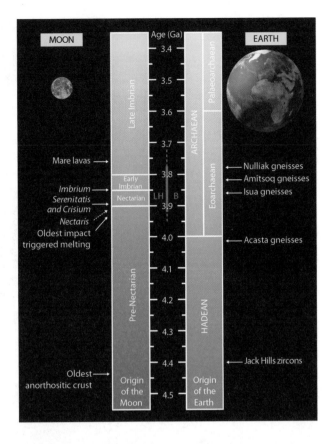

Nevertheless they may be several billion years old, such that their surface is ceaselessly remodeled by plate tectonics, which, among other things, builds mountain chains. Their surface is also subject to the action of water, which through the mechanisms of weathering and erosion destroys and erases any structures and any superficial landform.

It is certain, therefore, that the Earth, just like the other terrestrial planets, underwent a heavy meteoritic bombardment, but simply, no trace of it remains. Under these conditions, it seems to be very difficult, if not impossible, to evaluate the extent of this bombardment on our planet, except perhaps through the observation of another neighbor planetary body that would have undergone the same bombardment and of which it has recorded and retained the traces ... as is the case with the Moon.

Some 1700 craters larger than 20 km in diameter have been counted on our satellite, of which 15 have a diameter between 300 and 1200 km. All are dated between 4.0 and 3.85 Ga, which, in lunar chronology, corresponds to the Nectarian and to the beginning of the Imbrian periods (▶ Fig. 5.2). With its larger size, greater mass and, as a result, stronger gravitational attraction, the Earth should have received a more significant meteoritic flux than the Moon. It is estimated that, over the same period of time, our planet would have been subjected to a meteoritic flux that was between 13 and 500 times greater than that suffered by its satellite. According to this scenario, the Earth would have been collided by nearly 22 000 meteorites that would have left craters more than 20 km in diameter. Among these, between 40 and 200 would undoubtedly have had a diameter equal to, or exceeding, 1000 km, and some might even have reached 5000 km (the size of a continent).

Although all the scars of these terrestrial craters have been effaced, other traces and witnesses of the meteoritic bombardment have persisted, in particular in the Archaean sedimentary record. These include abnormally high contents in platinum-group elements, unusual

5

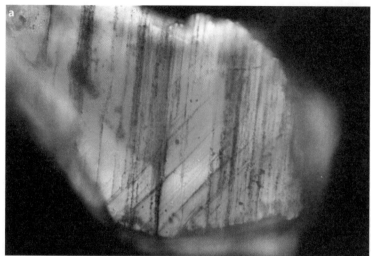

Fig. 5.3 Witnesses of meteoritic impacts on Earth: shocked minerals (a) and tektites (b). After a meteoritic crater has been destroyed by erosion it is often still possible to find signatures of the impact that produced it. This is the case, for example, with the presence of shocked minerals (that is, minerals that have been deformed by the high pressures induced by the shock waves), and with tektites (formerly molten silicates with a glassy texture resulting from the increase in temperature caused by the impact). (a: photo H. Leroux; b: photo P. Claeys.)

abundance of shocked minerals, or of tektites, but also isotopic anomalies. Let us examine each of these different indices.

The platinum group of elements (also called platinoids) is a family of rare chemical elements; it consists of ruthenium (Ru), rhodium (Rh), palladium (Pd), osmium (Os), iridium (Ir), platinum (Pt), and sometimes depending on the sources, rhenium (Re). Although they are relatively abundant in meteorites (for example, the concentration of Ir is 455 ppb in the C1 chondrites), the platinoids are very rare in the mantle (Ir ~ 3.2 ppb) or in the Earth's crust (Ir ~ 0.03 ppb). Regarding the shocked minerals, they were deformed by the very high pressures induced by the shock waves, whereas tektites are formerly molten silicates that are glassy in appearance and that result from the great increase in temperature caused by the impact (◘ Fig. 5.3).

To this day, these markers have still not been discovered in the oldest Archaean terranes. The explanation that comes immediately to mind is that, with the exception of the Acasta gneisses in Canada, which outcrop over an area of only 20 km² (*see* ▶ Chap. 6); no rock contemporary with the period of the heavy bombardment has been preserved. In addition the Acasta gneisses are plutonic rocks, which means that their parental magma emplaced and crystallized into the crust at a depth of several kilometers. The oldest known sedimentary rocks (deposited on the Earth's surface), are the Isua gneisses from Greenland, aged 3.872 ± 0.010 Ga. However, neither the Acasta gneisses, nor those from Isua, contain shocked minerals, tektites nor any excess of platinoids. That could mean that either the meteoritic bom-

■ **Fig. 5.4 Tungsten (^{182}W/^{183}W) isotopic ratio in terrestrial rocks and meteorites.** ε_w represents the value of the ^{182}W/^{183}W ratio, normalized to a reference value, which here is the terrestrial average. The diagram allows a comparison of the ε_w of sediments from the Early Archaean (Isua in Greenland and Nulliak in Labrador) with those of meteorites as well as with those of more recent terrestrial rocks. The light blue band between the two white dashed lines corresponds to the compositional field for the terrestrial samples. Four samples of Early Archaean sediments display ε_w that are significantly lower than the terrestrial range, and which are of the same order of magnitude as those in meteorites. This isotopic signature may be the proof that the Earth, like the other terrestrial planets, has undergone the Late Heavy Bombardment. (After data from Schoenberg, et al., 2002.)

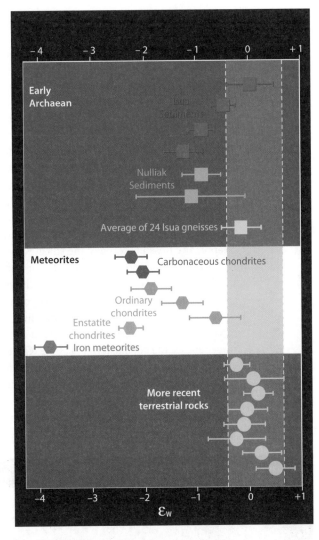

bardment had ceased by the time the rocks were emplaced, or else that these extraterrestrial markers have been subsequently greatly dispersed and diluted within the sediments.

And what about isotopic markers? Recent developments in analytical techniques allowed, in 2002, the investigation of the ^{182}W/^{183}W isotopic ratio (W = tungsten) in the ancient sediments from Isua in Greenland and Nulliak in Labrador. Among eight analyzed samples, four revealed an abnormal ^{182}W/^{183}W ratio. ■ Figure 5.4 shows that some ancient sediments (dated between 3.7 and 3.87Ga), have an ε_w – which represents the value of the ^{182}W/^{183}W ratio, normalized to a reference value (here the terrestrial average) – that is significantly different from the terrestrial average value and similar to that of extraterrestrial materials such as chondrites. If these results are corroborated, they provide the first tangible proof of the existence of an intense meteoritic bombardment on Earth! This enthusiastic conclusion should, however, be tempered by a study of the chromium isotopes (^{53}Cr/^{52}Cr), the results of which were published in 2005 and which did not detect, in the same Isua rocks, any isotopic signature of chromium that could be ascribed to an extraterrestrial component …

Despite these uncertainties, the existence of an intense meteoritic bombardment of the Earth is unanimously accepted by the scientific community. On the other hand, the frequency of collisions between 4.5 and 3.8 Ga remains a controversial subject: two contrasted scenarios are in contention.

Late, or Continuous, Bombardment? The Two Competing Scenarios

Two main theories have been proposed to account for the facts observed both on the Earth and on the Moon. The first one considers that the frequency of meteoritic impacts has declined slowly and regularly since the end of planetary accretion and that, as a consequence, the craters observed on the Moon, and dated between 4.0 and 3.85 Ga, mark the end of a period of continuous bombardment that lasted 600 Ma. The curve representing the evolution of the cratering rate of the lunar surface between 4.0 and 3.85 Ga (◘ Fig. 5.5) has been extrapolated, and from it the conclusion has been drawn that the meteoritic flux must have been greater and greater the further back in time one went. Such a flux would have had devastating consequences for our planet. The energy released by the collisions would have kept the crust in a quasi-permanent molten state, obviously turning the whole of the oceans into vapor, and possibly some of the silicates as well.

◘ Figure 5.5 provides a first factor contradicting the theory of a continuous bombardment since 4.5 Ga. Indeed, given that the size of craters is proportional to that of the "impactor" (and thus to its mass), it is thus possible to determine the mass of matter added to the planet solely through the meteoritic flux. If, as has been proposed, the measured fluxes are extrapolated backward over the period corresponding to the beginning of the Earth's history, one reaches the conclusion that at 4.1 Ga the contribution of meteoritic material was equal to the mass of our satellite. In other words following this scenario, accretion of the Moon would have lasted 100 Ma and not 600 Ma! A conclusion that is obviously completely unrealistic. In addition, the considerable volume of extraterrestrial material brought to the surface of the Earth and the Moon by such a heavy bombardment over such a period of more than 600 Ma

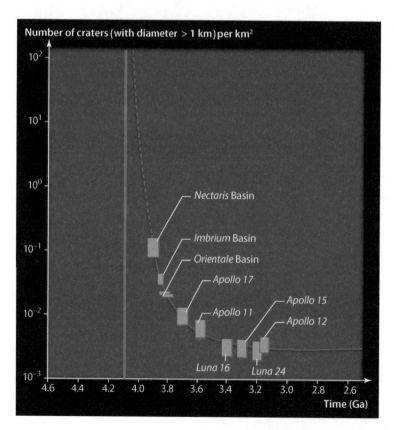

◘ Fig. 5.5 The evolution of the cratering rate for the lunar surface as a function of time: based on the theory of a continuous meteoritic bombardment. When the red curve is extrapolated back towards the beginning of the Moon's history (dashed line), it implies that the accretion of our satellite took place at 4.1 Ga, which is totally unrealistic. From this it has been concluded that no intense meteoritic bombardment took place before 3.92 Ga. The labels "Apollo" and "Luna" followed by a number, correspond to the landing sites of the American and Russian missions during which samples were collected and returned to Earth, where they were dated. (After Ryder, 2002.)

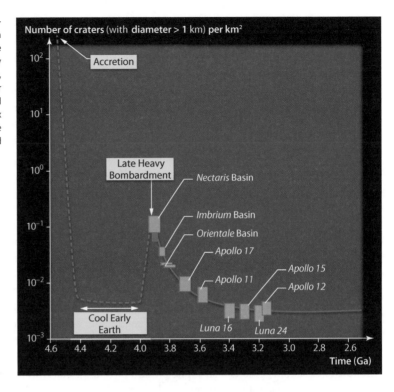

■ **Fig. 5.6 The evolution of the crater-ing rate for the lunar surface as a function of time: based on the theory of the Late Heavy Bombardment,** whose peak activity lies around 3.9 Ga. According to this theory, the meteoritic flux declined very rapidly after the formation of the Moon, reaching a level equivalent to twice that of the current flux (dashed line). This calm period, known in the case of the Earth as "Cool Early Earth", lasted more than 400 Ma.

5

should have provided and enriched those two bodies in siderophile elements such as the platinoids. Yet the surface of the Moon shows absolutely no trace of any such enrichment. Finally, among the meteorites of lunar origin that have been collected on Earth, those that correspond to tektites (rocks melted by the thermal effects of an impact) are never older than 3.92 Ga. From this it has been concluded that the period before 3.92 Ga was not subjected to any heavy meteoritic bombardment.

The Jack Hills zircons, which we have described in great detail in Chap. 3, also clearly contradict this model. Their analysis does, in fact, show that the continental crust has been continuously generated between 4.4 and 4.0 Ga, and that it was stable, such that it could be eroded and recycled. Those same zircons also prove the presence of liquid water on the sur-face of the planet from 4.4 Ga. All these data are inconsistent with the scenario of a continu-ous meteoritic bombardment throughout the Hadean. Finally, the presence on the Moon of an anorthositic crust dating back to 4.456 Ga (*see* ▶ Chap. 2) is, in addition, not compatible with an intense, continuous meteoritic bombardment lasting 600 Ma.

The result is that the curve of the cratering rate for the lunar surface cannot be extrapo-lated backwards. The beginning of the history of our planet – and, of course, of its satellite – was a calm period (the Cool Early Earth). From this comes the second theory, which today is favored by most of the scientific community: the meteoritic bombardment was only a limited catastrophic event that affected the Earth and the Moon around 4 Ga ago. This is the theory of the Late Heavy Bombardment (or LHB), shown in ■ Fig. 5.6. We shall now examine this theory in detail.

The Late Heavy Bombardment: a Cataclysmic Scenario

Until very recently, the theory of the Late Heavy Bombardment posed a problem that re-searchers were unable to resolve. Indeed, this implies, that somewhere in the Solar System, there must have been a large reservoir of asteroids, which had remained stable during the

600 Ma that followed the planetary accretion and that subsequently this reservoir was suddenly destabilized, giving rise to an intense, catastrophic bombardment. This scenario is in complete conflict with the commonly held idea that the Solar System is a stable system that had not undergone any significant alteration since its formation.

It was in 2005 that researchers developed, through theoretical modeling, a two-stage scenario that simply accounts for the triggering of the Late Heavy Bombardment. This model that has been initially developed in the Observatoire de la Côte d'Azur, in Nice, (France), is also known as the "Nice model". Currently, the orbits of the jovian planets (Jupiter, Saturn, Uranus, and Neptune) are slightly eccentric and lie between 5.2 and 30 AU (1 AU = astronomical unit = Earth–Sun distance = 150 million kilometers) (◘ Fig. 5.7). According to the Nice model, it was not the same, immediately after the accretion of the Solar System. These same planets had quasi-circular orbits, which, above all, were more "compact", that is, lying between 5.5 and 15 AU (◘ Fig. 5.8). Moreover, at that time the planet Uranus was the outermost of the Solar System, instead of Neptune today. In addition, a disk of ice and rock planetesimals (the Kuiper Belt) was located, outside the orbit of Uranus and extended from this orbit up to some 35 AU. At that time, this disk – equally referred to as the trans-Neptunian disk – was very massive (35 Earth masses). The gravitational interactions between the giant planets and the planetesimals in the trans-Neptunian disk have slowly and progressively modified the orbits of the planetesimals of the inner boundary of the disk. Consequently, after a certain time, the latter cut across the orbits of the giant planets. The resulting impacts of planetesimals on the jovian planets had the effect of progressively modifying their orbits and thus of inducing their migration. All the computed simulations predict that Jupiter would have migrated towards the inner region of the Solar System, while Saturn, Neptune and Uranus would have migrated towards the outside (◘ Fig. 5.8). This divergent motion accounts for the decrease in the "compactness" of the orbits of the giant planets. This is the first act of the play.

The ratio of the orbital periods of Jupiter and Saturn (the orbital period is the time for one revolution of the planet around the Sun) is currently slightly less than 2.5. Jupiter completes an orbit around the Sun in 11.86 years, whereas it takes 29.46 years for Saturn to do the same. The Nice model considers that this ratio was initially slightly less than 2.0. The simulations show that a certain time after Jupiter started to migrate towards the center of the Solar System and Saturn towards the outside, the ratio of their orbital periods reached the value of exactly 2.0, which corresponds to what astronomers call an orbital resonance. This sudden change provisionally destabilized the overall system consisting of the four giant planets: all acquired eccentric orbits, such that the orbit of Neptune crosscuts both the orbit of Uranus and the disk of planetesimals. This, then completely destabilized the latter, and some of the planetesimals migrated towards the center of the Solar System. This is the second act of the scenario, and one that provides a realistic explanation of the Late Heavy Bombardment. It has, in fact, been calculated that about one millionth of the planetesimals could have collided with the Earth and about one tenth of that number with the Moon. Only a thousandth of them would have remained within the disk and would form the current Kuiper Belt. The calculations predict, in addition, that the destabilization of the disk of planetesimals must have occurred less than 1200 Ma after the formation of the Solar System. The Late Heavy Bombardment, which occurred about 650 Ma after the start of accretion (that is at 3.95 Ga) fits perfectly within this large range of time.

The trans-Neptunian disk of planetesimals is not the only one which provided objects that collided with the Earth and the Moon during the Late Heavy Bombardment. The migration of Jupiter and Saturn did equally destabilize the asteroid belt located between Mars and Jupiter (◘ Fig. 5.8). It is estimated that more than 90 percent of these asteroids could have crossed the Earth's orbit and thus collided with our planet (◘ Fig. 5.9). Assuming that the mass of the asteroid belt corresponds to 5×10^{-3} the mass of Earth, it means that this belt certainly did feed the Late Heavy Bombardment in a substantial manner. However, calculations do not yet allow us to determine the respective roles played by nearby objects (from the asteroid belt)

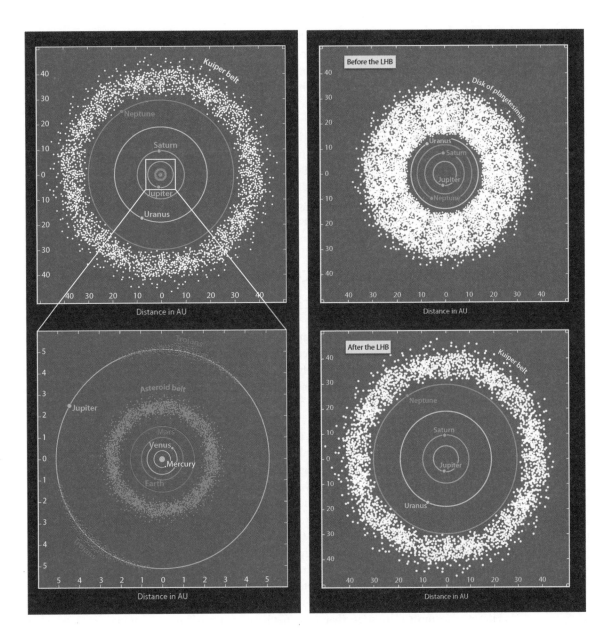

◘ **Fig. 5.7** A schematic representation of the present-day Solar System showing the relative positions of the various planets as well as the Asteroid belt, the Kuiper belt and the Trojans. Distances are expressed in astronomical units (AU); 1 AU represents the Earth–Sun distance, i.e., 150 million kilometers.

◘ **Fig. 5.8** A schematic representation of the relative positions of the giant planets and the belt of planetesimals, before (a) and after (b) the Late Heavy Bombardment. Just after its accretion, Jupiter slowly started to migrate towards the center of the Solar System, whilst the other three giant planets migrated outwards. A certain time after, the ratio of the orbital periods of Jupiter and Saturn reached the value of exactly 2.0, which corresponds to what astronomers call an orbital resonance. This gave rise to a sudden change that provisionally destabilized all of the giant planets: all acquired eccentric orbits, such that the orbit of Neptune cut across both the orbit of Uranus and the disk of planetesimals, thus causing a major destabilization, which led to the intense meteoritic bombardment. After Gomes *et al.* (2005).

and distant ones (from the Kuiper Belt) in this cataclysmic event. The only certainty comes from chemical analyses carried out on lunar samples collected in the immediate vicinity of major meteoritic craters. Their composition is identical to that of the ordinary chondrites or enstatite chondrites, bearing witness that this type of material did definitely contribute to the Late Heavy Bombardment.

The major strength of the Nice model, which we have just described, lies in the fact that it does not represent an *ad hoc* scenario, but a general model that takes into account all the prominent features of the Solar System. Indeed, it provides an explanation not only for all the characteristics of the Late Heavy Bombardment (intensity, duration, sudden onset, etc.), but also, and above all, for other properties of the Solar System, such as the characteristics of the orbits of the giant planets (spacing, eccentricity, inclination of the orbital planes) and the orbital distribution of the Trojans; (the Trojans are asteroids that revolve around the Sun in the same orbit as Jupiter; ◼ Fig. 5.7).

A Rain of Meteorites: The Consequences of the Late Heavy Bombardment

The new concept of Late Heavy Bombardment requires us to revise our conception of the first 600 Ma of the Earth's history. The theory of a continuous meteorite bombardment implied that the Earth's surface was completely molten, throughout the whole Hadean. The primitive continental crust would have been destroyed, and liquid water could have been totally vaporized. Under such extreme and chaotic conditions, the emergence of life and its development appeared utterly impossible. This environment was not favorable either for the development of prebiotic chemistry and less even for the preservation of its reaction products. In other words, this scenario of continuous bombardment, led to the logical conclusion that the Hadean Earth was hostile to the appearance and to the development of life and that its surface was utterly desert. The theory of a Late Heavy Bombardment reduces the duration of the cataclysm from 600 to about 100 Ma, that is, to an event that is far more limited (on a geological time scale). Consequently, between the end of the crystallization of the magma ocean (4.4 Ga) and the beginning of the Late Heavy Bombardment (3.95 Ga), the meteoritic flux remained low (twice that of the current rate), which is perfectly consistent with the message delivered by the Jack Hills zircons (Chap. 3) that during the Hadean, a stable continental crust and oceans of liquid water existed. In other words, all the required conditions were available for prebiotic chemistry to be able to operate, for life to appear and to develop. However, we must again stress that we do not have any trace, any proof or indication, whether direct or indirect, that life did actually appear during the Hadean. The only certainty is that, from 4.4 Ga, planet Earth was potentially habitable, which does not imply by any means that it was inhabited.

If life had not appeared before the Late Heavy Bombardment occurred, this cataclysm obviously would not have had any effect in the biological sphere. It would be different if one or more life forms had already arisen in the Hadean. One can then imagine two different situations.

The intensity of the bombardment could have been such that the oceans were totally vaporized, volcanism was triggered by the effect of the impacts, and the continental crust was destroyed. Under such conditions, the planet could well have been totally sterilized, both at the surface as well as at depth. Life would then have reappeared after the Late Heavy Bombardment, which leads to suppose that its emergence is not an exceptional event, but indicates instead an "unremarkable" evolution as soon as the environmental conditions and all *ad hoc* ingredients were available. An alternative scenario is, however, equally plausible. The Jack Hills zircons (4.4 Ga), the Acasta gneisses (4.03 Ga) as well as lunar anorthosites (4.456 Ga)

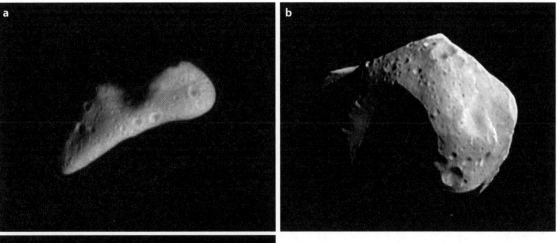

5

□ **Fig. 5.9 Photographs of three asteroids from the main asteroid belt (between Mars and Jupiter): a.** Eros (33 km long, 13 km wide), a photograph taken on 29 February 2000 by the NEAR probe; **b.** Mathilde (59 km long, 47 km wide), a photograph taken on 27 June 1997 by the NEAR probe; **c.** Ida (56 km long), a photograph taken on 28 August 1993 by the *Galileo* probe. One characteristic of all these objects is that their surface is riddled with impact craters. Some of these are very big, like the large crater (the black area) that fills the center and the bottom part of the image of Mathilde and whose estimated depth is more than 10 km.

show that the energy released during the Late Heavy Bombardment was not sufficient to cause the melting and complete destruction of the continental crust that had been generated before the event.

Similarly, several scientists now believe that even if a major impact could have vaporized a huge volume of oceanic water, it is highly unlikely that such a vaporization could have affected the whole of the oceans, so liquid water undoubtedly remained on the surface in certain areas of our planet. In such a scenario, the Late Heavy Bombardment does not necessarily imply the extinction of all forms of life on Earth. In addition, in the light of the recent discovery that bacteria can, today, live in the interior of rocks down to depths of 3000 meters, it is possible to envisage that forms of life confined to such "protected" environments could have survived the Late Heavy Bombardment.

All the theories, rapidly summarized here, are only possible scenarios, and we have no information that allows us to decide in favor of one or the other. The fundamental point is that the Hadean Earth was potentially habitable, and it is by no means absolutely certain that the Late Heavy Bombardment could have completely sterilized our planet.

The Messages from the Oldest Terrestrial Rocks

Chapter 6

From 3.87 to 2.50 Ga: A Habitable Planet Becomes Inhabited

Between 3.87 and 2.5 billion years before the present (3.87–2.5 Ga), significant continental masses were generated in an environment where, due to a greater internal heat production, the mechanisms that operated were different from those active nowadays. As far as the atmosphere was concerned, it was lacking in oxygen. And life? The most ancient, incontestable, fossil signatures have an age of 2.7 billion years. Nevertheless, a sheaf of arguments suggests that life might have been born by at least 3.5 Ga.

▣ **The Warrawoona region in Western Australia,** from where the oldest potential microfossils have been described. They have ages of 3.3 to 3.5 billion years. (Photo: P. López-García.)

M. Gargaud et al., *Young Sun, Early Earth and the Origins of Life*,
DOI 10.1007/978-3-642-22552-9_6, © Springer-Verlag Berlin Heidelberg 2012

The only remnants from which scientists attempt to reconstruct the face of the Earth before 4.031 Ga are minerals: the zircon crystals discovered at Jack Hills in Western Australia. These have taught us that, very early in its history, probably as early as 4.4 Ga, our planet was potentially habitable (*see* ▶ Chap. 3). Unfortunately, the zircons crystallized from a magma at high temperatures (around 700 °C) and high pressure. So they could not have been in contact with potential living beings, and are of no help in detecting the presence of life on Earth. The inquiry on the origins of life changes radically at 4.0 Ga, because from then on, geologists, geochemists and biologists have rocks at their disposal. As we shall see in this chapter, some of these rocks, which are among the oldest known on Earth, are of sedimentary origin: they were deposited at the bottom of an ocean or of a lake. If these environments sheltered life forms, then these rocks were likely to record and preserve any traces. So it is only starting from about 4.0 Ga that we may hope to find direct and tangible traces of life on the surface of the Earth, in the fossil record.

That age of 4.0 Ga marks the beginning of the eon known as the "Archaean". All Archaean rocks belong to the continental crust, because as explained in Chap. 3, due to plate tectonics, the oceanic crust is recycled (returns) into the mantle 180 Ma at the most after its formation. In other words, the oldest oceanic crust presently known has an age of 180 Ma. By contrast, the continental crust has a lower density such that, independently of its temperature, the continental lithosphere "floats" over the asthenosphere; therefore it cannot be recycled into the mantle in any significant amount. As a result, it has been able to undergo and record, several episodes of the Earth's history.

◘ Figure 6.1 shows that today, the Archaean terranes still cover a significant area of the planet's surface. They form what geologists call cratons or shields, and which are vast volumes of continental crust (together with the underlying lithospheric mantle) that have been stable for several billion years. The most ancient terranes (dated between 4.031 and 3.8 Ga) outcrop

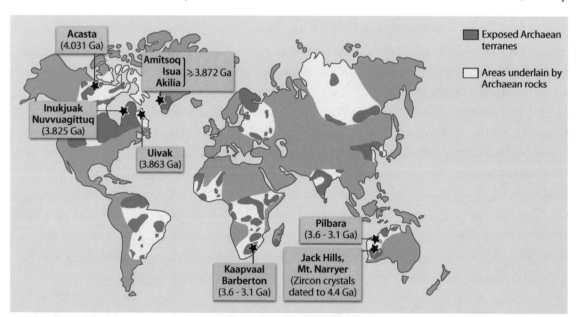

◘ **Fig. 6.1 Simplified map of the Earth, showing the main outcrops of Archaean continental crust.** The oldest rocks known today are the Acasta gneisses; they are of magmatic origin and have a TTG composition (*see* p. 175). They outcrop over about 20 km² in the Northwest Territories in Canada. Slightly more recent, the Amîtsoq gneisses in Greenland and the Uivak gneisses in Labrador (Canada) are also ancient TTG. The oldest volcanic and sedimentary rocks catalogued today, crop out at Isua and Akilia (Greenland) as well as at Nuvvuagittuq (east Quebec). With surface areas of 1.2 and 0.06 million km², respectively, the Kaapvaal craton (South Africa and Swaziland) and the Pilbara craton (Australia) are the oldest Archaean terranes of continental size. Finally, it is worth remembering that the oldest terrestrial materials yet known as the zircon crystals discovered in Western Australia, at Jack Hills and at Mount Narryer (*see* ▶ Chap. 3).

only on small areas, and somehow, they represent relics of the Archaean continents. The first cratons having truly continental dimensions are slightly more recent (about 3.5 Ga). We shall now examine these different terranes in more detail in order to draw as accurate a picture as possible of the Earth during the Archaean. And we shall see that a collection of arguments strongly pleads in favor of an already inhabited planet…

The Scattered Remnants of One of the Oldest Continents

The oldest rocks currently known are the Acasta gneisses, which crop out over an area of only 20 km² in the Northwest Territories, in Canada. These gneisses were previously magmatic rocks, known as TTG (for Tonalite-Trondhjemite-Granodiorite), from which zircon crystals have been extracted and dated at 4.031 ± 0.003 Ga. It may be noted that the tonalites, trondhjemites, granodiorites and granites, belong to the so called granitoid group. This latter consists of acid (silica-rich) plutonic rocks, which differ from one another in their relative abundance of plagioclase feldspar and alkali feldspar (*see* the classification of rocks, p. 267). A plutonic rock is a rock whose parental magma crystallized at depth in the Earth's crust. The TTGs are almost exclusively restricted to Archaean terranes, which means that they are at least 2.5 Ga old. In Acasta, the TTGs outcrop as foliated and banded rocks, which are associated with small amounts of amphibolites and ultrabasic rocks.

Rocks emplaced on the surface, which geologists term supracrustal rocks are either sediments or volcanic rocks. Canada also hosts supracrustal rocks that are among the oldest known to date. They come from the Nuvvuagittuq greenstone belt, on the eastern shores of Hudson Bay in Quebec, in the Inuit territory of Inukjuak. These outcrops contain, amongst others, beds of acid volcanics, whose zircon crystals have been dated to 3.825 ± 0.016 Ga. In September 2008, Canadian and American researchers dated an amphibolite in this greenstone belt: the Ujaraaluk unit, which before 2011, was referred to as the Faux amphibolite

6

☑ **Fig. 6.2 The Ujaraaluk unit, (previously referred to as Faux amphibolite), in the Nuvvuagittuq belt (East Quebec).** Although its age has not been established with certainty, this is one of the most ancient volcano-sedimentary (supracrustal) rocks known. Its age is at least 3.82 Ga. The white layers consist of cummingtonite (amphibole), the rounded red minerals are garnets, and the other minerals are plagioclase, quartz and biotite. (Photo: E. Thomassot.)

( Fig. 6.2). Depending on the method used, they obtained contradictory results. Dating based on the disintegration of ^{146}Sm (^{146}Sm → ^{142}Nd) gave a very ancient age of 4.28 ± 0.05 Ga, whereas that based on the ^{147}Sm → ^{143}Nd system gave a much more recent date of 3.82 ± 0.27 Ga. Nevertheless, these supracrustal rocks are still among the oldest in the world.

The most significant remnants of very ancient continental crust known to date are located on the southwestern coast of Greenland, in the Nuuk region. There, TTGs, termed the Amîtsoq gneisses ( Fig. 6.3), are exposed over a vast expanse (3000 km²); they have been dated to 3.872 ± 0.010 Ga. However, this part of Greenland harbors even older rocks. These are supracrustal rocks which crop out in the shape of large lenses and enclaves intruded by dykes of the Amîtsoq gneisses. They are thus older than the cross cutting Amîtsoq gneisses, which means that they were emplaced prior to 3.872 Ga. Sometimes, as at Isua, these enclaves may be very large (~35 km long), but in general as, for example, on the island of Akilia, their size does not exceed 1 kilometer ( Fig. 6.4). Unfortunately, it proves to be very difficult to obtain an accurate radiometric date for these materials. In fact, recent measurements of argon isotopes (^{40}Ar/^{39}Ar) carried out on minerals included in pyrite crystals from the Isua gneisses ( Table 6.1) have given an age of 3.86 ± 0.07 Ga. The error in the measured age is significant: indeed, the latter may lie anywhere between 3.93 Ga and 3.79 Ga, which does not allow us to conclude whether or not this age is consistent with the age, greater than 3.872 Ga that has been deduced from the intrusive relationships of Amîtsoq gneisses into the Isua supracrustals.

On the island of Akilia, Banded Iron Formations (BIF,  Fig. 6.4) contain carbonaceous (graphite) inclusions whose carbon-isotopic ratio (δ^{13}C = –30 to –35 ‰) has been interpreted by some authors as a possible biological signature. We shall discuss this point in detail again in the last part of this chapter, because these conclusions are still the subject of lively controversy. On the other hand, the fossil "bacteria" that were formerly described in the Isua formations – *Isuasphaera Isua* and *Appelella ferrifera* – are now unanimously considered to be artefacts.

◘ Fig. 6.3 Detailed view of the Amîtsoq gneisses in Greenland (3.872 ± 0.010 Ga). These remnants of very ancient continental crust outcrop over an area of 3000 km². The gneisses are grey in color with a well defined metamorphic banding. They are cut by white dykes of younger granite (the Qorqût granite dated to 2.55 Ga). (Photo: G. Gruau.)

Table 6.1 **Principal characteristics and ages of the oldest crustal remnants known at present.** The TTG (tonalites, trondhjemites and granodiorites) are ancient plutonic rocks whereas the BIF (Banded Iron Formation) and the Nuvvuagittuq gneisses are of sedimentary and volcano-sedimentary origin, respectively. *See also* the classification of rocks on p. 267

Name	Location	Lithology	Age in Ga
Acasta gneisses	Canada (Northwest Territory)	TTG	4.031 ± 0.003
Nuvvuagittuq gneisses	Canada (Quebec, Hudson Bay)	Volcano-sedimentary	> 3.825 ± 0.016
Akilia BIF and Isua gneisses	Greenland	BIF	> 3.872 ± 0.010
Amîtsoq gneisses	Greenland	TTG	3.872 ± 0.010
Uivak gneisses	Canada (Labrador)	TTG	3.863 ± 0.012

Fig. 6.4 **Banded Iron Formations (BIF) at Akilia in Greenland.** The age of these sedimentary rocks is greater than 3.872 Ga. They enclose carbonaceous (graphite) inclusions, which have a carbon isotopic composition that, according to some researchers, could be the signature of fossil biological activity. These conclusions are, however, highly controversial. (Photo.: B. Marty.)

This is also at Isua that, in 2007, researchers reported an association of pillow lavas and a dyke complex, which they interpreted as an evidence for the opening of an ocean. Such a conclusion would imply that plate tectonics was active on our planet since at least 3.8 Ga.

It may also be noted that along the Canadian coast of Labrador, at Uivak, there are exposures of geological formations that closely resemble the Amîtsoq and Isua gneisses. Zircon crystals, extracted from these, have been dated to 3.863 ± 0.012 Ga. The remnants of very ancient (> 3.8 Ga) continental crust therefore have a wide distribution geographically, which provides a weighty argument in favor of the existence of a vast continent (or of several smaller ones) from 3.8 Ga, and that is today located in the northern hemisphere. The main features of

6

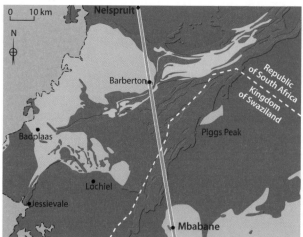

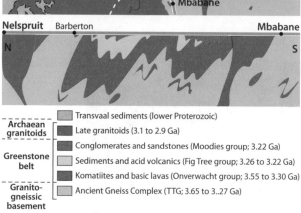

☐ **Fig. 6.5 Schematic map of the Kaapvaal craton (a) and a north–south section through the Barberton greenstone belt (b).** These two maps clearly show that the Barberton greenstone belt rests on the granito-gneissic basement (also known as the "Ancient Gneiss Complex" and having a TTG composition), which represents about 80 per cent of the volume of the craton. Both the basement and the greenstone belt are crosscut by late granitoids. These three lithological and chronological units are typical of Archaean terranes. Because the parental magmas of TTGs crystallized at depths ranging between 15 and 25 km, these rocks provide geologists with information on the composition and internal structure of the Archaean continental crust. Contrastingly, the greenstone belt consists of volcano-sedimentary rocks that were emplaced at the surface (probably at the bottom of a sea or ocean). This is the reason why these formations provide geologists with information about the conditions that prevailed on the planet's surface during the Archaean. (After Anhaeusser *et al.*, 1981.)

Legend:
- **Archaean granitoids**
  - Transvaal sediments (lower Proterozoic)
  - Late granitoids (3.1 to 2.9 Ga)
- **Greenstone belt**
  - Conglomerates and sandstones (Moodies group; 3.22 Ga)
  - Sediments and acid volcanics (Fig Tree group; 3.26 to 3.22 Ga)
  - Komatiites and basic lavas (Onverwacht group; 3.55 to 3.30 Ga)
- **Granito-gneissic basement**
  - Ancient Gneiss Complex (TTG; 3.65 to 3..27 Ga)

these continental remnants are summarized in ☐ Table 6.1. These remnants are of fundamental importance to scientists for at least three reasons: they show that the principal components of the Archaean continental crust and in particular the TTGs, existed by 4.0–3.86 Ga; they provide arguments suggesting that plate tectonics was active by 3.8 Ga; and they contain potential "indications" of life, which, even if these are at the heart of heated polemics and controversies, are key elements in the discussion about the origin of life.

3.4 Billion Years Ago, in the Heart of a Vast Archaean Continent ...

The oldest Archaean terranes of continental size are presently exposed in the southern hemisphere: the Kaapvaal craton (1.2 million km²), in South Africa and Swaziland, and the Pilbara craton (0.06 million km²) in Australia (☐ Fig. 6.1). Both of these cratons were formed between 3.6 and 3.1 Ga. They share several structural and lithological similarities, such that geologists consider that they initially formed a single supercontinent, called the Vaalbara, and which was finally dislocated around 2.8 Ga ago. These two shields provide outcrops of an exhaustive variety of rocks, both of the deep continental crust and of formations that were deposited on the surface. They therefore allow us to attempt a reconstruction of the terrestrial environment and the mechanisms that operated on Earth about 3.4 Ga ago.

Just like all the Archaean terranes, the Kaapvaal and Pilbara cratons consist of three main lithological and chronological units (☐ Fig. 6.5), specifically, from oldest to youngest:

— a granito-gneissic basement, which forms most of the volume of the craton (about 80 per cent);

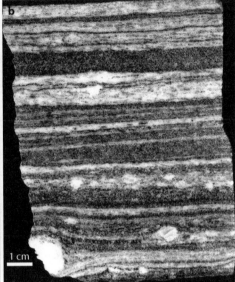

Fig. 6.6 a. An outcrop of TTG gneisses from the Ancient Gneiss Complex at Piggs Peak, in Swaziland; b. a detailed view of these gneisses. The two photographs clearly show the banded and foliated nature of these rocks, which have been dated to 3.644 ± 0.004 Ga. (Photos: H. Martin.)

- volcanic and sedimentary rocks (supracrustal rocks) known as greenstone belts;
- late granitoids that were intruded into the two earlier formations.

The "Spine" of Cratons: an Old Basement of Granite and Gneiss

In the Kaapvaal craton, the granito-gneissic basement is known as the "Ancient Gneiss Complex" (Fig. 6.6). It consists in magmatic rocks, TTG in composition and very similar to those that we have already described at Acasta and Amîtsoq. They correspond to a juvenile continental crust, which means that they have been directly or indirectly extracted from the mantle. The Ancient Gneiss Complex was emplaced in several episodes, over a long period of time, between 3.644 ± 0.004 Ga and 3.263 ± 0.097 Ga. In the Pilbara craton the TTG ages range from 3.65 Ga to 3.17 Ga. The parental magmas of TTGs crystallized at depth, between 15 and 25 km and, as a result, these rocks provide information about the internal composition and structure of the continental crust. It should be noted that TTGs, although extremely widespread in the Archaean, remain practically absent from terranes that are younger than 2.5 Ga. This means that the mechanisms that governed and controlled the genesis of the continental crust on the Archaean Earth were different from those that are active today.

At the Surface: the Greenstone Belts

The greenstone belts of the Kaapvaal and Pilbara cratons share many lithological and structural characteristics. They appear as elongated units, like the Barberton greenstone belt (in South Africa), that is some 130 km long by only 30 km wide on average (Fig. 6.5), and which we will take as an example. It consists of volcanic and sedimentary rocks, which were emplaced at the surface, above the TTGs of the Ancient Gneiss Complex. These supracrustal formations were deposited in three successive episodes.

First, between 3.55 and 3.30 Ga, the Onverwacht group was emplaced. This mostly consists of basic and ultrabasic lavas (komatiites). These lavas are associated with small volumes of detrital sediments that were deposited in a deep marine environment, as well as with some cherts, which are chemical deposits formed by the direct precipitation of silicon ions in solution in water, which bears witness to significant hydrothermal activity. Second, the Fig Tree group was emplaced between 3.26 and 3.226 Ga. It is characterized, not only by less significant volcanic activity but also by the nature of the erupted lavas whose composition is no longer ultrabasic, but intermediate to acid. The dominant rock types are thus detrital sediments (sandstones, schists, clays) as well as cherts (◘ Fig. 6.15), which are associated with Banded Iron Formations (BIF) consisting of alternating beds, generally centimeter-sized, of minerals rich in iron (as oxides, carbonates, etc.) and beds of silica (◘ Fig. 6.16). Finally, around 3.22 Ga, we observe primarily shallow, fluvial and marine detrital deposits, which are also associated with BIF. This is the Moodie group.

Unlike the TTGs which provide information about the internal composition and structure of the continental crust, the greenstone belts informs us about the conditions that prevailed at the surface of the planet.

The Late Granitoids

The last event that affected the Kaapvaal craton, during the Archaean, consisted of the intrusion of granitic plutons of calc-alkaline affinity. They are intrusive into both the granito-gneissic basement and the greenstone belt. They were emplaced over a long period of time, from 3.201 to 2.61 Ga. In the Pilbara craton, this magmatic event started at 3.02 Ga and finished at 2.93 Ga.

The Saga of the Oldest Archaean Continents

Each of the geological formations that we have just described – the granito-gneissic basement, greenstone belt and late granitoids – recorded a lot of information about the composition, structure and environments in the Archaean Earth. We shall, therefore, once again try to explore and summarize the clues available to us. The collected information will be all the more precious and original in that three lithologies among those described in the Kaapvaal and Pilbara cratons are only known in any abundance in Archaean terrains: these are the komatiites, the TTGs and the BIFs.

A Hot Interior to the Earth

Let us first turn our attention to the komatiites (◘ Fig. 6.7). These ultrabasic lavas that are very abundant before 2.5 Ga are, with one exception, completely absent after that date. They consist of an association of olivine and pyroxene, which crystallized as acicular crystals (like tiny and very fine needles), with a characteristic dendritic texture (called spinifex texture), which bears witness to sudden cooling.

These characteristics, as well as laboratory experiments on melting and crystallization, have shown that komatiites were the result of the crystallization of a magma generated through partial melting at depth (> 120 km, as is confirmed by the occasional presence of diamonds) and at very high temperatures. Indeed, it has been estimated that the lavas erupted at temperatures between 1525 and 1650 °C. By way of comparison, the lavas currently erupted at the mid-ocean ridges or at hot spots are basalts, whose temperature ranges between 1250 and 1350 °C. These high temperatures resulted in a high degree of partial melt-

6

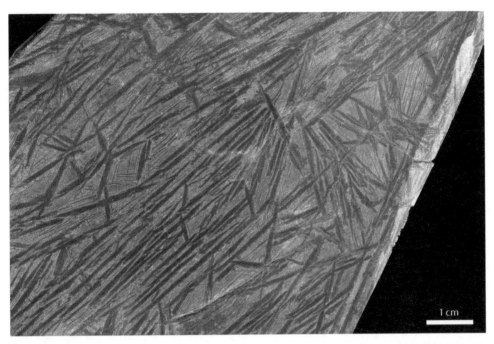

■ **Fig. 6.7 A komatiite sample from Pyke Hill (Canada).** Komatiites are ultrabasic lavas typical of the Archaean Earth: with one exception they are found exclusively in terranes older than 2.5 Ga. They result from partial melting of a mantle peridotite at very high temperatures. It is estimated that komatiitic magma erupted at temperatures of between 1525 and 1650 °C. When reaching the surface, it cooled very abruptly such that the minerals (olivine and pyroxene) crystallized as acicular crystals (tiny, very fine needles) with a characteristic dendritic texture, called "spinifex texture". (Sample: C. Nicollet; photo.: H. Martin.)

ing of the mantle, which is estimated to be between 50 and 60%, whence the ultrabasic nature of Archaean komatiites (■ Box 6.1).

Nowadays, the lower temperature of the mantle does not permit degrees of melting greater than 25 to 30%, which give rise to basic lavas (basalts). We know nothing about the geodynamical environment in which the komatiitic magmas were generated and emplaced. It does seem, however, that they result from the ascent of a deep mantle diapir, which nowadays, would probably correspond to a hot spot geodynamic environment. However that may be, the komatiites bear witness to the fact that the Archaean mantle was hotter than nowadays. They are thus the most spectacular witnesses of how our planet has cooled.

The Message from the TTGs

Let us now turn our attention to the component that, by volume, forms the largest part of Archaean terranes: the granito-gneissic basement. It consists, as we have seen, of an association of tonalites, trondhjemites and granodiorites, soberly described by the acronym TTG (*see* the rock classification on p. 267). As with the komatiites, the geodynamic environment in which they were generated has long since disappeared. It was therefore by analyzing their mineralogical and chemical compositions and then by developing both quantitative modeling and melting/crystallization experiments that is has been possible to constrain and reconstruct the conditions that prevailed during their genesis.

Regardless of their age and the location where they were formed, the TTGs have constant and very homogeneous mineralogical and chemical characteristics. They are granitoids which typically consist of an association of quartz + plagioclase feldspar + biotite (black mica), together with (in the less differentiated facies), hornblende (amphibole). In this they differ from the rocks of the modern continental crust, which have granodioritic to granitic

◨ Box 6.1   **Partial Melting and Fractional Crystallization**

Changes of state do not occur in the same way depending on whether one is dealing with a pure phase (consisting of a single type of component) or a mixture, such as a rock. For pure phases, changes take place almost instantaneously. For example, at a pressure of 1015 mbar, pure water turns to ice at 0 °C, and to vapor at 100 °C. Rocks are not pure phases and changes of state are progressive. For example, if one heats a rock from the terrestrial mantle (peridotite), in the absence of water and at a pressure of 1015 mbars, the first drop of liquid will appear at 1200 °C, but the rock will not be completely liquid until 1850 °C. The curve that marks the appearance of the first drop of liquid on a pressure-temperature (or depth-temperature) diagram is known as the *solidus*, while the one that marks the end of melting is known as the *liquidus* (◨ Fig. 6.8).

For temperatures lying between the temperatures of the solidus and liquidus, a magma will consist of a mixture of crystals and liquid. ◨ Figure 6.8 shows that in the case of the terrestrial mantle (that is to say of an anhydrous peridotite); this temperature interval is 650 °C.

If we consider a magma chamber located at a depth of a few tens of kilometers in the terrestrial crust, the time it requires to cool by 650 °C is of the order of several hundreds of thousands of years. Put another way, in this magma chamber, crystals and liquid will coexist for several hundred thousand years. Taking account of the contrasting physical properties of liquids and crystalline solids, this coexistence will be able to lead to a separation of these phases. For instance, ◨ Table 6.2 shows the density of several mineral phases as well as that of a basaltic magma. It is perfectly clear that all the mineral phases, with the exception of plagioclase, have

◨ **Table 6.2 Comparative densities of the main magmatic minerals and of a basaltic liquid.** With the exception of plagioclase, all the mineral phases are denser than the basic magma (basalt).

Mineral / magmatic liquid	Density
Magnetite	5.2
Ilmenite	4.7
Olivine	3.32
Orthopyroxene	3.55
Clinopyroxene	3.4
Hornblende (amphibole)	3.3
Plagioclase (feldspar)	2.65
Basic magma = basalt	2.85

densities greater than that of the basaltic liquid. Consequently, due to their greater density, these minerals will sink in the basaltic magma and settle to the bottom of the magma chamber, where they will accumulate (◨ Fig. 6.9a and b). Geologists term this accumulation of crystals "a cumulate". As for the plagioclase crystals, being less dense than the basaltic magma, they will ascend and accumulate in the upper portion of the magma reservoir. In the case of a melting process, when the amount of generated liquid is sufficient, this latter, given its low density, will tend to rise towards the surface, most frequently by means of fractures (◨ Fig. 6.9c). It is also possible that, in a compressive tectonic environment, the liquid would be expelled, while the crystals remain in place (filter press effect). Whatever physical mechanism is involved, it results in a complete or partial separation of the crystals from the liquid. In a crystallization context, one speaks of "fractional crystallization" whereas in the case of fusion, one uses the term "partial melting".

Regardless of whether partial fusion or fractional crystallization is involved, separation of crystals and liquid will result in a progressive change in the composition of the liquid. For example, in a basaltic liquid containing 9% MgO, if pyroxene crystals containing 20% MgO are extracted, it is easy to understand that the pyroxene crystallization will consume a lot of MgO, the content of which will progressively decrease in the magmatic liquid. The mass-balance equation governing the behavior of major elements during fractional crystallization is as follows:

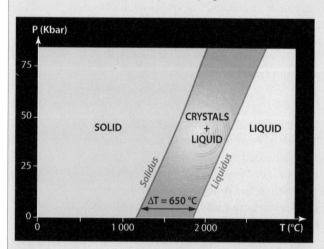

◨ **Fig. 6.8 A pressure-temperature diagram showing the *solidus* and *liquidus* curves for an anhydrous mantle peridotite.** As we can see, when there is a change of state, either solid → liquid (melting), or liquid → solid (crystallization), crystals and liquid may coexist over a temperature range of about 650 °C.

6

☐ **Fig. 6.9 a. An illustration of the sedimentation of amphibole crystals (black)** during the crystallization of a dioritic magma (Island of Guernsey, United Kingdom). The crystals sank into the magma and accumulated at the bottom of the magma chamber, forming a darker layer, called a cumulate. **b. Accumulation of potassium feldspar crystals** (white) during crystallization of the La Margeride granite (Massif Central, France). **c. Accumulation of a liquid** (white, now solidified), originating in the melting of a gneiss (grey foliated rock), in a vertical shear zone (Kainuu, Finland). (Photos: H. Martin.)

$$C_0 = (1 - X)C_l + XC_s$$

where C_0 = concentration of an element in the parent liquid; C_l = concentration of an element in the residual liquid; C_s = concentration of an element in the cumulate; X = degree of crystallization. This equation shows that the composition of the liquid changes progressively as a function of the X value, that is as the crystallization progresses. Taking the example of the fractional crystallization of pyroxene within a basaltic magma, one can calculate that after 10% of crystallization (X = 0.1), the liquid will contain 7.8% of MgO (C_l), whereas for X = 0.3, its MgO content will only be 4.3%. All that we need to remember of all this is that fractional crystallization or partial melting are mechanisms that change the composition of a magma; geologists call this: magmatic differentiation.

Another parameter that plays an important role in the processes of fusion and crystallization is the presence of water. Water does, in fact, have the effect of considerably lowering the temperature of the *solidus*. ☐ Figure 6.10 shows that at a depth of about 100 km (30 kbars), a mantle peridotite will start to melt at 900 °C in the presence of water as against 1500 °C under anhydrous conditions. To put it another way, the presence of water has lowered the temperature of the beginning of melting by 600 °C, which is considerable. This factor is so important, as we shall see later, that many authors refuse to consider that the continental crust could have been generated under anhydrous conditions.

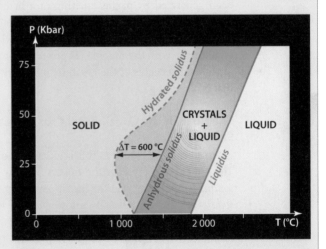

☐ **Fig. 6.10 A pressure-temperature diagram showing the anhydrous** *solidus* **and the hydrated** *solidus* **for a mantle peridotite.** The presence of water considerably lowers the temperature of the *solidus*. For example, for a pressure of 30 kbars (100 km depth) a mantle peridotite would start to melt at 900 °C in the presence of water, as against 1500 °C under anhydrous conditions.

compositions, that is, typically a quartz + alkali feldspar + plagioclase feldspar + biotite association. From a geochemical point of view, the Archaean TTGs are sodium-rich rocks (☐ Fig. 6.11), unlike the modern continental crust which is potassium-rich. Another difference: the modern continental crust is rich in the heavy rare earth elements (Yb ~ 3 ppm), whereas the TTGs have very low abundances of these elements (Yb < 1 ppm).

6

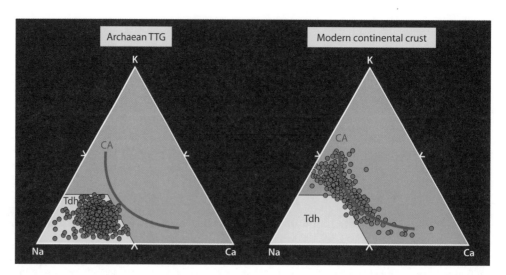

■ **Fig. 6.11 K-Na-Ca triangle illustrating the differences between Archaean TTGs and the present-day continental crust.** Each apex of the triangle corresponds to a pure pole (100% of the element concerned). The points representing the TTGs plot near the sodium apex in the trondhjemite region (Tdh, pale yellow), whereas their modern equivalents are enriched in potassium and follow a classical calc-alkaline differentiation trend (CA). This shows that the Archaean and modern continental crusts were generated through the melting of different sources, which was a direct consequence of the progressive cooling of our planet. (After Martin, 1995.)

All the research and geochemical modeling that have been carried out on the Archaean TTGs led to the conclusion that their parental magma (from which they crystallized), has been generated by partial melting of a hydrated basalt. In addition, experimental melting of hydrated basalts carried out in the laboratory has shown that all the geochemical characteristics of TTGs, can be reproduced only when the residue of melting which is in chemical equilibrium with the magma (■ Box 6.1), contains both garnet and hornblende. The presence of residual garnet is of the greatest significance, indeed, it requires pressures above 12 kbars, that is, a melting depth of at least 40 km.

The Birth of the Continental Crust: the "Cold" Scenario *Versus* the "Hot" Scenario

Let us analyze ■ Fig. 6.12. This is a diagram of pressure (and thus depth) *versus* temperature. The red curve represents the *solidus* for a hydrated basalt (5% water): it is only beyond this *solidus* temperature that the basalt partial melting will start. The orange curve represents the *solidus* for the same basalt, but anhydrous. It is clear that for any given pressure, the temperature at which a basalt will start to melt closely depends on its degree of hydration. For example, at a pressure of 10 kbars (i.e., a depth of 35 km), the temperature at which basalt melting begins is slightly more than 600 °C in the presence of 5% water, whereas it is above 1200 °C if the basalt is anhydrous. Generally, water significantly lowers the temperature at which a rock starts to melt. (■ Box 6.1).

At present, the basalts of the ocean floor take, on average, 60 Ma (180 Ma at most) to complete their trip from the mid-oceanic ridge where they are generated to the subduction zone where they sink into the mantle. In the mid-oceanic ridge systems, the basalts are generated through partial melting of an anhydrous mantle, consequently, are themselves also anhydrous (~0.3% water); which results in the crystallization of anhydrous minerals such as olivine (Mg_2SiO_4), orthopyroxene ($Mg_2Si_2O_6$), clinopyroxene ($CaMgSi_2O_6$) and plagioclase ($CaAl_2Si_2O_8$). In addition, the ridge is generally intensively faulted, which allows water to penetrate deep into the newly formed oceanic crust, and to react with its anhydrous minerals, which are thus transformed into hydrated minerals. These latter are typically the serpentine antigorite [$Mg_3Si_2O_5(OH)_4$], chlorite [$Mg_6Si_4O_{10}(OH)_8$], talc [$Mg_3Si_4O_{10}(OH)_2$], and hornblende amphibole, [Ca$_2$Mg$_5$Si$_8$O$_{22}$(OH)$_2$], thus the water content of this hydrated basalt ranges between 1.5 and 7%. The effect of ocean water also consists of cooling down the oceanic crust. When it enters a subduction zone, the cold oceanic crust cools the "mantle wedge" (the piece of mantle located above the sinking plate), and the temperature of this subducted crust increases only slowly with depth. In other words, the geothermal gradients along the Benioff plane (the interface between the subducted oceanic crust and the mantle) are low: the temperature increases slowly with depth (◘ Fig. 6.12a, blue arrow).

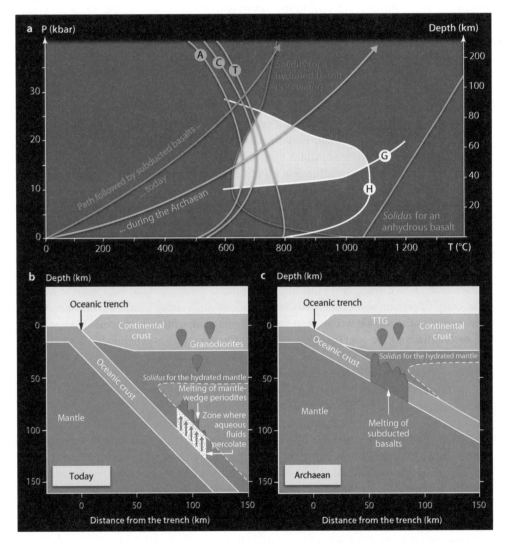

◘ **Fig. 6.12 A pressure (depth) versus temperature diagram (a) and schematic sections of a subduction zone, showing the conditions under which the continental crust has been generated today (b) and during the Archaean (c).** During the Archaean, the geothermal gradients along the Benioff plane were high (pink arrow), such that the subducted oceanic crust reached its *solidus* temperature before it could have been dehydrated. It was therefore able to melt at a relatively shallow depth, in the region where hornblende and garnet are stable (yellow region), giving rise to the TTG magmas, typical of the Archaean continental crust. Today, the geothermal gradients along the Benioff plane are low (blue arrow) and the subducted oceanic crust becomes dehydrated before it can start to melt. The fluids released by the dehydration of the subducted slab ascent through the "mantle wedge", causing its rehydration and metasomatism. While hydrated, the mantle wedge melts and gives rise to calc-alkaline magmas typical of the modern continental crust. The orange and red curves respectively represent the *solidus* for an anhydrous and a hydrated (5% water) basalt. The green curves represent the dehydration reactions that occur in the basalt when hydrous mineral phases are destabilized: (A) = antigorite (serpentine), (C) = chlorite and (T) = talc. The other curves are those for the stability of hornblende (H) and garnet (G). In the schematic sections, the red areas are those where magma is found, and the white area is the region where fluids percolate. (After Martin and Moyen, 2002.)

When following such a geothermal gradient from the diagram's origin (temperature = 0 °C, depth = 0 km), in fact, one follows step by step, the evolution of oceanic basalts as they progressively sink into the mantle. Around a depth of 80 km, this gradient intersects the destabilization curve for antigorite (curve A), then that of chlorite (curve C), then that of talc (curve T) and of hornblende (curve H). This means that at the high temperatures and pressures that have been reached, these minerals are no longer stable and react to give rise to new minerals, which will be stable under the new pressure and temperature conditions. For example, talc and olivine react giving a pyroxene as reaction product; the reaction is as follows:

$$Mg_3[Si_4O_{10}(OH)_2] \quad + \quad Mg_2[SiO_4] \quad \rightarrow \quad 5(Mg[SiO_3]) \quad + \quad H_2O$$

Talc   Olivine   Pyroxene   Water

All these reactions are dehydration reactions, which means that they release the water contained in the basalts of the ocean floor, which thus progressively become anhydrous. At a depth of 120 km, the subducted basalts reach a temperature of 750 °C, that is, the *solidus* temperature for a hydrated basalt. However, by that depth they have long been dehydrated. Consequently, the temperature at which their melting could start is thus that of the anhydrous *solidus*, which, under these conditions is 1400 °C. Because the temperature of the anhydrous *solidus* increases with pressure, the oceanic crust never reaches this *solidus*, given the local geothermal gradient: it is unable to melt and returns into the mantle as a dry and solid slab.

Compared to the surrounding rocks, the water released by the dehydration reactions has a low density, so it rises towards the surface across the "mantle wedge". ◙ Fig. 6.12b shows that, although the latter is hot, it is unable to melt because it is anhydrous. The water that percolates through the mantle wedge rehydrates it, and considerably lowers the temperature of its *solidus*, which triggers its melting. In addition, the fluids released from the oceanic crust carry with them some dissolved chemical elements, among which is potassium, which they deliver to the mantle wedge, thus modifying its composition (a process that geologists call metasomatism). To sum up, the modern continental crust is generated in a subduction environment, through the partial melting of the hydrated mantle wedge, whose composition has been modified by the fluids released by the dehydration of the subducted oceanic crust. During the fusion of the mantle wedge, the residue of melting consists of olivine + pyroxenes; it does not contain any garnet or hornblende.

Let us now return to the TTGs from the Kaapvaal and Pilbara cratons. We have seen that the magma that gave rise to the TTGs was in equilibrium with a melt residue containing garnet and hornblende. In ◙ Fig. 6.12a, the region where these two mineral phases coexist with a TTG magma is colored in yellow. If we assume that the Archaean continental crust was also generated in a subduction-like environment (a concept, the validity of which we shall discuss later), then the Archaean geothermal gradients had to pass through the yellow region, which implies that they would have been higher than nowadays. When, in an Archaean subduction zone, a basalt followed such a gradient (pink arrow), at a depth of 40 km, it reached the temperature of its hydrated solidus, that is about 620 °C (where the gradient crosses the red curve of the hydrated *solidus*). At this depth, basalt dehydration can only take place for temperatures higher than 650 °C. Because the oceanic basalts were still hydrated, they were able to melt at a low temperature, and thus generating TTG-like magmas (◙ Fig. 6.12c).

As a result, unlike the present-day continental crust, which comes from melting of the mantle wedge, the Archaean continental crust resulted from the melting at a relatively shallow depth of a hydrated oceanic crust (transformed into garnet-bearing amphibolite). In other words, the source and the mechanisms of genesis of the Earth's continental crust have changed between the Archaean and today.

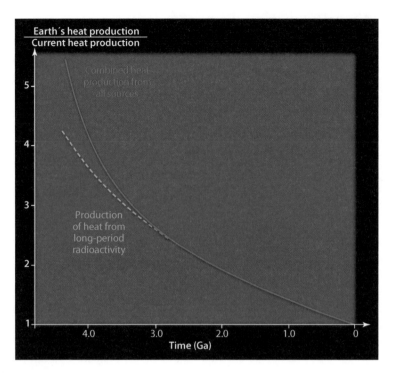

Fig. 6.13 The evolution of the Earth's internal heat production since 4.55 Ga. The dashed curve represents the heat resulting from just radioactive decay. Our planet is slowly consuming its store of energy: it progressively cools. (After Brown, 1986.)

A Hotter Earth: by How Much and Why?

Both the komatiites and the TTGs naturally lead to the conclusion that the interior of the Earth was much hotter in the Archaean than it is today. Estimates of the temperature of the Archaean mantle vary in a wide range from one author to another, but the most recent models indicate that it was no more than 200 °C (between 100 and 200 °C) higher than the present-day mantle. Why was the Archaean Earth hotter than the Earth today?

On Earth the internal heat flux is a result of the dissipation of the energy that it accumulated at the period of its accretion. This internal heat has several sources, the main ones being: the residual heat from the planet's accretion, the heat resulting from the exothermic differentiation of the core and mantle, the latent heat released by the crystallization of the liquid core and above all, the heat produced by the disintegration of radioactive elements such as ^{235}U, ^{238}U, ^{232}Th, and ^{40}K, as well as of extinct radioactivity (i.e., from radioactive elements with short half-lives, now fully decayed). Ever since 4.55 Ga this stock of energy has been consumed bit by bit, which has resulted in the progressive cooling of the planet (■ Fig. 6.13). For instance, 4 Ga ago, the planet produced 4 times as much energy as today, and 2.5 Ga ago, it was still generating more than twice as much as nowadays. Currently, the Earth's internal heat production amounts to 42 TW (1 TW = 1 Terawatt = 10^{12} W), of which more than 32 TW comes from radioactive decay.

A Thick Continental Crust and Emerged Continents?

The study of the TTGs and komatiites has led us to the conclusion that the Archaean upper mantle was hotter than today by at least 100 °C. As a result its viscosity would have been at least 10 times lower, than that of the present-day mantle. We may recall that the continental crust "floats", so to speak, on the subjacent mantle, which is denser. A less viscous mantle must therefore be less resistant to the load formed by the continental crust – a significant load because the Kaapvaal and Pilbara cratons represent very large volumes – consequently, the former would tend to sink more easily. Some researchers have theorized that the continental crust itself was hotter, softer, and more ductile than nowadays, such that flowing under its own weight, it could not thicken in any lasting manner. From this they have concluded that mountain ranges or significant relief could not exist during the Archaean. Is this vision of the Archaean Earth correct?

This time, it is the detailed study of the mineralogical associations within Archaean rocks that allows us to tackle this question. Indeed, every mineral or, more widely, every mineralogical association is stable within a very specific range of temperature and pressure. If these conditions change, the mineralogical assemblage is no longer stable; it is thus destabilized and new minerals crystallize that are stable at the new pressure and new temperature. For example, let us consider a sedimentary rock (thus deposited at the surface) containing a mineralogical assemblage of muscovite (white mica) and quartz, which is buried until it reaches a pressure of 5 kbars (at a depth of 17 km) and a temperature of 700 °C. Under such new conditions, both the muscovite and quartz become unstable and recrystallize in the form of an association of potassium feldspar and aluminosilicate (sillimanite), in a reaction of this type:

$$KAl_2[Si_3AlO_{10}(OH)_2] \;+\; SiO_2 \;\rightarrow\; K[AlSi_3O_8] \;+\; Al_2SiO_5 \;+\; H_2O$$

| Muscovite | | Quartz | | Potassium feldspar | Sillimanite | Water |

This process, which does not modify the overall chemical composition of the rock, but simply redistributes the chemical elements into new mineral phases, is known as metamorphism. Geologists use the metamorphic mineralogical assemblages to deduce the pressure and temperature conditions of the metamorphic event to which the studied rock has been subjected. Such studies applied to the Kaapvaal and Pilbara cratons as well as to several other Archaean cratons have shown that the continental geothermal gradients were of the same order of magnitude as current gradients, namely about 30 °C. km^{-1}. In other words, the Archaean continental crust was not significantly hotter than its present-day equivalent.

This result, which may at first seem surprising, is not really so. Today, the Earth's internal heat is not released in a uniform manner from the surface of the globe: there are localized areas where the heat flux is high – at the mid-oceanic ridges – whereas everywhere else, this flux is weak. In fact, these two kinds of area correspond to two different processes of heat dissipation. At the mid-ocean ridges, the heat is released through an effective mechanism: convection. Indeed, the ridges are located above the ascending branches of the mantle convection cells, and the internal heat is removed by volcanism and by intense hydrothermal activity. Everywhere else, the heat is dissipated by conduction, which is a slow and relatively inefficient mechanism. Nothing gives us any reason to believe that anything was different in the Archaean, where, just like today, the higher geothermal gradients were equally associated with mid-oceanic ridges.

The study of the mineralogical associations in Archaean rocks also provides an estimate of the thickness of the continental crust at that time. In the Kaapvaal craton, for example, or more precisely within the Barberton greenstone belt, basalts and sediments that have undergone metamorphism contain garnet crystals, which have experienced pressures of 15 kbars. This result, published in 2006, means that these rocks have been buried to depths of about 45 km. The Kaapvaal continental crust may therefore have been as thick as its present-day equivalent, and this thickness was maintained over a sufficiently long period of time, such that it could have been recorded by the mineral systems. Even better: the rocks that have preserved the traces of these high pressures were first deposited at the surface – we may recall that they are lavas and sediments – and thus reveal the existence of mechanisms capable of drawing them down to considerable depths, or in other words to thicken the crust. At the present time, the abnormally great thickness of the continental crust is only locally observed in mountain ranges. There, the crustal thickening mainly results from the processes of subduction and collision, that is, from convergent horizontal movement of the lithospheric plates. The question now arises of knowing whether, such horizontal movements were already operating during the Archaean, and whether this continental crustal thickening has led, or not, to the emergence of continents. In accordance with Archimedes' principle (isostasy) the altitude reached by a continent depends on its thickness. Today, the continental crust has an average thickness of 30 km (occasionally more than 70 km below recent mountain ranges) and its average altitude is 300 meters above sea level. Assuming that the volume – and thus the level – of the Archaean oceans was of the same order of magnitude as nowadays, it is possible to estimate that, at that period, the same thickness of crust would produce similar relief. In the Barberton area, if the crustal thickness really reached 45 km, this obviously implies that, in Archaean times, the Kaapvaal craton was an emerged continent.

These conclusions are corroborated by the kind of sedimentation observed in the greenstone belts. This time, it is the structure of the rocks that provides precious indications. Indeed, in both the Pilbara craton and the Kaapvaal craton, detrital sediments are very abundant (■ Fig. 6.14a to d). This is particularly the case with conglomerates, which may form impressive series (up to 3000 m in thickness) and that share many characteristics with the orogenic molasse, which is a frequently coarse sediment generated by the rapid destruction of young mountain chains. These conglomerates contain rounded pebbles that originate not only from the greenstone belts (cherts, basic lavas), but also from the TTG basement, which bears witness to the existence of emerged reliefs in the vicinity of the sedimentation basin. Indeed, we have seen that the parent magma of the TTGs crystallized at depth in the continental crust (15 to 25 km). Yet the conglomerates formed on the surface of the planet. Consequently, the presence of pebbles from the TTG basement in the Pilbara and Kaapvaal conglomerates implies that these TTGs have been displaced vertically by several tens of kilometers. Such a path is most frequently the result of a collision between two continental masses, a process which in our days, is the origin of mountain ranges such are the Alps or the Himalaya. The Archaean landscape must therefore also have included mountainous massifs. In addition, in the same greenstone belts, sandstones with cross-bedding textures are widespread (■ Fig. 6.14c), these sedimentary textures are typical of fluviatile or deltaic deposits. Mud cracks (■ Fig. 6.14e) are also observed in these beds. These features, which are identical to the structures that appear today in the mud of dried lagoons, lakes and puddles (■ Fig. 6.14f), indicate periods of emersion and of drying, and show, without any ambiguity, the existence of emerged lands in the Archaean.

In the Barberton greenstone belt, the fine structure of the sediments, associated with the exceptional quality of their preservation, is such that it has allowed researchers to go even further in describing the environment of the Archaean Earth. In the sandstones, they have been able to demonstrate cyclicity in detrital deposits that they interpret as resulting from the effect of the tides. These cycles of some twenty days indicate that the duration of the Archaean lunar month was shorter than that of the present-day lunar month (29.5 days), which is consistent with the fact that during the Archaean, the Earth-Moon distance was less than today.

By way of conclusion it may be noted that these proofs of emerged continents are not the oldest known. Part at least of the sediments at Isua in Greenland were formed through the alteration and erosion of acid and basic magmatic rocks, which researchers have interpreted as bearing witness to the existence, at 3.87 Ga, of an emerged continent.

Significant Hydrothermal Activity and a Reducing Atmosphere

A hot internal Earth, thick and emerged continents, as well as mountain chains: the message delivered by the rocks from the Kaapvaal and Pilbara cratons is already very rich. We shall now attempt to complete this picture by looking at two other kinds of Archaean sedimentary rock: the cherts and the Banded Iron Formations or BIF.

6

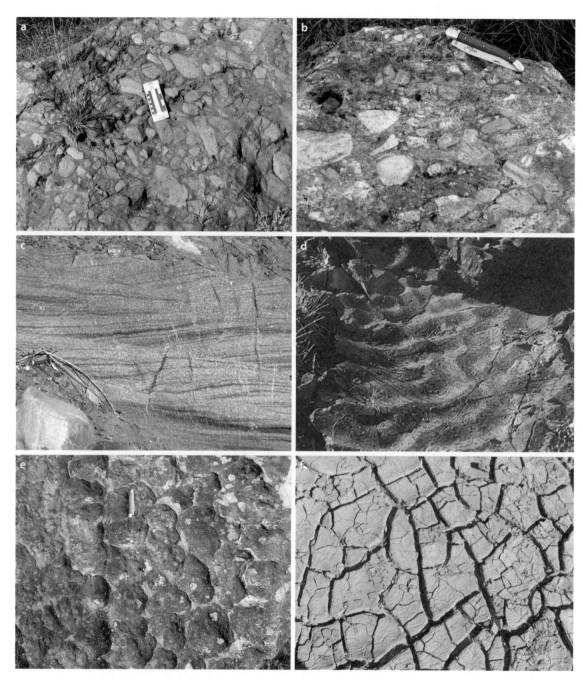

■ **Fig. 6.14 Some sediments from the Kaapvaal and Pilbara Archaean cratons. a and b.** Polymict conglomerates of the molasse type consisting of rounded pebbles of chert (white), basalt (black) and granitoid gneiss (light grey). These pebbles are not just derived from the greenstone belt but also from the TTG basement, which argues in favor of the existence of mountains in the Archaean. The TTG magmas, in fact, crystallized at depth in the crust. The fact that they could be eroded and integrated into the conglomerates proves that they have been displaced vertically by several tens of kilometers. **c.** Sandstone, showing cross-beddings typical of fluvial or deltaic deposits. **d.** Ancient wave ripple marks in sandstone. **e.** Polygonal mud cracks indicating periods of emergence and of drought (**f.** present-day mud cracks in a dried silt, by way of comparison). Photo a. (H. Martin) is from the Onverwacht group, and the photos b, c, e (H. Martin) from the Moodies group in the Barberton greenstone belt in South Africa. Photo d (J.-F. Moyen) from the Dresser formation in the Pilbara craton in Australia.

Fig. 6.15 Cherts: witnesses to an intense hydrothermal activity. a. Landscape in the North Pole region of the Pilbara craton, showing the abundance of cherts (all the brown rocks generally outcropping as dykes and veins). **b.** Chert from Marble Bar (Pilbara) consisting of alternating layers of white, grey and red silica. **c.** Chert from the Barberton greenstone belt. **d.** Modern hydrothermal vent (a black smoker) at a mid-oceanic ridge. The Archaean cherts are believed to have formed in a similar environment. The abundance of cherts indicates that hydrothermal activity was particularly widespread during the Archaean. (Photos a to c: H. Martin.)

Cherts, which are chemical deposits of silica, are very abundant in the Kaapvaal and Pilbara greenstone belts. They are associated with basic or ultrabasic (komatiite) volcanic rocks, which they alter according to a process known as silicification or into which they form intrusive dykes (Fig. 6.15). Researchers generally consider that this association is characteristic of an active oceanic environment, either a hot-spot type or a mid-oceanic ridge. The cherts would then be the result of intense hydrothermal activity, probably similar to that which, today, is associated with black and white smokers, on mid-ocean ridges.

Another sedimentary lithology that is abundant in the Pilbara and Kaapvaal cratons consists in the banded iron formations. Typically they are made up of an alternation of beds of iron-rich minerals (oxides, carbonates, etc.) with chert beds (mostly siliceous) (Fig. 6.16). In the early Archaean, they were generally modest-sized deposits associated with volcanic formations and hydrothermal deposits. On a global scale, no BIF is known after 2.2 Ga. Researchers have wondered, on the one hand what was the origin of the iron, and on the other why these sediments did not form later than 2.2 Ga.

The iron comes, basically, from the weathering of surface rocks and from the leaching of the continents. For example, when a granite is weathered, the biotite (black mica with the formula $K(Fe, Mg)_3[Si_3AlO_{10}(OH)_2]$ becomes unstable and the iron that it contains is released. The mobility of iron subsequently depends closely on its oxidation state: in its reduced ferrous state (Fe^{2+}) it is easily soluble in water, whereas it is not at all soluble in its ferric oxide

6

◘ **Fig. 6.16 The Banded Iron Formations (BIF).** The red layers are iron oxide-rich and the black layers correspond to siliceous cherts. The iron is derived primarily from the weathering and leaching of the emerged continent surfaces. Because this element is mobile (that is to say soluble in water) only in its reduced state Fe^{2+}, the abundance of the Archaean BIF proves that, at that period, the atmosphere and the oceans were reducing (devoid of dioxygen). Photo a was taken in the Barberton greenstone belt in the Kaapvaal craton (South Africa), while photo b is a detailed view of a folded BIF at Gopping Gap, in the Pilbara craton (Australia). (Photos: H. Martin.)

state (Fe^{3+}). Today, the Earth's atmosphere is rich in oxygen, the iron is oxidized (Fe^{3+}), it is thus "immobile" and remains on the continents in the form of oxides or hydroxides, which form, for example, the laterite crust (ferruginous crust) that is found in tropical countries, and which originates in the weathering and leaching of the underlying rocks. The existence of the Archaean BIFs testifies that, at this time, iron was in solution in the oceans and consequently that it was under its reduced state (Fe^{2+}). Accordingly iron was mobile, which means that the atmosphere was reducing, that is, devoid of dioxygen (O_2). However, in the BIFs, the iron is in the ferric Fe^{3+} state, it originates from the oxidation of the ferrous iron (Fe^{2+}) dissolved in the ocean water. This fact implies that, if the atmosphere and the ocean were reducing, there existed small, localized islands of oxidation, which fixed the iron in the BIF.

This conclusion may be of particular significance in our enquiry into the traces of Archaean life. Indeed, certain scientists have put forward the hypothesis that colonies of cyanobacteria (that perform oxygenic photosynthesis) could have produced a localized environment that was sufficiently rich in oxygen to establish the oxidizing conditions required for iron precipitation. After the sedimentary formations at Isua (3.86 Ga), the BIFs therefore provide a new example of rocks that might contain traces of life on Earth during the Early Archaean. However, as with all the potential signatures of life earlier than 2.7 Ga, other interpretations can be envisaged. For example, the acid hydrothermal plumes, associated with volcanism could have locally increased the oceanic pH, which would have caused the iron dissolved in the ocean to precipitate.

However that may be, in both cases, the precipitation of iron in a oxidized state that enabled the BIF to form, was a very local phenomenon, confined to the immediate surroundings of hydrothermal vents or volcanic centers. The abundance of these sedimentary formations in the greenstone belts proves the ubiquity of the hydrothermal activity on the Archaean Earth. In the rest of this book we shall attempt to explain why.

Modeling the Terrestrial Atmosphere at 3.8 Ga

At this stage of our tale, we intend to abandon observations and lowly "down-to-earth" considerations to take a higher view. We shall provisionally leave the old Kaapvaal and Pilbara cratons to consider the nature of the atmosphere on the Archaean Earth. This question remains problematic and often speculative because, although rocks and minerals have recorded

Fig. 6.17 Mean relative abundance of gases released today by volcanoes. These gases mainly consist of water and, to a lesser extent, of carbon dioxide.

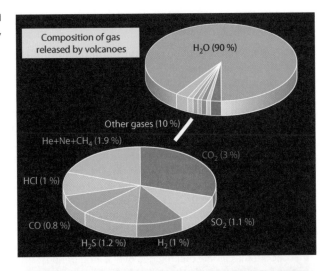

Fig. 6.18 Evolution of the atmospheric gas content from 4.5 Ga to the present. Whether it concerns the Hadean (*see* ▶ Chap. 3, Fig. 3.13) or the Archaean, deciphering the composition of the atmosphere is problematic and speculative, because no direct record of it has been preserved. It is therefore sensible to keep in mind that our knowledge in this field is based on theoretical models, which, themselves, are based on a certain number of theories that are liable to be modified as a result of progress in the science. (After Kasting, 1993.)

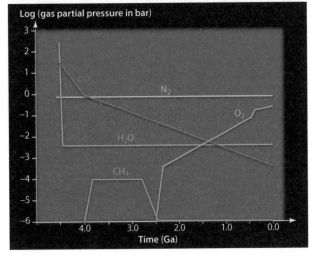

and preserved the memory of a large part of the history of the terrestrial continental crust, there is no direct record of the composition of the Archaean atmosphere. Our knowledge in this field is therefore founded on theoretical models, which, themselves, are based on a certain number of assumptions.

In 1993 James Fraser Kasting, a planetologist at Pennsylvania State University in the United States, put forwards a relatively robust reconstruction of the atmospheric composition at 3.8 Ga, that is, just after the Late Heavy Bombardment (*see* ▶ Chap. 5). It is based on four primary hypotheses:

- the atmosphere's partial pressure of molecular nitrogen (N_2) was already the same as measured today (i.e., 0.8 bar). The models assume, in fact, that degassing of the N_2 contained in the terrestrial mantle had been completed less than 230 Ma after the beginning of the Earth's accretion. The partial pressure of atmospheric N_2 would therefore not have varied since then;
- the composition of the gases released by volcanoes was essentially the same as today: mainly water, and in lesser amounts, carbon dioxide (Fig. 6.17);
- the surface temperature was already regulated by the silicate–carbonate cycle (*see* ▶ Chap. 3). This particular hypothesis requires the weathering of emerged continents, as well as a CO_2 atmospheric pressure of at least 0.2 bar (between 0.2 and 1 bar);

▣ Box 6.2   The Faint Young Sun Paradox

Models of the Sun's evolution predict that the luminosity of our star was 20% and 27% weaker 2.8 and 4.0 Ga ago, respectively, than it is today. Under these conditions, if we assume as a working hypothesis, that at that epoch the Earth atmosphere was the same as today, it means that it would have been necessary to wait 2 Ga for its surface temperature to exceed 0 °C (▣ Figure 6.19 opposite). However, Archaean rocks have not recorded even the slightest worldwide glaciation. On the contrary, they show that up to 2.0 Ga, the surface of the planet was hot (above 0 °C), the sole exceptions being the local glacial deposits dated to 2.9 Ga (Pongola), around 2.4 Ga (Huronian) and 2.2 Ga (Makganyene).

This faint young Sun paradox raises a major questionmark about the climate that reigned over a long period of the Earth's history that included both the Hadean and the Archaean. The only way of resolving this apparent contradiction (with which we are equally faced if we consider that the atmospheric CO_2 pump that is the silicate weathering – a powerful climatic brake – was as effective as today; see ▶ Chap. 3, p. 74), consists of assuming that the atmospheric composition was different and, in particular, that it was far richer in some greenhouse gas than today (the atmospheric partial pressure of CO_2 today is 3.5×10^{-4} bar), such that the surface temperature was always above 0 °C.

This greenhouse gas could have been CO_2. As today, its abundance was regulated by a balance between the production through volcanism and absorption by the silicate weathering followed by the precipitation of carbonates (see ▶ Chap. 3). However, such regulation could have been strongly perturbed by certain characteristics of the primitive Earth's environment. The silicate weathering was perhaps less effective than supposed because of the limited area of emerged continents, such that CO_2 would not have been efficiently removed from the atmosphere, where it could have accumulated, which would have led to temperatures above 0 °C. But the composition of minerals found in sedimentary Archaean rocks suggest that the partial pressure of CO_2 in the atmosphere was not sufficient to maintain high temperatures, as is shown, for example, by the limited presence of siderite ($FeCO_3$). On the other hand, from the Hadean onward, and independently of the emerged continental area, a possible, very efficient mechanism for absorption of dissolved CO_2 consists of the direct transformation of the ocean floor silicates into carbonates. This would have led to cooling to temperatures well below 0 °C.

The other potential candidate is methane (CH_4), a powerful greenhouse gas, whose presence in the atmosphere would have heated the surface of the planet. In the absence of dioxygen (O_2), its lifetime in the atmosphere was greatly increased and its content could have been, for an identical production rate, 100 to 1000 times greater than the value currently measured. But the origin of this gas nowadays is primarily biological. A high CH_4 atmospheric content in the past therefore implies either an increased efficiency of abiotic mechanisms for the production of CH_4, or an abundance of methanogenic organisms just after the emergence of life (but the presence of methanogenic *archaea* at the beginning of the Archaean would imply an extremely early diversification of life). Another point to be taken

▬ life was absent, or if it had already emerged, it was not involved in the atmospheric CO_2 and O_2 exchange cycles, except in a very localized and minor fashion.

Under such conditions, what was the composition of the Early Archaean terrestrial atmosphere? ▣ Figure 6.18 schematically summarizes the results published by Kasting. We shall see that some points still remain subjects for discussion.

In his model, Kasting estimated the molecular hydrogen (H_2) content by calculating the balance of the volcanic degassing and of the thermal escape (Jeans escape) of atomic hydrogen (H) to interplanetary space (see ▶ Chap. 3). Assuming that the thermal escape in the Archaean was comparable with that measured today, he calculated that the partial pressure of H_2 was 10^{-3} bars. This value is a minimum estimate because, today, due to the significant O_2 content in the atmosphere, the temperature of the exosphere (the outermost layer of the atmosphere) is high, which favors thermal escape (see ▶ Chap. 3). Such a process was not definitely operating 3.8 Ga ago, indeed, the atmosphere was an anoxic atmosphere, which implies a cold exosphere, and consequently a less efficient thermal escape of atomic hydrogen, resulting in a partial pressure of H_2 that could have reached 10^{-1} bar.

In addition, the Kasting modeling assumes that the amounts of gases released by terrestrial volcanoes were identical to those today. This hypothesis is minimalist, because it is very likely that, due to greater internal heat production during the Archaean, magmatic activity was more

into account is that a higher CH_4 content in the atmosphere would have resulted in the appearance of photochemical hazes, like those found today on Titan, increasing the Earth's albedo, and contributing to a decrease in its surface temperature. In addition, high concentrations of CH_4 would have absorbed solar radiation in the visible and infrared regions, which would have had the effect of heating the stratosphere, and, again, of cooling the surface of the planet. All these processes imply that in a CH_4-rich atmosphere and with a given CO_2 content, a limiting value for the surface temperature cannot be exceeded. This may be estimated as 30 °C at 2.9 Ga.

The appearance of photosynthetic organisms, producing O_2 – preceding the oxygenation of the surface which began at 2.45 Ga – must have corresponded to drastic modifications, not only of the atmospheric composition, but also of the climate. Indeed the presence of this gas progressively caused the disappearance of the anoxic conditions required for the development of methanogens. This would have resulted in a decrease in the amount of atmospheric CH_4 and consequently in the lowering of the temperature, which again could have led to a global planetary glaciation. In a glacial period, oxygen photosynthesis decreases greatly, such that the concentrations of CH_4 and CO_2 can increase again, thus enhancing the greenhouse effect and consequently augmenting the surface temperature. It is generally believed that this new type of regulation could have maintained the surface temperature slightly above 0 °C. However, the localized glacial episodes mentioned earlier (at 2.9, 2.4 and 2.2 Ga) bear witness to the relative instability of this regulating system.

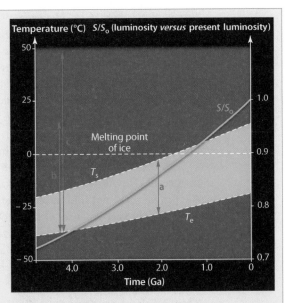

■ **Fig. 6.19 Evolution of the relative luminosity versus the current value (S/S_0) as a function of time for a star of one solar mass (for instance the Sun) and of the theoretical surface temperature for a planet like the Earth.** The radiative equilibrium of the Earth on its own gives rise to a surface temperature T_e well below 0 °C. The contribution from the greenhouse effect corresponding to the present-day atmosphere composition is represented by the pale blue area. At 2 Ga (**a**), it is just enough to raise the surface temperature T_s to a point near the melting point of ice. At 4 Ga, taking the surface temperature to 15 °C (**b**) or to 50 °C (**c**) would require a very significant change of the intensity of the greenhouse effect. (After Kasting and Catling, 2003.)

6

significant than at the present day, and that, as a result, the emission of gases by volcanoes was also more abundant. It is, therefore, reasonable to believe that the atmospheric concentration of the gases released by volcanoes was higher than the value proposed by Kasting.

In the absence of photosynthetic living organisms, the dioxygen (O_2) is produced only by photochemical reactions: by photolysis of H_2O, followed by the thermal escape of the released hydrogen. Because of the reaction of this molecule with all the reducing volcanic gases, the partial pressure of O_2 at 3.8 Ga therefore remained extremely low (about 10^{-10} bar). As a consequence, the Archaean atmosphere was reducing ... as already shown by the study of the BIFs, typical of Archaean cratons. At the end of the Hadean, methane (CH_4) was exclusively of abiotic origin: it could have been produced in small amounts by serpentinization of the ultrabasic rocks that covered the magma ocean (*see* ▶ Chap. 2). During the Archaean, it mainly resulted from hydrothermal activity that altered the basic and ultrabasic rocks of the oceanic crust. In the previous pages, it has been emphasized that during the Early Archaean, in the Kaapvaal and Pilbara cratons, hydrothermal activity was very significant. If, today, the partial pressure of methane of abiotic origin is about 10^{-6} bar, it is possible to estimate that at that period it ranged between 10^{-5} and 10^{-4} bar. Equally, some scientists made the assumption that if life appeared and diversified very early in the Archaean, it is highly likely that the development of methanogenic *Achaea* strongly contributed to the increase in the partial pressure of atmospheric CH_4. However, prevalence of methanogenic *Achaea* at this epoch is not proven, and their possible role thus remains largely speculative.

Other gases such as NO, HCN and CO could have formed from meteorite impacts or from the action of lightning, but there is no reliable model that allows us today to quantify their concentration in the Archaean atmosphere, especially as this latter depends closely on the concentration of other atmospheric components, such as CH_4, CO_2 and H_2.

Obviously, Kasting's model should be taken for what it is, that is ... a model – and not as an established truth – that has the merit of sketching the basic outline of the geological face of the superficial layers of the Earth during the Early Archaean.

However this outline would not be complete without invoking other important players – the oceans. The Australian Jack-Hills zircons have told us that the Earth harbored oceans at 4.4 Ga, but have not told us anything about the composition, the temperature, or even the volume of the vast expanses of water. During the Archaean, the situation is different, because geologists and geochemists can analyze the rocks and, in particular, sedimentary rocks. The latter were formed and deposited in an oceanic environment, so they thus represent ideal objects for attempting to reconstruct some of the characteristic of Archaean oceans.

The Archaean Oceans: Saline and Hot?

The supracrustal formation at Isua (3.86 Ga) and the greenstone belts of the Kaapvaal and Pilbara cratons (3.6–3.1 Ga) contain great amounts of sedimentary rocks. These latter will serve as a basis for our outline of the nature of the Archaean oceans.

The Composition of the Archaean Oceans: a Lot of Iron, but How Much Salt?

Two complementary approaches allow us to discuss the composition of these oceans. The first of these is indirect; it consists of analyzing the composition of Archaean sedimentary rocks in order to decipher the conditions that prevailed when they were deposited. We have already seen that the presence of BIF bears witness to the great scarcity of oxygen in the atmosphere and in the oceans: overall, the latter were reducing. Carbonate rocks are formed by the precipitation of carbonates, even if they are rare in the Archaean terrains, they can provide important information. The Archaean carbonate rocks mainly consist of calcite, aragonite (two minerals whose chemical formula is $CaCO_3$) and dolomite ($[Ca,Mg]CO_3$), whereas siderite ($FeCO_3$) is rare. This shows that the oceans were then saturated with Ca^{2+} and Mg^{2+} ions. In addition, it is estimated that the Ca^{2+}/Fe^{2+} ratio in Archaean seawater was about 250, while today it is about 10^7. This considerable difference is interpreted as due to the high content of dissolved Fe^{2+} in the Archaean ocean, which itself results from the low content of atmospheric and oceanic dioxygen. We may recall that, today, in the presence of O_2, ferrous iron (Fe^{2+}) oxidizes into ferric iron (Fe^{3+}). The latter precipitates in the form of ferric hydroxide $Fe(OH)_3$, which is insoluble in the ocean, whence the very high Ca^{2+}/Fe^{2+} ratio of the oceanic water.

In addition, by analyzing the fractionation of sulfur isotopes (in particular of ^{34}S) in Achaean marine sediments, geochemists have concluded that at that epoch, the concentration of sulfates in the ocean was about 0.2 mmol.L^{-1}, that is, one-hundredth of its value today (28.7 mmol.L^{-1}). This result is not really surprising, because nowadays, a large proportion of the dissolved sulfates originates from the oxidation of the pyrite (FeS_2) that is contained in the rocks. This oxidation takes place in the presence of water and dioxygen, in accordance with a reaction of the type:

Fig. 6.20 A fluid inclusion within a mineral. When minerals crystallize, they develop a certain number of crystal defects in the form of small cavities whose volume ranges from a few μm³ to 100 μm³. In the case of sedimentary minerals, these cavities trap and retain fluids (most frequently water) present in the medium from which they crystallize. Analysis of the fluid inclusions therefore provides researchers with information about the composition of the medium and, in the case of Archaean sediments, of the composition of the Archaean ocean. The inclusion photographed here contains not only a fluid (saline water) but also vapor and a crystal of salt (halite). (Photo: F. Gibert.)

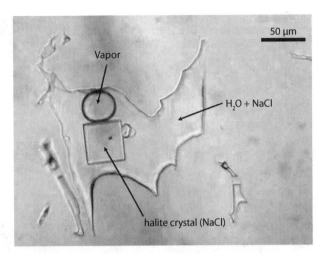

6

$$FeS_2 + 7/2\ O_2 + H_2O \rightarrow Fe^{2+} + 2\ SO_4^{2-} + 2\ H^+.$$

During the Archaean, in the absence of O_2, pyrite could not be oxidized and so was not transformed into sulfate, whence their low concentration in the Archaean oceans.

The second method of determining the composition of these Archaean oceans is more direct, indeed, it consists of studying tiny volumes of fluids (gas or liquid, or both), which during the Archaean, were trapped in minerals. When they crystallize and precipitate in the ocean, the minerals of sedimentary rocks, alike all other crystals, develop a certain number of crystal defects in the form of tiny cavities going from a few μm³ to 100 μm³, and into which the sedimentation medium, namely water, may be trapped (Fig. 6.20). The composition of these fluids therefore corresponds, in theory, to that of the ocean at the time when they were trapped within the inclusion. At Isua quartz globules preserved in pillow lavas dated to 3.75 Ga, contain the remnants of two independent fluid systems. One is almost pure methane, whereas the other is a highly saline aqueous solution: 25 wt% equivalent NaCl (the weight obtained assuming all the Cl⁻ ions are in the form NaCl), whereas the present-day ocean contains 3.5 wt% equivalent NaCl. The researchers have interpreted these fluids as being representative, not of the composition of the ocean, but of just the water circulating in the hydrothermal systems on the ocean floor. However, this interpretation is open to discussion, because the Cl⁻ content of hydrothermal fluids nowadays is not very different from that of the ocean; indeed, it ranges from 0.4 to 7 wt% equivalent NaCl. The controversy is all the more intense because the Isua rocks that have been analyzed have undergone significant metamorphism: their mineralogical composition has recorded traces of pressures of 4 kbar and temperatures of about 480 °C, which could have modified the composition of the fluids.

Other Archaean sedimentary rocks have been brought into this discussion. In the Pilbara craton, researchers have studied a sedimentary formation dated to 3.49 Ga, that has been subjected to low-grade metamorphism (its temperature has never exceeded 200 °C), which contains inclusions, where several fluids have been detected. One of these is a saline solution, which is interpreted as being seawater. It contains 12 wt% equivalent NaCl, which is 4 times more than present-day seawater (Fig. 6.21 and Table 6.3). Analyses have, however, given a comparable result for Br⁻ and K⁺ ions. As a result, these high ionic concentrations have been interpreted as resulting of an intense evaporation of seawater, and the geologists have concluded that the host minerals of these inclusions have probably crystallized in an evaporitic environment. From these examples, it would seem that the fluids trapped in the sedimentary minerals, if they can provide information about local

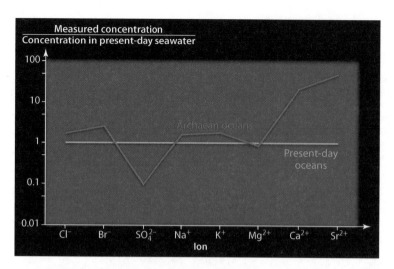

Fig. 6.21 The concentration of dissolved ions in the Archaean seas (red curve) relative to that in present-day seawater (green line). The data are derived from the analysis of fluid inclusions discovered in a sedimentary formation in the Pilbara craton, dated to 3.23 Ga. It is worth bearing in mind that some uncertainty remains. It is not absolutely established that the water trapped locally in the inclusions is representative of the average composition of Archaean seawater. It may also have sampled a very local environment such as an evaporatic environment or a hydrothermal system. (After de Ronde *et al.*, 1997 and Gutzmer *et al.*, 2001.)

Table 6.3 Concentration of the principal ions dissolved in the Archaean ocean and in the present-day ocean. *See* the remarks in the caption to Fig. 6.21. (After de Ronde, *et al.*, 1997.)

Ions (mmol.L⁻¹)	Cl⁻	Br⁻	I⁻	SO₄²⁻	Na⁺	K⁺	Mg²⁺	Ca²⁺	Sr²⁺
Archaean ocean	920	2.25	0.037	2.3	789	18.9	50.9	232	4.52
Modern ocean	556	0.86	0.0005	28.7	477	10.1	54.2	10.5	0.09

conditions, are useless for deciphering the overall composition of the Archaean ocean. So the debate remains unresolved...

The Temperature of the Archaean Oceans: a Controversial Issue

The temperature of the Archaean oceans is also the subject of controversy. Although all the researchers agree that the oceans were hotter than nowadays, they disagree on how much they were hotter. In some Archaean cherts, for example, geochemists have analyzed two stable isotopes of oxygen: ^{16}O and ^{18}O, the fractionation of which is temperature dependent. They have deduced that throughout the Archaean the temperature of the surface waters ranged between 55 and 85 °C. It was only at the beginning of the Proterozoic that it decreased progressively until it reached its current value (Fig. 6.22).

However, similar measurements carried out on fluid inclusions, 3.2 Ga old, have yielded very different results; indeed, the temperature of surface waters has been calculated to be 39 °C. In 2006, French workers studied the fractionation of the stable isotopes of silicon – ^{28}Si and ^{30}Si, a fractionation that is also temperature dependent – in the cherts and flint of different geological periods. They calculated the surface temperature of the ocean at 3.4 Ga as being about 70 °C, in perfect agreement with the data obtained from oxygen isotopes (Fig. 6.22). Nevertheless, the debate is not over, because one cannot preclude that the fractionation of the oxygen and silicon isotopes does not reflect the average temperature of the ocean, but rather that of the hydrothermal systems in which the cherts were deposited.

This question of the temperature of the surface waters in the Archaean is particularly important. Indeed, if really the Archaean oceans were hot, then there is a contradiction with the models that consider that during the Hadean, the Earth was, at least sporadically, cold (*see* ▶ Chap. 3). In addition, as the solubility of O_2 decreases strongly when the temperature and salinity increase, it is highly probable that a hot, salty ocean would have remained anoxic even when, from 2.2 Ga, the O_2 content became significant. If this was indeed the case, it was

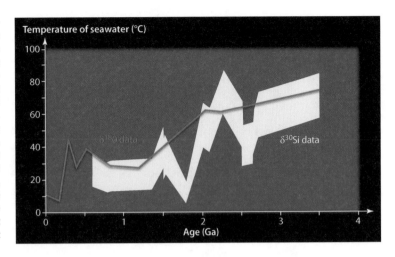

Fig. 6.22 Evolution of the temperature of ocean water over the course of the planet's history. Measurements have been performed on cherts of different ages. The pale yellow area corresponds to the temperatures computed from silicon isotopes ($\delta^{30}Si$), while the red curve has been derived from the oxygen isotopic signature ($\delta^{18}O$). The temperatures calculated for Archaean surface waters range between 55 and 85 °C. If these temperatures do really represent those of the ocean at the beginning of the Archaean, then, they contradict the models, which suggest that during the Hadean, the terrestrial climate, governed by the silicate-carbonate relationship, was cold, at least episodically (see ► Chap. 3). (After Robert and Chaussidon, 2006.)

6

certainly not devoid of consequences for any forms of life that might have already developed in the aquatic environment.

The Earth's Machinery During the Archaean: Plate Tectonics Between 3.8 and 2.5 Ga

During our trip to the heart of the remnants of the Archaean Earth, we have so far primarily considered small-scale structures: rock samples, minerals, and even the micrometric inclusions that the latter may enclose. But geologists excel in the practice of changing scale: they search for the indications and information that they need, from mountain ranges, considered as a whole, to microscopic inclusions in minerals. In this last section of our geological quest, we shall focus on a range of scales from outcrops to continents. That will lead us to linger over another vital question: whether or not plate tectonics operated on Earth during the Archaean.

Plate tectonics describes the present-day displacements of lithospheric plates over the asthenospheric mantle as well as their relative (horizontal) motion, the whole being integrated into an overall dynamical model of our planet. The convective motions that drive the mantle – and efficiently participate in dissipating the Earth's initial stock of energy – are also the main engine driving plate tectonics (■ Box 3.1, and ■ Fig. 3.3 in Chap. 3).

First let us remind ourselves that the notions of lithosphere and asthenosphere rely on criteria that are neither geochemical nor mineralogical, but rheological. The lithosphere is rigid. It consists of the crust (continental or oceanic) and the rigid and cold portion of the upper mantle. The asthenosphere corresponds to the deeper mantle which is ductile (it deforms by flowing, a bit like modeling clay) and hot (■ Fig. 3.2, Chap. 3).

The Debate

In the last twenty years, within the geoscientist community, two major schools of thought have developed and confronted one another. One considered that, given the significant production of heat within the Earth during the Archaean, the basaltic oceanic crust was not rigid enough, nor dense enough to be able to sink and to be recycled into the mantle through the subduction mechanisms. Similarly, the continental crust was too "soft" to be able to support the weight of mountain ranges. According to this theory, plates such as those we know today

6

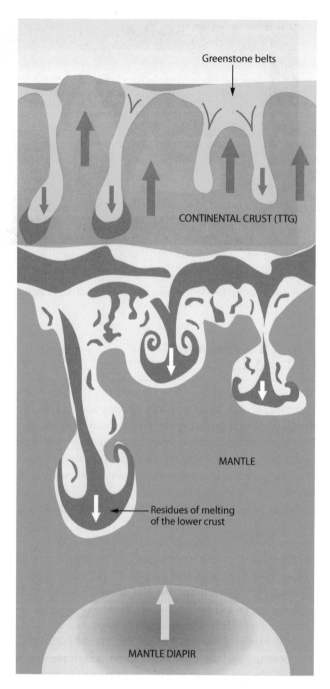

Fig. 6.23 Schematic section of the continental crust and upper mantle, illustrating the mechanisms operating on an Archaean Earth, where plate tectonics would not have existed. In such a model, the TTG constituting the continental crust (orange) is generated by the partial melting of basalts delaminated and detached from the base of a thick oceanic crust. All the movements are vertical, and gravity-driven. In the mantle, hot (and thus less dense) diapirs rise, while the residues of melting of the lower, basic crust (dark green) sink. Similarly, in the continental crust, the dense greenstone belts (pale green) sink into the TTG basement (orange), which is less dense. (After Bédard, 2006.)

could not have existed and the Earth's surface was little (or not) affected by horizontal movements. It was dominated by "hot spot" type environments, corresponding to the ascent of diapirs from the deep and hot asthenospheric mantle (like the modern hot spots: Hawaii, La Réunion, etc.). Such a geodynamic setting would have favored the genesis of komatiites. In this environment the TTGs are assumed to have been produced by partial melting of basalts that had been detached from the base of a thick oceanic crust, perhaps similar to a modern oceanic plateau (▣ Fig. 6.23). (An oceanic plateau consists in an over-thickened oceanic crust, whose thickening is caused by the emission of huge volumes of magma linked to the activity of a hot spot. In a continental setting, the hot spot magmatic activity leads to the emplacement of traps).

The other school of thought assumed that plate tectonics was active on the Archaean Earth, but that the details of its processes differed from what is known today. The lithospheric plates today are subject to horizontal displacements, such that the form of tectonics itself is primarily characterized by horizontal structures, like the great thrust faults that are typical of collision zones such as the Alps or the Himalaya. The Archaean cratons exhibit such horizontal structures, even in the oldest terranes (Amîtsoq, Pilbara, Kaapvaal, etc.). This argument in favor of Archaean plate tectonics has recently been reinforced by the discovery that we have already mentioned at the beginning of this chapter. In 2007, an ophiolitic complex, dated to 3.8 Ga, was found in Greenland, in the Isua formation. The dyke complex of this ophiolite, resembling present-day complexes, is a proof of seafloor spreading, which itself, is characteristic of plate tectonics.

As a result, a consensus has gradually been established around the idea that plate tectonics has taken place at least since the beginning of the Archaean. However, the two schools of thought do not exclude each other, and as we shall see, although plate tectonics seems to have predominated, it by no means precludes the existence of numerous hot spots, thick oceanic plateaus, and vertical tectonics (as on the present-day Earth – where horizontal movements of the lithospheric plates predominate – but where, locally, there are tectonics and magmatic activity associated with the rise of hot, mantle diapirs, and thus vertical motions),

6

Smaller and "Faster" Plates

We shall now see how the Archaean global tectonics differed from its modern equivalent. The study of typical Archaean rocks (TTG and komatiites) has shown that the Earth's internal heat production was between twice and four times greater during the Archaean than today. This thermal energy had, perforce, been released, failing which its accumulation would have resulted in the melting of at least a part of the planet, an event absolutely no trace of which is found in the geological record. During the Archaean, as today, convection was by far the most efficient mechanism for dissipating the Earth's internal heat. The latter is released at the mid-oceanic ridges (located above the ascending branches of the convection cells), through magmatic and hydrothermal activity. In this context, the amount of removed heat is a function of the cube root of the length of the ridge. As a result, if there was more heat to be released, the overall length of the mid-oceanic ridges should also have been greater. Given that the Earth has kept a constant surface area, it can logically be concluded that the lithospheric plates demarcated by the oceanic ridges were smaller than nowadays (◘ Fig. 6.24). The surfeit of energy to be released could also have resulted in a more vigorous convection and, therefore, it is rationale to assume that the plates moved more rapidly than they do today.

The present-day Earth offers us the possibility of testing these conclusions "full scale". Indeed, in the North Fiji Basin (◘ Fig. 6.25), the overall measured heat flux is abnormally high and can reach 240 mW.m^{-2}, or about 4 times as much as the average oceanic flux today, which is about 70 mW.m^{-2}. To a certain extent, this area of the planet can be considered as an analogue of the Archaean Earth; at least as far as the heat flux is concerned. This area of the world is characterized by the abundance of active ridge segments which demarcate very small plates. In the North Fiji Basin, the length of ridge relative to the surface is 20 times as great as that measured for the rest of the Pacific Ocean. There is, therefore, an obvious correlation between the heat flux, which is the amount of internal heat that is being released, the length of ridges and the size of the lithospheric plates.

The mid-oceanic ridges today are the site of significant hydrothermal activity, which is revealed, in particular, by the presence of black smokers and white smokers. If the overall length of the ridges was greater during the Archaean, the associated hydrothermal activity should also have been more vigorous. This prediction is corroborated by the great abundance of cherts and

6

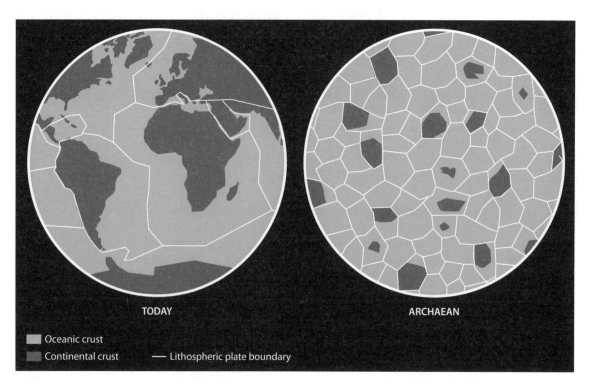

TODAY   ARCHAEAN

■ Oceanic crust
■ Continental crust   —— Lithospheric plate boundary

■ **Fig. 6.24 Comparison of the size of present-day lithospheric plates with the assumed size of Archaean lithospheric plates.** During the Archaean, Earth's internal heat production was greater than today. This heat was released through the mid-oceanic ridges, whose overall length, as a consequence, was also greater, which resulted in a mosaic of much smaller lithospheric plates than those found on the Earth today. (After De Wit and Hart, 1993.)

rocks of hydrothermal origin in all the terranes older than 2.5 Ga. In addition, if the Archaean plates were smaller and moving faster, the oceanic crust should have entered into subduction far younger than today. It has been calculated that the average age of the oceanic crust when it entered in subduction was about 10 Ma in the Archaean, against 60 Ma at present.

The subduction of a young, hot oceanic crust would result in high geothermal gradients along the Benioff plane. It is this characteristic of the Archaean Earth that, as we have explained earlier, enabled the hydrous melting of the subducted basalts, before they could be dehydrated, thus generating the TTG magmas, which constitute most of the continental crust at that period. Because density decreases as temperature increases, a young subducted crust was hotter and consequently less dense than its present-day equivalent. Therefore, the angle at which the oceanic slab sank into the mantle would be lower (resulting in a flat subduction, ■ Fig. 6.26). In addition, as melting typically occurs at depths of 50 to 80 km, a flat subduction would favor the formation of a wide volcanic arc.

This last hypothesis may also be tested at full scale on the Earth today. For example, a low angle (almost flat) subduction has been reported in Ecuador, where the Nazca Plate is being subducted beneath the South-American Plate. The Nazca Plate carries the young aseismic Carnegie Ridge, which has formed above the Galapagos hot spot. The Carnegie Ridge, being younger, is also hotter and less dense than the Nazca Plate supporting it and it contributes to the buoyancy of the whole. At the place where the Carnegie Ridge is subducted, the angle of subduction is only some twenty degrees, the associated volcanism is very active and, above all, the volcanic arc is three times as wide as in other subduction zones of this type (150 km as against 50 km generally). This situation, which is exceptional today, must have been the rule in the Archaean.

Obviously, an oceanic lithosphere, young, hot and thus of low density, would be buoyant and could not spontaneously enter into subduction. It is this finding that has led certain researchers to put forward the theory that subduction could not operate in the Archaean. How-

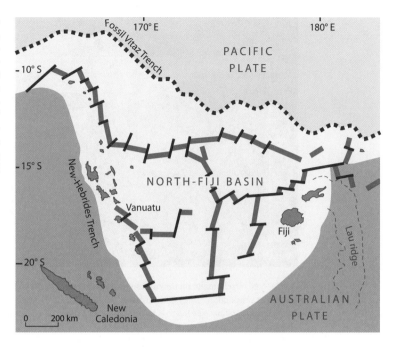

◻ Fig. 6.25 A simplified geological map of the North Fiji Basin. This map shows that, in this area of the world where the present-day geothermal flux is very high, and consequently where Earth releases significant amounts of heat, the length of the active mid-oceanic ridges (in red) is also abnormally great. This has resulted in the creation of individual micro-plates, which are undoubtedly similar to those that existed on the Archaean Earth. Transform faults are shown in black. (After Lagabrielle *et al.*, 1997, and Lagabrielle, 2005.)

ever, even if subduction is not spontaneous, it may be forced, as for example, what is occurring today in Ecuador. There, the oceanic lithosphere is sinking, less because its density has become greater than that of the mantle, but rather because it is obliged to do so by the forces that compel it to move horizontally. Indeed, the production of new oceanic crust induces an increase in the Earth's surface area. Given that the surface area of our planet is constant, as much oceanic crust must disappear as is generated in the mid-oceanic ridge systems. The constraints produced at these ridges may therefore force young oceanic crust to be subducted.

Another factor to be taken into account is the possible presence of komatiites associated with Archaean basalts. These could significantly increase the density of the oceanic crust, such that it could become denser than the underlying asthenospheric mantle and spontaneously sink. In this context, it is worth noting that, unlike hydrated basalts, the solidus temperature of komatiites is far too high, such that these rocks cannot melt under subduction conditions, even when hot. They could not, then, play any part in the genesis of the TTGs of the continental crust.

A Different Kind of Tectonics: Sagduction

Although the Archaean cratons display horizontal structures symptomatic of plate tectonics, they also exhibit evidences for another form of tectonics, which is specific to them: sagduction. This tectonic form, omnipresent before 2.5 Ga, is characterized by vertical movements. It consists of broad structures where the TTG basement forms large domes between which the greenstone belts are wedged in the form of narrow bands (◻ Fig. 6.27). Sagduction is a gravity-driven mechanism. In fact, deposition of high-density lavas such as komatiites ($d = 3.3$), or even of sediments such as the banded iron formations (BIF), onto a continental crust consisting of TTGs with a lower density ($d = 2.7$), creates a strong, inverse density gradient, which leads to the sinking of the denser rocks (the greenstone belt) into the lower density rocks (the TTG). Once it has started, the phenomenon evolves resulting in some kind of inverse diapirism, caused not only by the descent of the denser rocks, but equally by the concomitant ascent of the lower-density rocks. Subsidence of the greenstone belts is able to create a topographic depression where sediments can deposit (◻ Fig. 6.27).

6

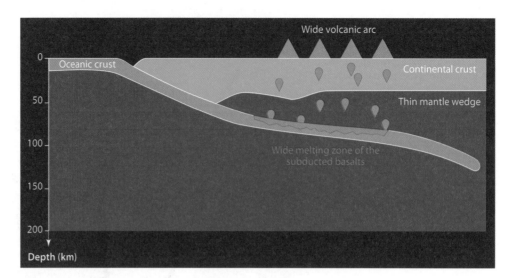

■ **Fig. 6.26 A schematic section of a subduction zone, where a young and hot oceanic crust subducts with a low angle (flat subduction).** This type of subduction – which is only exceptionally observed today, (in Ecuador, for example, where the young Carnegie Ridge is being subducted under the South-American Plate) – was probably the general rule on the Archaean Earth.

In such a configuration, the subducted oceanic lithosphere is young, hot and less dense. It then tends to resist the forces that are dragging it down, such that the subduction angle is low, or even nearly zero: it is known as a flat subduction. Consequently, the zone where the subducted slab can melt is wider, which results in a large volcanic arc.

Komatiites are abundant in Archaean terranes and, with one exception, completely absent after 2.5 Ga. Indeed, after 2.5 Ga, the cooling of the Earth is such that the mantle is no longer able to reach degrees of melting of at least 50%, required for the genesis of komatiitic magmas Today, the degrees of mantle melting, ranging between 25 and 30%, are only able to give rise to basalts, whose density does not exceed 2.9 or 3, which does not generate an inverse density gradient, sufficient to initiate sagduction. So, as komatiites are necessary to trigger sagduction and that they were no longer generated after 2.5 Ga, sagduction and vertical tectonics also logically ended and disappeared after 2.5 Ga. During the Archaean times two contrasting tectonic styles were active concomitantly: a horizontal tectonics which, like today, predominated at rigid plate edges, and a vertical tectonics, localized to the center of the continental plates. Only the horizontal tectonics survived the Archaean-Proterozoic transition.

To sum up, we may draw up a list of the specific features of Archaean global tectonics:

- plate tectonics was in operation, with the genesis of oceanic crust at mid-oceanic ridge systems and its subsequent destruction at subduction zones;
- relative to the current plate tectonics, the Archaean plates were smaller, moved more rapidly, were subducted earlier and with a lower angle;
- the central part of continental plates was affected by a gravity-driven vertical tectonics: sagduction;
- hydrothermal activity was far more significant than today;
- it was the hydrated oceanic crust, drawn into subduction that melted, (thus generating TTG of the continental crust) and not as today the mantle itself.

A Newly Habitable and Already Inhabited Planet

Our voyage in time has taken us back into a past that has often been more than three billion years old. We have gathered a whole mass of clues, which after interpretation enable us to draw up a "geological portrait", as exhaustive as possible, of the Archaean Earth.

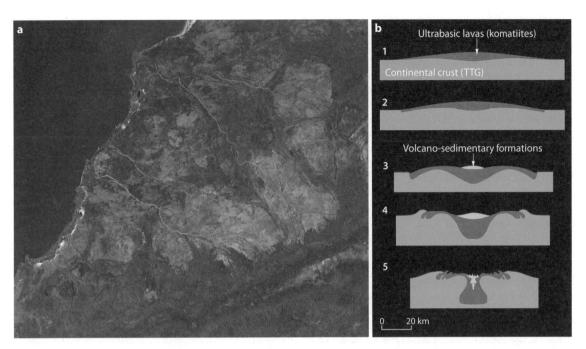

6

□ **Fig. 6.27 A form of tectonics typical of the Archaean Earth: sagduction. a.** Satellite image of the Pilbara region, where the terranes, with ages of 3.5 to 3.2 Ga, exhibit sagduction structures: the greenstone belts, dark in color, are wedged between domes of TTG that are in a lighter color. The image covers an area that is about 450 km across. **b.** Schematic representation of the sagduction process. Very high-density ultrabasic rocks such as komatiites (in green) are deposited on a low-density (TTG) continental crust (in orange). Under the effect of their weight, the komatiites sink into the TTG, creating a depression in the center in which sediments and volcanic rocks (in yellow) can deposit. (b: after Gorman *et al.*, 1978.)

Although, in certain aspects, it resembles the present-day Earth, it differs from it in many other ways. All these differences, in fact, are the result of a single and simple process: the primitive internal Earth was hotter, and since its accretion, it has progressively cooled. Indeed, it is the progressive cooling of the planet that explains the disappearance of komatiites after 2.5 Ga; the new way in which the continental crust was formed – the TTG produced by the melting of hydrated basalts giving way to the granodiorites generated by melting of the "mantle wedge" – the end of vertical tectonics (sagduction); the increase in the size of lithospheric plates; the decrease in hydrothermal activity, etc. Detailed geochemical work (□ Box 6.3) has even revealed the way in which this cooling progressed throughout the Archaean. For example, it has been possible to show that between 4.0 and 2.5 Ga, as a result of the decrease in geothermal gradients, the depth at which TTG magmas were produced by melting of hydrated basalts steadily increased.

On this Archaean Earth, which was more active geologically than the present-day Earth, continents emerged in the middle of oceans of liquid water, which was probably hot (about 70 °C). Overall, both the atmosphere and the ocean were reducing, that is to say devoid of oxygen, and they formed environments that may have been suitable for prebiotic organic synthesis (*see* ▶ Chap. 4). The climate, and more precisely, the greenhouse effect, was regulated by the interactions between the weathering of the emerged continental crust and the gases released by the volcanoes, such that despite a "cold" Sun (□ Box 6.2), global glaciations ("Snowball Earth") were probably avoided. The emerged continents also provided a source of elements – including oligo elements – necessary for the synthesis of organic molecules and which, via rivers, were conveyed into the oceans.

Now, what about life? At the very beginning of the Archaean (4.0 Ga), all the conditions were present not only for life to appear – if it had not already appeared in the Hadean and had

▣ Box 6.3   The Source of the Continental Crust Becomes Deeper and Deeper

Over the whole of the Archaean period (from 4.0 and 2.5 Ga), it is estimated that the Earth's heat production decreased by one half (▣ Fig. 6.13). To what extent did this progressive cooling affect the growth of the Archaean continental crust, i.e., the melting of hydrated basalts of the oceanic crust at subduction zones? Advanced geochemical analyses have recently provided a clear answer to this question.

In 2002, geologists compiled 1100 chemical analyses of TTG, whose ages ranged from 3.86 to 2.5 Ga. They came to the conclusion that, in the TTG parent magmas, the content of some chemical elements had evolved over time (▣ Fig. 6.28): MgO, Ni, (Na_2O + CaO) and Sr contents progressively increased.

Let us first consider the behavior of MgO and Ni. Recent investigations have shown that the TTG liquids obtained by experimental melting of basalts were systematically poorer in MgO and in Ni than natural TTG. The high content of MgO and Ni in the TTG parental magmas has therefore been interpreted as the result of interactions between these TTG magmas and the overlying mantle. Indeed, during their ascent towards the surface, these magmas, which are acid, percolated through the "mantle wedge" located above the subducted oceanic slab and reacted with the mantle peridotites, which are ultrabasic rocks very rich in magnesium and in transition elements such as nickel. Consequently, during its ascent the TTG melt was contaminated by mantle peridotites, from which it took small amounts of Mg and Ni.

The temporal increase of MgO and Ni contents of the TTG parental magmas therefore reveals an augmentation of the efficiency of interactions of these TTG liquids with the "mantle wedge".

How should we now interpret the progressive enrichment of the TTG parental magmas in (Na_2O + CaO) and Sr? Plagioclase, which is a widespread mineral phase in basalts, is rich in these three elements. If the residue of basalt melting contains plagioclase, the latter will therefore retain Na_2O, CaO and Sr, and the magmatic liquid, in equilibrium with this residue, will be poor in those elements. So the increase in time of the (Na_2O + CaO) and Sr contents in the TTG parental magmas reflects the progressive decrease in the abundance of residual plagioclase during basalt melting. ▣ Figure 6.29 shows that pagioclase is only stable at low pressures (below 15 kbar). In other words, as the depth of melting increases, it progressively disappears from the residue of melting. The conclusion therefore is that over the course of time, melting of the basalts that are the source of the TTG took place at greater and greater depths.

The chemical evolution of the parental magmas of the Archaean continental crust therefore appears as a logical consequence of the gradual cooling of the Earth. During the early Archaean (> 3.4 Ga), the geothermal gradients along the subduction plane (the Benioff plane) were high, so that the melting of the basalt took place at shallow depths, in the domain where plagioclase was stable (▣ Fig. 6.29). The lat-

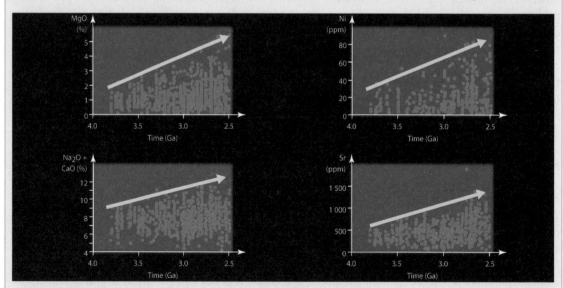

▣ **Fig. 6.28 Diagrams showing the temporal evolution of MgO, Ni, (Na_2O + CaO) and Sr contents, in the TTG parental magmas, from 4.0 to 2.5 Ga.** In all these diagrams, the upper envelope of the group of points represents the composition of the less-differentiated parental magmas, that it to say those whose composition has not been modified by fractional crystallization. The four diagrams clearly show that the composition of the parent magmas has evolved with time (yellow arrows). The content of all the elements studied increases from 4.0 to 2.5 Ga. These changes are interpreted in terms of increase in the depth of basalt melting in the Archaean subduction zones, which results of the gradual cooling of the Earth. (After Martin and Moyen, 2002.)

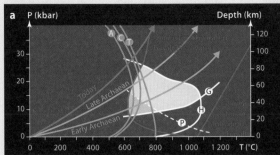

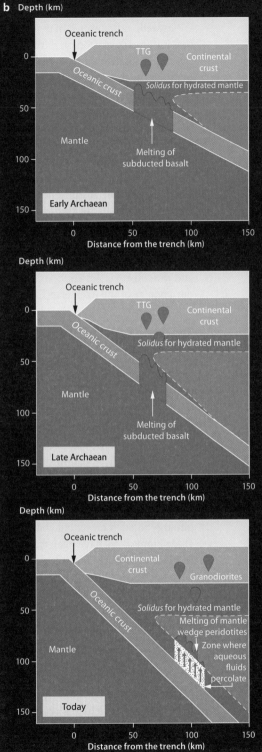

□ **Fig. 6.29 A pressure (depth) *versus* temperature diagram (a) and schematic sections of subduction zones, showing the conditions under which continental crust was generated during the Early Archaean (*T* > 3.4 Ga), Late Archaean (*T* < 3.4 Ga), and today (b).** During the Early Archaean, the geothermal gradients along the Benioff plane were high (orange arrow), in such a way that the subducted oceanic crust reached its hydrated *solidus* temperature before being dehydrated. It was thus able to melt at a relatively shallow depth, in the domain where hornblende, garnet and plagioclase were stable (dark orange field). During the Late Archaean, the geothermal gradients along the Benioff plane were weaker (pink arrow), but still sufficient for the subducted oceanic crust to reach its hydrated *solidus* temperature before becoming dehydrated. It then melted at a greater depth, in the region where just hornblende and garnet were stable (pale yellow field), but outside the plagioclase stability domain. Today, the geothermal gradients along the Benioff plane are low (blue arrow) and the subducted crust dehydrates before it can melt. The fluids released by this dehydration ascend and percolate through the mantle wedge, which, at the same time, they metasomatize and rehydrate. Consequently, this mantle wedge melts and gives rise to calc-alkaline magmas typical of the present-day continental crust. The red curves represent the anhydrous and hydrated (5% water) *solidus* for a basalt. The green curves correspond to the dehydration reactions that occur in the basalt when hydrous minerals are destabilized: (A) = antigorite (serpentine); (C) = chlorite and (T) = talc. The other curves are those for the stability of hornblende (H), garnet (G) and plagioclase (P). (After Martin and Moyen, 2002.)

ter remained in the residue, such that, the magmatic liquid was poor in (Na_2O + CaO) and Sr. During its ascent towards the surface, the TTG magma only had to percolate through a small thickness of the mantle wedge and it had little (or no) chance of reacting with the mantle peridotites. On the other hand, during the Late Archaean (< 3.4 Ga), the Earth had already cooled and the geothermal gradients had decreased. Consequently, the melting of the subducted basalts occurred at progressively greater and greater depths, outside the domain of plagioclase stability. The TTG parental magma was then rich in (Na_2O + CaO) and Sr. In addition, before beeing emplaced at the surface, it had to pass through a considerable thickness of the mantle wedge, causing even stronger interactions with the peridotites of the mantle, thus resulting in high MgO and Ni contents in TTG.

The evolution of the TTG composition, between 4.0 to 2.5 Ga, thus reveals a progressive increase in the depth of melting of the subducted oceanic crust, as the Earth became less and less hot.

not partially survived the Late Heavy Bombardment between 4.0 and 3.9 Ga (*see* ▶ Chap. 5) – but also, and above all, for it to evolve and maintain.

Indeed, although we may not know exactly when life arose, we can assert with certainty that life already existed on Earth 2.7 Ga ago, and perhaps even well before, by 3.5 Ga. In fact, the oldest fossils that bear witness of an Archaean microbial life are, indisputably, the stromatolites in the Fortescue formation, in northwestern Australia, dated to 2.7 Ga, and, with less certainty, macrofossil structures dated to 3.5 Ga ago, discovered at Pilbara (in Australia) and at Barberton (in South Africa).

Stromatolites are structures resulting of the activity of particular microbial mat communities that are quite diversified and complex. We must keep in mind that the Fortescue stromatolites are necessarily the outcome of an evolutionary process that took place before 2.7 Ga. However, the traces of this evolution are very tenuous and very difficult to identify in an unequivocal manner. Admittedly, several signatures of fossil life have been described from rocks far older than 2.7 Ga (the current record being 3.87 Ga) and the number of publications devoted to them increases every day, thanks to a scientific exploration that becomes more and more sophisticated and methodical. However, part of the scientific community still remains skeptical as to the fact that these fossil traces provide undoubted proof of past life. These data and the controversies concerning the traces of ancient life are the subject of the last section of our enquiry of the Archaean Earth.

Traces of Ancient Life: Data and Controversies

All potential life traces older than 2.7 Ga discovered in our planet's record are controversial. There are two main reasons for that. The first is that, as we have seen, the geology of the Earth before 2.5–2.7 Ga (that is during the Hadean and Archaean) has features that are not found later: ultrabasic volcanism of high temperature (komatiites); widespread hydrothermal activity; intense chemical sedimentation with the banded iron formations (BIF) and the cherts. Under these conditions, because of the lack of modern points of reference, it is very difficult to reconstruct the Archaean palaeoenvironments. The second reason concerns the degree of metamorphism undergone by the Early Archaean rocks. Most often the metamorphic grade is so high that the possible microfossil structures have lost their original morphology to such an extent that they are no longer identifiable. Organic molecules that include biomarkers have been transformed into kerogen – a solid organic substance, insoluble in organic solvents and often very complex – or into graphite. The isotopic signatures may have been altered by exchanges with hydrothermal fluids. Even worse, several metamorphic reactions may generate organic matter abiotically which, when metamorphosed, may give rise to kerogen or graphite without the intervention of biology. As a result, the possibility that a number of potential signatures are, in fact, the result of abiotic processes cannot be excluded.

What precisely is the nature of these possible traces of life recorded in rocks older than 2.7 Ga? At the beginning of the 20th century, the fossil record was the result of macroscopic observations and the oldest proof of life only dated back at the beginning of the Cambrian (544 Ma). Those fossils represented traces of fairly complex organisms (metazoans). This implied that other, simpler, forms of life preceded them. Indeed, as the Precambrian rocks were explored, an increasing number of fossils of various types were discovered (◘ Fig. 6.30): macrofossils, microfossils (fossilized microorganisms), molecular fossils and indicators of metabolic activity (isotopic markers) (◘ Box 6.4).

The only macrofossils that have been found in terrains older than about 2.1 Ga (before the Neoproterozoic) are stromatolites (◘ Fig. 6.31). These are laminated sedimentary structures formed by the precipitation (sometimes also accretion) of carbonates caused by complex mi-

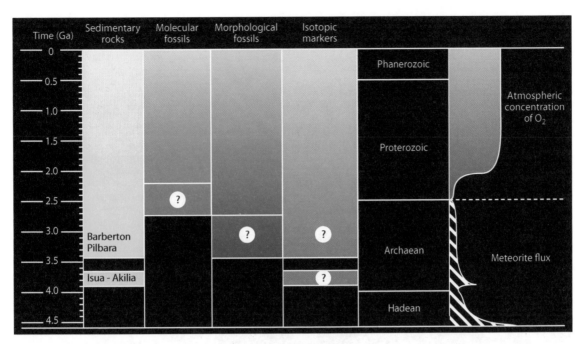

Fig. 6.30 Simplified scheme of the fossil record of life on Earth from its origin to the present. The first unambiguous morphological fossils are dated to 2.7 Ga; these are large expanses of stromatolites at Fortescue in Australia. Between 3.5 and 2.7 Ga, several structures that may correspond to macrofossils (stromatolites, but far less massive) or to microfossils have been identified, though they could also be the result of abiotic processes. "Molecular fossils" (fossil lipids) were identified in rocks aged 2.7 Ga, but it has recently been shown that they were the result of late contamination. So, the oldest, unambiguous molecular fossils have an age of about 2.15 Ga. Finally, with regard to the isotopic markers, isotopic fractionations resembling those generated by biological activity have been detected in Greenland, in BIFs on the island of Akilia (3.87 Ga) and the Isua greenstone belt (3.86 Ga), but it is by no means certain that these are indeed the result of biological activity. The first isotopic markers that appear to be unambiguous are dated to 3.45 Ga and correspond to layers of potential stromatolite fossils (North Pole, Australia). However, the intense hydrothermal alteration undergone by these rocks allows some doubt to persist about the biogenic nature of the isotopic fractionation.

crobial communities consisting of a large variety of bacteria, photosynthetic ones – such as cyanobacteria, which play an important part as primary producers – but also heterotrophs. Fossil stromatolites rarely contain microfossils because, in general, the fine structure of the microbial community's layers is destroyed when the sedimentary rocks are compacted (diagenesis) and subsequently, through the mineralogical transformations that these sediments undergo (metamorphism).

Structures that may be interpreted as fossil stromatolites have been described in the Australian Pilbara craton (in the cherts of the Warrawoona group, dated to 3.5 Ga) and in the Kaapvaal craton in South Africa (in the cherts of the Onverwacht group, in the Barberton greenstone belt, dated to 3.23 Ga). But this interpretation is controversial: these Archaean "stromatolites" could also be the result of abiotic processes. Indeed, because of the physical properties of viscous materials, organized structures with laminations and with a morphology even more complex than stromatolites may be produced naturally under certain conditions; for instance, when a chaotic regime prevails during sediment deposition.

The oldest stromatolites that are indisputable evidence of past microbial life are those in the Fortescue formation, in northwestern Australia, whose age is 2.7 Ga. Given the massive character of these stromatolites and the start of the oxygenation of the atmosphere towards 2.4 Ga, we may assume that cyanobacteria, whose oxygenic photosynthesis is far more efficient than that of anoxygenic photosynthetic organisms, already formed part of these structures. However the most ancient diagnostic biomarkers of cyanobacteria that have been discovered so far, date to only 2.15 Ga.

6

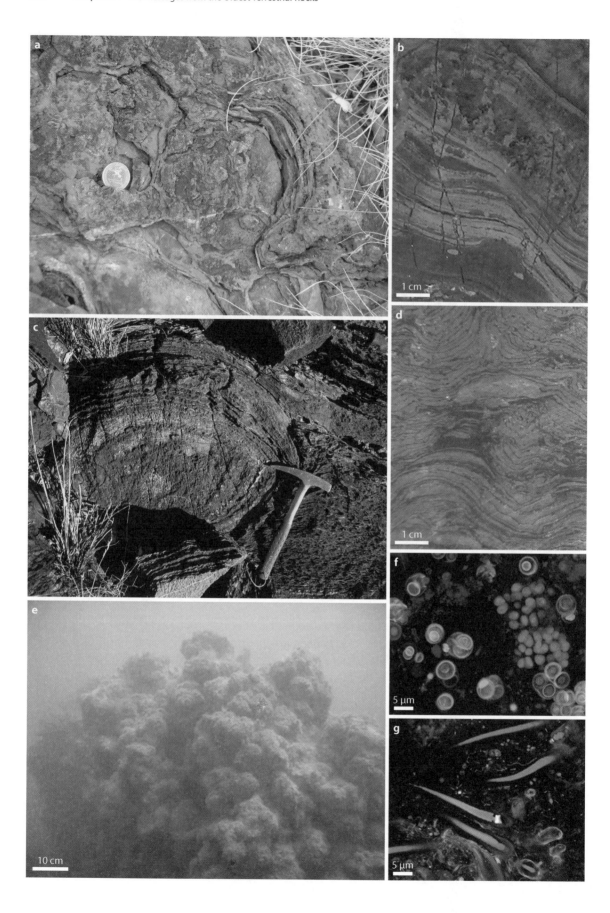

□ **Fig. 6.31 Fossil Archaean stromatolites and modern stromato-lites. a.** The oldest likely stromatolites; they are found at North Pole, in the Pilbara craton in Australia; their age is 3.45 Ga. **b.** Section of a drilling core sample through a layer of well-preserved stromatolites, aged 3.45 Ga; the gold-tinted layers enriched in pyrite (FeS_2) bear witness to an atmosphere devoid of oxygen. **c.** Confirmed stromatolites from the Tumbiana formation, in the Pilbara craton in Australia; they have an age of 2.72 Ga. **d.** Section of a drilling core sample through the well-preserved stromatolites in this formation; this core was taken in 2004, during a Franco-Australian expedition (the Pilbara Drilling Project). **e.** Present-day stromatolite from Alchichica (Mexico). **f.** and **g.** Cyanobacteria from the Alchichica stromatolites observed with a confocal microscope. The red color indicates auto-fluorescence of chlorophyll. Colonies of coccoid cyanobacteria surrounded by a sheath are visible as well as filaments of endolytic cyanobacteria are clearly visible. (Photos: P. López-García and E. Gérard.)

When it comes to microfossils, the work of the geologist William Schopf caused a sensation at the beginning of the 1990s. He announced the discovery, in the Pilbara craton (in the Apex chert formation), of apparently very well preserved microfossils, resembling present-day cyanobacteria, and with ages of 3.5 Ga. But here, once again, strong doubts were voiced at the beginning of the 21st century as to their biological origin. In fact, the geological formations that contain these "microfossils" have been affected by significant hydrothermal activity, such that these structures could just as well have resulted from abiotic processes that took place within hydrothermal veins (□ Box 6.5). Similarly, microstructures discovered in certain rocks at Isua in Greenland (3.8 Ga) and in the Barberton greenstone belt, in South Africa (3.5–3.2 Ga), have been interpreted as microfossils of bacteria. However these microstructures sometimes form almost perfect spheres, which seems suspect: given their very ancient nature and the fragility of microbial structures, any original biological morphologies would have been altered by metamorphism. Indeed, almost identical structures may result from the precipitation of liquid inclusions inside cavities and in a hydrothermal environment.

To avoid artifacts from being interpreted as microfossils on just the basis of morphological considerations, several scientists have suggested that a second criterion should be taken into account, namely that the structure of a microfossil should be intimately associated with organic matter, detectable by Raman spectroscopy *in situ*. Although this suggestion has the merit of clarifying the situation, it is far from being a miracle solution, because organic matter may be produced abiotically at high temperatures (□ Fig. 6.34). So, during the intense hydrothermal alteration to which the Archaean rocks were subjected, it is quite possible that the precipitates that formed into cavities could have absorbed abiotic organic compounds.

Numerous microfossils have been described within the range of time between 3.5 and 1.9 Ga. This last date corresponds to the Gunflint Chert formation in Canada, where the quality of preservation of the microfossils leaves no room for doubt as to their biological origin. As regards the other microstructures, some could have resulted from abiotic processes, whereas others could indeed correspond to genuine microfossils, but it is impossible to decide without taking the local environmental context into account, as well as their association with other fossil biological markers.

More reliable than organic matter detected by Raman spectroscopy, the abiotic origin of which cannot be excluded, there are the "molecular fossils", that is, derivatives of biological macromolecules. In fact, although most biological organic matter is rapidly degraded after cell death, certain macromolecules, notably membrane lipids, are transformed in more stable molecules, which, during fossilization, may persist over time. Sometimes, the "fossil" molecules retain signatures that provide information about the kind of organism that produced them. This is the case with fossil hopanoids, derivatives of bacterial lipids, and steranes, derivatives of the sterols that are typical of eukaryotic cells. However, Archaean metamorphic rocks potentially contain just infinitesimal quantities of these geolipids, and as a result, the risks of contamination are very significant, whether they are of human origin (during coring or during the handling of the samples), or natural (the deposit of lipids resulting from post-Archaean microbial activity). Although molecular fossils derived from biological macromolecules are the only "life traces" that are not ambiguous, it is unfortunately very difficult to avoid these problems of contamination. Indeed, the analyses are carried out on large amounts of rocks, whose preservation is by

▢ Box 6.4   Traces of Life

The search for the oldest signatures of life is a very active research field. These traces may be ambiguous when they may also result from abiotic processes. They may be also diagnostic when they can definitely be attributed to life or to a particular sub-set of living beings (the cyanobacteria, for example). In this case, they are called biomarkers (or bioindicators).

Three types of traces may be left by living organisms, each of which has advantages and/or disadvantages when it comes to interpreting them to infer their true origin.

1. Morphological Structures and Traces (Morphological Fossils)

– Fossils and microfossils derived from tissues and cells
– Imprints produced by the physical alteration of the environment (for example, imprints left by certain soft-bodied animals of the Edicarian fauna when they moved over a marine sediment).
➢ **Disadvantages:** morphological traces of microorganisms are difficult to detect, and they are not very reliable (▢ Box 6.5).

2. Indicators of Metabolic Activity

– **Excreted products**, particularly gaseous ones. For example, the content of oxygen (O_2) in the present-day terrestrial atmosphere results from the accumulation of this gas over millions of years through the activity of bacteria that perform oxygenic photosynthesis (cyanobacteria), where oxygen is produced by the photolysis of water. Before 2.1 Ga, the atmosphere of the Earth did not include significant quantities of oxygen. It started to become oxygenated from about 2.4 Ga.
– **Bio-minerals.** Examples are pyrite (FeS_2) and magnetite (Fe_3O_4). Pyrite is a secondary product of the metabolic activity of some bacteria, which either reduce the sulfate in an environment rich in ferrous iron (Fe^{2+}) or reduce iron oxide (Fe^{3+}) in the presence of sulfur. Magnetite is often the result of an active precipitation within the cells of certain bacteria living in redox transition zones. These magnetite precipitates form intracellular organelle-like structures called magnetosomes, thanks to which cells may orient themselves in a magnetic field. The presence of magnetite in the Martian meteorite ALH84001 was interpreted as a potential proof of the presence of fossil microorganisms in this extraterrestrial object, but it was subsequently shown that identical crystals could be formed under particular abiotic conditions.

– **Isotopic** fractionation of carbon (C), sulfur (S), nitrogen (N), and iron (Fe). They reflect the fact that light isotopes are preferentially mobilized in metabolic reactions, whence their accumulation in organic matter produced by living organisms. These differences are quantified by isotopic ratios ($\delta^{13}C$, $\delta^{34}S$, $\delta^{15}N$, and $\delta^{56}Fe$).
➢ **Disadvantages:** All these indicators, taken individually, could equally well be the result of abiotic processes.

3. Biological Macromolecules (Molecular Fossils)

– **Nucleic acids (DNA or RNA).** RNAs degrade quickly, while DNA may be preserved to several thousands of years, or even more.
– **Proteins.** Proteins are generally considered to be more stable in the long term than nucleic acids, but they do eventually degrade. Few studies of their persistence in the fossil record have been carried out. They have not, in any case, been detected in stromatolites, dated to 2.7 Ga.
– **Polysaccharides,** in particular those that are secreted outside the cell (EPS = Extracellular Polymeric Substances or ExoPolymeric Substances). EPS may persist for a very long time in the fossil record. Their presence has been detected in drill-cores of fossil stromatolites (dated to 2.7 Ga), thanks to X-ray microscopy observations. This method eliminates the risk of contamination by exogenous organic matter; because observations and measures are made *in situ* and at a microscale within the best-preserved portions of the fossil.
– **Lipids.** Some of these are also among the molecules that are best preserved in the fossil record (they may be several billion years old), in particular the hopanoids (derivatives of bacterial lipids) and the steranes (derivatives of sterols typically produced by eukaryotes). Their advantage over the EPS is that sometimes they allow the diagnosis of sub-groups of organisms.
➢ **Advantages:** Biological macromolecules constitute true biomarkers.
➢ **Disadvantages:** The labile nature of most macromolecules, notably the nucleic acids and the proteins, and their lack of preservation in the fossil record; the problems of contamination by extraneous organic matter or by later products.

❏ Box 6.5   The Most Ancient Microfossils: Reality or Artifact?

At the beginning of the 1990s, the geologist William Schopf caused a sensation in the scientific world by describing (in the journal *Science*) extraordinarily well-preserved microfossils, 3.465 Ga years old. These microfossils (❏ Fig. 6.32a), which come from the Apex Chert formation in Warrawoona (in northwestern Australia), strongly resembled the filaments of present-day cyanobacteria (❏ Fig. 6.32b). They were considered proof of the existence of photosynthetic bacteria 3.5 Ga ago.

In 2002, Martin D. Brasier and his collaborators contradicted (in the journal *Nature*) this view of things, by asserting that morphology was not a sufficiently reliable feature to draw conclusions about the biological origin of the observed microfossils. Indeed, in re-examining the same structures that had been analyzed by Schopf and new structures from the Apex Chert, they observed several anomalies; irregular arrangements that are atypical of biological cells, changes in diameter, and branching structures. In addition, Brasier maintained that the fossiliferous samples came from a hydrothermal vein and not from a sedimentary deposit. As a result, all the features assumed to be of biological origin could only be the result of hydrothermal activity. The organic matter associated with the samples ana-

lyzed by Schopf could then have been synthesized by Fischer-Tropsch reactions, which occur at high temperatures (❏ Fig. 6.34). The isotopic carbon fractionation of these samples, which was close to that typical of autotrophic organisms, could be as well the consequence of exchange with hydrothermal fluids. So, according to Brasier, all the fossil traces older than 2.9 Ga are suspect and the default hypothesis that ought to prevail in the absence of conclusive proof is that these structures and the associated traces were produced abiotically.

Juan-Manuel García-Ruiz and his collaborators also stressed the weakness of conclusions based solely on morphological features. In an article published in the journal *Science* in 2003, they demonstrated that it is possible to produce, in the laboratory and in an abiotic manner, mineral structures that resemble amazingly the microfossils described in the scientific literature (for example Schopf's filamentous structures). These deceptive structures are known as bacteriomorphs or biomorphs (❏ Fig. 6.33).

6

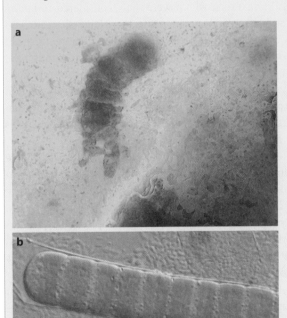

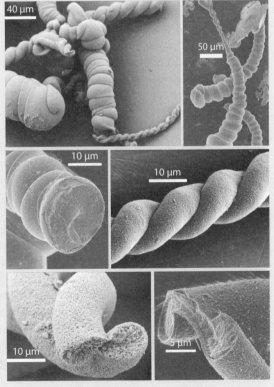

❏ **Fig. 6.32 a. One of the alleged most ancient microfossils; b. a modern filamentous cyanobacteria.** The structure in (a), 3.465 Ga old, has been interpreted as a microfossil of cyanobacteria. This interpretation is controversial. (Photos: M. Brasier (a) and P. López-Gárcia (b).)

❏ **Fig. 6.33 Misleading bacteriomorphs.** These are mineral structures produced in the laboratory in an abiotic manner; they resemble, in an astonishing fashion, recognized microfossils described in the scientific literature. (Photos: J.M. Garcia-Ruiz.)

6

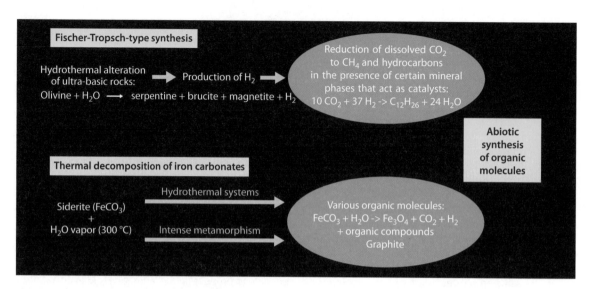

□ **Fig. 6.34 Some reactions that lead to the abiotic formation of organic matter.** The oldest microfossils are ambiguous: they may quite well be, either the structural trace of ancient organisms or the result of an abiotic process. To decide between the two, one can try to determine whether the microfossil is, or is not, associated with organic matter. However, even if the answer is in the affirmative, ambiguity is not totally resolved to the extent that several abiotic reactions may lead to the production of organic matter.

no means homogeneous (presence of fractures, etc.). Thus, scientists are expecting a lot from development of techniques that will allow the study of molecular fossils at a microscopic scale.

The severity of the problem of contamination was recently illustrated in the scientific literature. In 1999, Jochen Brocks and his collaborators published, in the prestigious journal *Science*, the discovery of unambiguous biomarkers, which were then, considered to be the oldest known. These were hopanoids and steranes from the Australian Hamersley (2.6 Ga) and Fortescue (2.715 Ga) formations, and which had been interpreted as genuine traces of cyanobacteria and the first eukaryotes, respectively. Nevertheless, these conclusions have been recently invalidated in the no less distinguished journal *Nature*. Study of the same samples by *in situ* isotopic analysis methods and on a microscopic scale had, in effect, shown that the biomarkers that they contained resulted from contamination that was more recent than 2.7–2.6 Ga. So, at present, the oldest diagnostic biomarkers for cyanobacteria and eukaryotes in the fossil record, go back only as far as 2.15 Ga and 1.7 Ga, respectively.

However, despite these observations and the impossibility of concluding in any reliable manner about the presence of lipids in the Fortescue formation, the fossil stromatolites in this formation are still considered to be the result of microbial activity. First, they are massive and extensive structures, whose origin cannot be easily explained by purely abiotic processes. In addition, very high resolution analysis by spectroscopic and microscopic techniques has revealed, in very well preserved parts, the presence of crystals, nanometric in size, associated with globules of organic matter (more precisely, EPS), which suggests that microorganisms were definitely involved in the formation of these stromatolites.

With regard to isotopes, the ones that have been studied most as "tracers of potential life" are those of carbon, sulphur, nitrogen and, more recently, iron. As far as carbon is concerned, the most informative reservoirs in this context are the carbonates ($\delta^{13}C$ of 0 ‰), the CO_2 released by the mantle degassing ($\delta^{13}C$ of -5 ‰), and the derivatives of organic matter of biological origin ($\delta^{13}C$ about -25 ‰). The $\delta^{13}C$ of carbonaceous material present in the Archaean cherts that has been analyzed is about -35 to -30 ‰, and it has been suggested that photosynthetic and methanogenic organisms might be the source of this fractionation. In the case of sulfur, sulfate-reducing bacteria produce sulfides with a $\delta^{34}S$ of -10 to -40 ‰, whereas in volcanic rocks the $\delta^{34}S$ is 0 ‰ ± 5. Several authors considered that the low values of $\delta^{34}S$ in some Archaean sedimentary sulfide deposits would have formed by biological reduction of sulphur or sulfates. More recently, values of $\delta^{34}S$ measured in the well-preserved sedimentary rocks of the Dresser formation, aged 3.5 Ga and located at Pilbara, have been interpreted as the result of the disproportionation of S^0 by bacteria. This process consists of using the S^0 simultaneously as a donor and acceptor of electrons. The result is that, starting with two molecules of S^0, reduced (S^{2-}, sulfide) and oxidized (SO_4^{2-}, sulfate) sulphur species are formed liberating energy that may be utilized by the cell. This interpretation is, nevertheless, disputed, at least as far as the nature of the microbial activity leading to this isotopic fractionation is concerned. Regarding nitrogen, the value of $\delta^{15}N$ measured in the kerogen in certain Archaean sedimentary rocks is lower (about 5 ‰) than the average for the present-day biosphere. Some consider this isotopic signature as a clue of the activity of nitrogen-fixing bacteria, nitrifying or denitrifying bacteria, or yet again that of deep-sea vent chemoautotrophic microorganisms using the NH_3 dissolved in hydrothermal fluids. Other researchers, however, explain these results by abiotic processes, they invoke nitrogen originating in the mantle and subsequently affected by metamorphic processes, which modified the overall chemical composition of the rocks (metasomatism).

In fact, all the isotopic signatures that we have just mentioned could quite well be the result of abiotic processes that have affected Archaean sediments, such as, in particular, their alteration by hydrothermal fluids and high-temperature metamorphism. There is also the risk of contamination by more recent bio-organic material. The interpretation of these isotopic indicators is therefore the subject of bitter discussion. Let us give one example here.

The oldest clues of potential life correspond to the fractionation of carbon isotopes that have been measured from highly metamorphosed sedimentary deposits from southwestern Greenland. Graphite inclusions characterized by very low $\delta^{13}C$ values have been discovered in BIFs on the island of Akilia (aged of 3.87 Ga; ◘ Fig. 6.35b) as well as in several outcrops in the Isua belt (dated to around 3.86 Ga).

Geologists do not yet agree on the origin of these graphite-bearing rocks. Some put forward arguments in favor of a sedimentary origin, while others think that, on the contrary, they are volcanic rocks, which were subsequently, strongly modified by metasomatic processes. The origin of the graphite is equally at the heart of the debate. The $\delta^{13}C$ of the graphite inclusions from Akilia and, above all, from Isua, could be the isotopic signature of a biological activity. The intense metamorphism undergone by the rocks, with exposure to temperatures of about 500–600 °C, could have led to the transformation of all the organic matter of biological origin into crystalline graphite with a very low $\delta^{13}C$ value. But although this explanation is very seductive, it is equally perfectly possible that at least a large part of these graphite inclusions with low $\delta^{13}C$, are the result of contamination by younger (post-metamorphic) organic matter or synthesized by thermal decomposition of iron carbonates during high-grade metamorphism, or by both means. Only graphite inclusions in rocks devoid of iron carbonates might possibly bear witness to genuine biological activity around 3.8 Ga.

In fact, most of the rocks dated between 2 and 3.5 Ga, display isotopic anomalies (at least for carbon, the most investigated) that are consistent with a biological origin. But they are also equally affected by a significant hydrothermal metamorphism, such that these isotopic anomalies could also have an abiotic origin. Once again, the more often one observes, in a rock, different kinds of fossil traces that are potentially biological, the greater the probability that these traces are really of biological origin. But this remains a probability. What a lot of uncertainties there are, therefore, about whether our planet was inhabited between 3.8 and 2.7 Ga!

Can we really not risk a prediction? To do so, before finishing our trip to the Archaean Earth, let us turn once more to the old continental basement of the Pilbara craton, in Australia. There we find, in the cherts of the Warrawoona group, aged 3.5 Ga, several potential traces of life: fossil stromatolites; fossils of microorganisms, of which some were perhaps photosynthetic; isotopic markers of C, S, and N, suggesting the presence of microorganism with various metabolisms. Even if each of these signatures is arguable and may be ex-

⬛ **Fig. 6.35 Outcrops of Archaean sedimentary rocks from which potential traces of life have been described: in Greenland (left) and in Australia (right). a.** *Iron Mountain*, at the extreme northeast of the Isua supracrustal belt (3.8 Ga). This hill, consisting of BIF is one of the oldest sedimentary deposits known on Earth. **b.** Rock rich in quartz and pyroxene from the island of Akilia (3.8–3.9 Ga). **c.** Metacarbonate veins in a basic rock from the Isua belt. They contain graphite inclusions, whose isotopic anomalies are regarded as the oldest known evidences of life (3.8 Ga); this interpretation is very controversial. **d.** The Warrawoona region (Pilbara craton) from which the oldest potential microfossils have been described (3.3–3.5 Ga). **e.** and **f.** Stromatolite-type formations aged 3.45 Ga at North Pole (Pilbara craton). (Photos: M. van Zuilen and P. López-García.)

plained, as we have seen, by abiotic processes, the frequency and variety of these traces (and in particular the diversity in the morphology of the stromatolites) form a bundle of arguments that speak in favor of the probable presence of life, even though it may be difficult to demonstrate formally.

A Planet Where Life Diversifies

Chapter 7

From 2.50 to 0.50 Ga

It is from 2.5 billion years before the present, that the Earth progressively lost all its archaic features and that it became totally modern. In particular, the atmosphere became enriched in oxygen (O_2) and life spread into every ecological niche. Despite some limited perturbing events (glaciations, meteorite falls, etc.), life diversified until the "Cambrian explosion", which, 540 million years ago, marked the appearance of "modern" groups of animals.

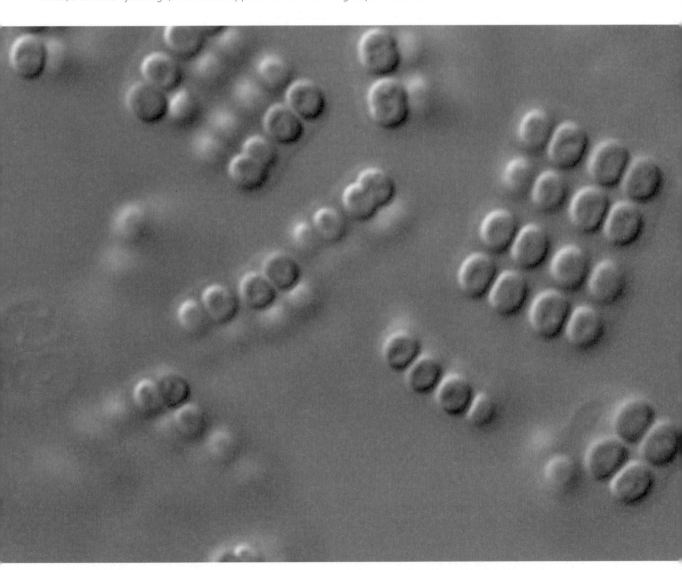

▣ Colonial coccoid cyanobacteria observed through a microscope. It was photosynthesis carried out by cyanobacteria from 2.7 Ga (and perhaps even earlier) that enriched the Earth atmosphere in oxygen between 2.4 and 2.0 Ga. (Photo: López-García.)

M. Gargaud et al., *Young Sun, Early Earth and the Origins of Life*,
DOI 10.1007/978-3-642-22552-9_7, © Springer-Verlag Berlin Heidelberg 2012

In the preceding chapters we have seen how the Archaean Earth differed significantly from the planet that we know today, in particular in many fields of the geological record: the nature of the rocks, the thermal regime, the type of tectonics, the geodynamics, etc. Between the Archaean and today, the atmosphere has also undergone a genuine revolution. It was rapidly enriched in one component of fundamental importance for the evolution of living beings, molecular oxygen (O_2). One of the questions that then arises is to know when and how this change to our present-day Earth came about. This chapter, which is the last stage in our temporal exploration of the planet, will provide some elements of a response to that question. In doing so, it will also allow us to discover the major stages in the evolution of life between 2.5 Ga and the dawn of the Palaeozoic era with, in particular, the sudden appearance of the first animals in the fossil record. The tour will stop 540 million years ago, just before the Cambrian explosion, a major evolutionary radiation, after which all the main present-day branches of life will have emerged.

7

From the Primitive Earth to the Modern Earth

With regard to terrestrial geodynamics, it is very difficult to state exactly the date when a major change occurred. Indeed, in this field, changes are rarely sudden; on the contrary, they operate progressively over several hundreds of millions of years and, above all, they are not synchronous all over the Earth. For example, we have seen that a feature of the Archaean was the production of enormous volumes of TTG magma. After 2.5 Ga, this activity decreased dramatically, but it did not cease entirely, and even today, still, under exceptional conditions, such as the subduction of a mid-ocean ridge, very small volumes of magma similar to TTGs are formed (the adakites). In other words, what geologist call "change" is not a single event when one geological mechanism is replaced by another, but a period when, without ceasing completely, the mechanism becomes less important, then rare, and finally exceptional.

On our planet, most of the petrogenetic and geodynamic changes are the result of the progressive, and inexorable, cooling of the internal Earth. For example, because it cools, our planet is not hot enough to allow high degrees of mantle melting. So komatiite volcanism has disappeared. And in the absence of these very dense ultrabasic lavas, the vertical tectonics – the sagduction – typical of the Archaean Earth has also disappeared. Similarly the "hot subduction", which allowed the melting of the subducted basalts before they start dehydrating, thus giving rise to the TTG, cedes its place to "cold subduction" where the calc-alkaline magmatism (granodioritic) arises from the melting of the metasomatized mantle wedge. Because the Earth had less internal heat to release, the length of the mid-ocean ridges decreased, resulting not only in an increase in the size of the lithospheric plates, but also in a decrease in hydrothermal activity.

These changes were not synchronous, far from it, but all took place at about 2.5 Ga, a date that marks the end of the Archaean and the beginning of the Proterozoic. The same applies to the Wilson cycle and to the supercontinent cycle: it is very probable that they started to be operative in the Archaean, perhaps from 3 Ga.

However, although it seems reasonably easy to assemble a single supercontinent from a small number of large plates, it appears much more difficult to achieve this with a large number of small plates. This is why it seems rationale to think that, even if large size continents could have formed in the Archaean, the cycle of single, vast and global supercontinents really started either at the very end of the Archaean, or at the beginning of the Proterozoic.

Birth, Life and Death of an Ocean: the Wilson Cycle

The present-day Earth consists, as we may remind ourselves, of rigid lithospheric plates that move above the ductile portion of the mantle known as the asthenosphere. Today, the rate of plate motion is typically of about 1 to 15 centimeters per year. We have mentioned several times that one of the features that distinguishes the oceanic crust from the continental crust is its relatively short lifetime (< 180 Ma) and the fact that it is recycled into the mantle, on average 60 Ma after its genesis in a mid-ocean ridge system. This is from that observation that the notion of a cycle has been elaborated: the oceanic crust is extracted from the mantle, by melting at divergent plate boundaries (mid-ocean ridges), and returns into the mantle in subduction zones (convergent plate boundaries). Thus, it sounds perfectly logical that, on a planet the surface of which must be considered as being constant, if the surface of the ocean floor increases at one point, it must decrease at another. Things become a bit more complicated when the continental crust is taken into account. Its low density ($d = 2.75$) endows it with great buoyancy, which prevents any significant recycling into the mantle. So, when two continental blocks confront one another at a subduction zone, they overthrust. A thickening of continental crust results, that is to say, the erection of a mountain chain and also, most often, a halt to subduction.

The Wilson cycle describes, in five major stages, this succession of divergent and then convergent motions (◘ Fig. 7.1).

Fragmentation of a Continent

The first stage of the Wilson cycle is the fragmentation of a continent. It starts with a thermal anomaly (that is, an increase in the thermal flux), the origin of which is still controversial. It involves the ascent of hot asthenosphere beneath the continent; either as a hot spot or along the ascending part of a convection cell (◘ Box 3.1).

The continental crust can only transfer this heat from its base to its surface by conduction, which is a highly inefficient mechanism. In other words, the continental crust acts as a thermal screen, leading to the heat accumulation beneath the continent. The increase in temperature has the effect that a part of the lithospheric mantle, which was rigid, becomes ductile, and transforms *de facto* into asthenosphere, decreasing the thickness of the lithosphere by that amount.

In parallel, the rising asthenosphere causes a camber and uplift of the continental crust, which stretches and thins through the effect of a series of normal faults (◘ Fig. 7.1a). The latter form the boundaries of blocks that collapse, thus giving rise to a valley known as a rift. The continental crust becomes thinner and thinner and the tilting of the blocks causes the progressive divergence of the rift edges. The adiabatic ascent (without the addition of heat) of hot peridotites from the asthenospheric mantle induces their melting, thus generating alkali basaltic magmas. The latter are generally injected into the faults, which also plays a part in widening the rift and starting the separation of the plates.

This situation is currently observed in the Great East-African Rift Valley, which extends for nearly 10 000 km from Ethiopia to Mozambique.

Birth of an Ocean and Sea Floor Spreading

The continental crust thins and parts to such an extent that it finally disappears. The bottom of the rift then consists solely of the basalts resulting from the melting of the asthenosphere

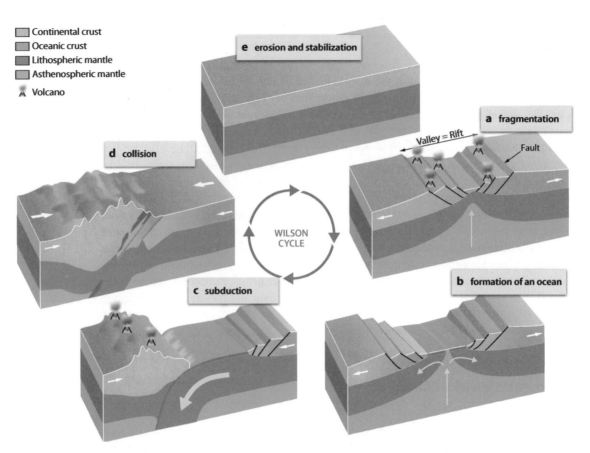

Continental crust
Oceanic crust
Lithospheric mantle
Asthenospheric mantle
🜂 **Volcano**

e erosion and stabilization

a fragmentation

Valley = Rift

Fault

d collision

WILSON
CYCLE

c subduction

b formation of an ocean

◘ **Fig. 7.1 A schematic summary of the Wilson cycle. a. Fragmentation of the continental crust** is caused by the ascent of a hot asthenospheric diapir: the crust arches upwards and thins through the effect of a series of normal faults. Partial melting of the asthenosphere generates alkaline basaltic magmas. **b. Birth of an ocean and sea floor spreading:** the continental crust thins so much that it disappears. It is replaced by basalts resulting from melting of the asthenospheric mantle, which forms a new oceanic crust. A new ocean is born. **c. Subduction, disappearance of an ocean:** the oceanic crust cools as it moves away from the mid-ocean ridge. Its density in- creases to the point where it becomes greater than that of the asthenosphere into which it then sinks, giving rise to a subduction zone. **d. Collision, formation of mountain ranges:** once the oceanic crust has completely disappeared in the subduction zone, the continental blocks collide. Because of their low density they cannot be recycled into the mantle. They stack up on top of one another, thus forming a mountain chain. **e. Erosion and stabilization:** once relief has been established, erosion starts to destroy it, giving birth to a new, stable, continental block.

(◘ Fig. 7.1b). These new basic magmatic rocks correspond to a new oceanic crust. Being hot and therefore with a relatively low density, it "floats" on the asthenosphere. The persistence of the rising asthenosphere (the ascending branch of a mantle convection cell) results in an on-going volcanism and thus in the continuous formation of oceanic crust. The rift has become a mid-ocean ridge and the ocean expands bit by bit.

The present-day Red Sea corresponds to such a setting: it is a very young ocean.

Disappearance of an Ocean: Subduction

The oceanic crust formed at the mid-ocean ridges gradually cools in contact with cold sea- water. Its density thus increases progressively and the oceanic lithosphere gradually sinks into the asthenospheric mantle. The depth of the oceans is approximately 2500 meters at the mid-ocean ridges but reaches 5500 meters well away from them (◘ Fig. 7.3a). Inevitably, the density of the oceanic lithosphere becomes greater than that of the asthenospheric mantle on which it floats. It then starts to plunge into the mantle: a subduction zone has formed (■ Fig. 7.1c). From being a passive margin, the ocean-continent boundary has become an active margin where seismic and magmatic activity are very significant. Here, the oceanic lithosphere is dragged under the continental lithosphere and the movement of the plates is convergent, which may lead to the erection of high relief on the continental block.

Today, the Andes, as well as the whole edge of the Pacific Ocean (Pacific Ring of Fire), correspond to very active subduction zones.

The Collision: Formation of Mountain Ranges

After some time, the oceanic crust will be completely consumed by subduction and will have disappeared, being totally recycled into the mantle. The two continental blocks previously separated by this oceanic crust, will now collide (■ Fig. 7.1d). Because of its low density, the continental crust cannot be recycled into the mantle. So the pieces of continental crust will stack up, thus resulting in a thickening of the continental crust and the formation of mountain ranges.

The Alps and Himalaya are present-day examples of mountain chains erected by the collision of two continental blocs.

Erosion and Stabilization

Once relief has formed, erosion starts to destroy it. So if the movements of compression that were active disappear, mountain chains such as the Alps will be totally eroded (becoming a peneplain) in less than 5 Ma. A new, continental block will eventually be stabilized (■ Fig. 7.1e), under which heat may accumulate, causing it to fracture and beginning a new cycle.

Ephemeral Giants: the Supercontinents and their Cycle

One of the consequences of the Wilson cycle is the formation of supercontinents. Indeed, our planet being spherical, it is inevitable that two plates that move away from one another on one side, must approach one another on the other, and if both bear a continental crust, the situation will inevitably lead to a collision. So, all the continental blocks scattered across the planet will get progressively closer to one another and collide, accreting together, thus resulting in a single and unique continental block, known as a supercontinent. On average between 300 and 500 Ma elapses, from the fragmentation of one supercontinent until the end of the accretion of a new supercontinent.

The aggregation of the last supercontinent ended 280 Ma ago: it is known as Pangaea (■ Fig. 7.2). Around 245 Ma ago, it began to break up, starting with the opening of the Atlantic Ocean and the migration of India from Madagascar towards Asia. The process is still continuing today with the opening of the Red Sea and the great East-African Rift valley.

The preceding supercontinent was Pannotia, also referred to as the Vendian supercontinent, which existed between 600 and 540 Ma. Before it, Rodinia formed about 1.1 Ga and started to break up around 900 Ma. And if we go even farther back in time, we find the supercontinents of Columbia (1.8 to 1.5 Ga), Kernorland (2.7 to 2.1 Ga) and even, in the Archaean, Vaalbara (3.1 to 2.8 Ga). It is obvious that the farther we go back to ancient periods, the harder it is to specify the shape, the number and the size of supercontinents. By way of example, the Vaalbara was probably smaller in size than Australia today: so it would be more a continent than a supercontinent.

7

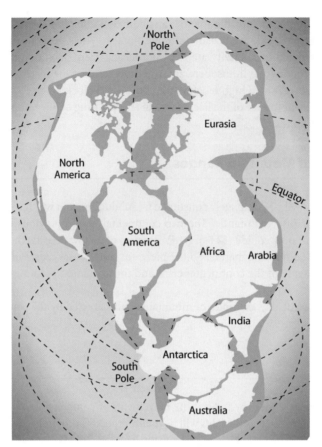

▣ **Fig. 7.2 The reconstruction of the Pangaea supercontinent, 280 Ma ago (during the Permian).** Pangaea started to break up about 245 Ma ago. The distribution of the continents that we know today resulted from that event. The area in a khaki color represents the continental shelf (the submerged portions of the continents). (After Windley, 1984.)

The Crucial Consequences of the Supercontinent Cycle for the Earth's Environment

Whether the continental masses are grouped together in a single block (supercontinent) or spread across the surface of the globe will have important consequences for the environment and the conditions that will prevail on the surface of the Earth.

The supercontinent cycle does indeed have a major influence on sea level. In the Wilson cycle, the accretion of the supercontinents takes place at the end of a cycle, when the oceanic crust is old, cold, and thus dense, which decreases its buoyancy. Consequently, the oceans are deeper over an old oceanic crust (▣ Fig. 7.3a); and therefore, for the same surface area, deeper oceans will contain more water. In other words, the global level of the seas will decrease, leading to the emersion of most of the continental shelf. The surface of emerged lands will thus increase considerably. On the other hand, during a period of continental break-up, the oceanic crust is young and hot such that the oceans are shallower, and the sea level is higher, which drowns a large portion of the low-lying areas of the continental masses. The land surface is reduced and the continental shelf then covers very large areas (▣ Fig. 7.3b).

Currently, more than 85% of the oceans' biomass is concentrated on the continental shelf. The periods when supercontinents exist, which correspond to a lower sea level and to an almost completely emerged continental shelf, are thus potentially less favorable for the proliferation and expansion of life. This effect is further exacerbated by the climatic consequences resulting from the accretion of the supercontinents, which favors the development of con-

◘ **Fig. 7.3 a. The depth of the oceans as a function of the age of the ocean floor; b. the link between the depth of the oceans and the surface area of emerged land.** A young oceanic lithosphere is hot, light and buoyant, it cannot sink deep into the asthenosphere. This results in shallow oceans whose mass of water covers the continental shelves, correspondingly reducing the emerged surface of the continents. On the other hand, an old oceanic lithosphere is cold and dense, and it sinks into the asthenosphere. The oceans are then deeper; the overall sea level decreases, leading to the emersion of most of the continental shelf: the emerged surface area of the continents increases. (a: After data by Parson and Sclater, 1977.)

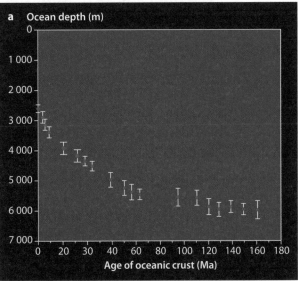

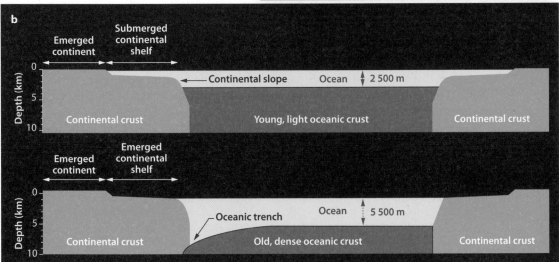

tinental climates, with, in particular, very cold winters that promote the formation of polar caps. The latter trap and fix part of the water on the continents as ice and so contribute to the lowering of sea level. Conversely, during splitting of the continents, maritime climates, which are milder, predominate, thus causing melting of the ice sheets and raising sea level.

The supercontinent cycle will also play a very important role in the evolution of living beings. Indeed, the selection pressure is greater in isolated environments. When all the continents are gathered into a single supercontinent, the possibilities of isolation are low, which results is a lower degree of diversity. On the other hand, when the number of independent continental masses is greater, each of them forms an isolated environment where both the evolution and the diversification can occur regardless of what is happening on the other continents. In addition, the displacement of the continental masses on the surface of the globe is able to cause climatic changes – for example when a continent migrates from a pole towards the equator – which induces an additional selection pressure. The work of palaeontologists has shown that there is a close correlation between the number of families living on the surface of our planet and the supercontinent cycle.

The Appearance of Atmospheric Oxygen: a Revolutionary Event!

The events associated with the Wilson cycle and the cycle of supercontinents are, inherently ... cyclic! However, there are others that have occurred only once throughout the history of our planet. This is particularly the case with the appearance of oxygen in the atmosphere. This event, also known as the Great Oxidation Event (GOE) took place between 2.4 and 2.0 Ga. It corresponds to an overall change in the composition of the Earth's atmosphere, which, from being reducing, became oxidizing ... a genuine revolution whose consequences were crucial for life!

The Great Changes

From a geological point of view, the Great Oxidation Event is vouched for by four major events: the appearance of oxidized palaeosols; the disappearance of the banded iron formations (BIF) and of the uraninite deposits, as well as the appearance of a mass-dependent fractionation of sulfur isotopes..

The Disappearance of BIFs

The last event of massive BIF production took place between 2.2 and 2.0 Ga, even though a few of these formations were sporadically able to form until 1.8 Ga. We saw, in the last chapter, that the BIFs arose from the local precipitation of iron dissolved in water. And yet this iron is mainly released by the alteration of surface rocks and leaching of the emerged continents. In order to reach the oceans, the iron must therefore be mobile. As it is soluble in its reduced form Fe^{2+} and absolutely insoluble in its oxidized form Fe^{3+}, the BIFs bear witness to the reducing nature of the atmosphere and the oceans in the Archaean. Reciprocally, their disappearance testifies that both the atmosphere and the ocean became oxidizing.

The Appearance of Oxidized Palaeosols

Red palaeosols, similar to the lateritic soils of tropical regions today (◘ Fig. 7.4) appeared on a massive scale at 2.2 Ga. Their color is caused by the presence of iron oxides and hydroxides

◘ **Fig. 7.4 General view (a) and detail (b) of the area near Uluru (central Australia).** The red color is due to iron oxides and hydroxides, minerals where the iron is in the Fe^{3+} state, because of its oxidation by the atmospheric oxygen. In this oxidized state, iron is not soluble in water, it is therefore not very mobile or is immobile and remains in place. (Photos: H. Martin.)

(haematite Fe_2O_3, goethite $FeO(OH)$, ferrihydrite $5Fe_2O_3 \cdot 9H_2O$, etc.). In all these minerals, the oxidation state of iron is Fe^{3+}, which proves, once again, the oxidizing nature of the atmosphere and the presence of atmospheric dioxygen. The 2.2 Ga old red palaeosols bear witness to a partial pressure of atmospheric dioxygen that ranged between 1 and 10 mbar as a minimum (the current value being 210 mbar).

The Disappearance of Uraninite Deposits

The most important oxidation states of uranium are U^{4+} and U^{6+}. When it is in its reduced state (UO_2), uranium is poorly soluble in water and precipitates, then forming deposits of a mineral known as uraninite (UO_2). Contrastingly, in its oxidized state (UO_3), uranium is very soluble, such that it remains dissolved in water. Uraninite deposits, abundant before 2.2 Ga, disappeared almost completely after that date, which, once again, corroborates that a rapid change in the oxidation state of the atmosphere and the ocean took place at that time.

The Modification to the Sulfur Isotopic Fractionation

Today, in an oxygen-rich atmosphere, the isotopes 33 and 34 of sulfur (^{33}S and ^{34}S) undergo a mass-dependent fractionation, that takes place during unidirectional kinetic processes or during equilibrium processes. The mass-dependent isotope fractionation law is: $\delta^{33}S = 0.515 \times \delta^{34}S$. However, the reactions for the sulfur dioxide (SO_2) photolysis under ultraviolet radiation with a wavelength less than 310 nm are likely to induce a mass-independent isotopic fractionation. Except for major volcanic eruptions, which eject gas as far as the stratosphere, the bulk of atmospheric SO_2 is found in the troposphere. Today, stratospheric ozone (O_3) protects the troposphere from UV radiation, such that photolysis of SO_2 cannot be achieved. Consequently, the only possible fractionation for sulfur isotopes is thus linked to kinetic and equilibrium processes that are mass-dependent. Conversely, in the absence of the ozone layer, the lower atmosphere is not protected from UV radiation and the latter performs SO_2 photolysis; which fractionates sulfur isotopes regardless of their mass. The formation of ozone requires partial pressures of dioxygen higher than 5×10^{-6} mbar.

Researchers measure the difference between an analyzed sample and the mass-dependent fractionation law by the parameter $\Delta^{33}S = \delta^{33}S - 0.515 \times \delta^{34}S$; the value of $\Delta^{33}S$ is equal to zero when the fractionation is mass-dependent, and differs from it in case of mass-independent fractionation. The sulfur analyzed in all sediments whose ages range between 3.8 and 2.32 Ga shows a mass-independent fractionation ($\Delta^{33}S \neq 0$; ◩ Fig. 7.5), which implies the absence of stratospheric ozone and a partial pressure of dioxygen below 5×10^{-6} mbar. After 2.32 Ga, the sulfur fractionation becomes mass-dependent ($\Delta^{33}S \approx 0$; ◩ Fig. 7.5), indicating the existence of an oxidizing (oxygen-bearing) atmosphere.

The Delay in the Oxygenation of the Atmosphere: Some Possible Explanations

The question that naturally arises is that concerning the origin of the dioxygen that caused the drastic compositional change of the atmosphere at 2.4 Ga. The only mechanism capable of producing large volumes of dioxygen is oxygenic photosynthesis, according to a reaction as follows:

$$2H_2O \;+\; CO_2 \;+\; 8h\nu \;\rightarrow\; CH_2O \;+\; H_2O \;+\; O_2$$

water   carbon   light   organic   water   dioxygen
            dioxide   matter

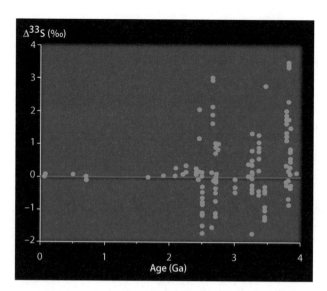

■ Fig. 7.5 Diagram showing the evolution of $\Delta^{33}S$ (= $\delta^{33}S$ − 0.515 × $\delta^{34}S$) in sediments as a function of time. Before 2.2 Ga, $\Delta^{33}S$ is very clearly different from zero, which indicates a mass-independent fractionation, a marker of the absence of stratospheric ozone and thus of an oxygen-depleted atmosphere (partial pressure of $O_2 < 5 \times 10^{-6}$ mbar). After 2.2 Ga, $\Delta^{33}S$ is equal to zero, a value typical of a mass-dependent fractionation, thus showing the existence of an ozone layer and therefore of an oxygen-rich atmosphere. (After Farquhar and Wing, 2003.)

However, this process is closely linked to the existence of living beings such as cyanobacteria and eukaryotes. In other words, it was the oxygenic photosynthesis carried out by these organisms that progressively enriched the atmosphere in dioxygen. This apparently, poses a small chronological problem…

Indeed, as we have explained in the previous chapter, life appeared on Earth well before 2.4 Ga, perhaps even as early as 3.8 Ga. Although eukaryotes only emerged around about 2.0 Ga, the Archaean traces of fossil life include stromatolites, whose complex structures attest, at least with some of them, the presence of cyanobacteria at 2.7 Ga. We have also mentioned in Chap. 6 the fact that the Archaean BIFs vouch for the existence of oxidizing oases in an overall reducing ocean that allowed the precipitation of dissolved iron and which could be linked to the presence of cyanobacteria. So everything appears to indicate that there is not a direct temporal relationship between the appearance of living organisms producers of oxygen and the Great Oxidation Event. How can we explain this delay in the atmospheric oxygenation? Here again, several mechanisms may be invoked.

Organic Carbon Should be Efficiently Sequestrated After the Death of Organisms

Living beings that produce oxygen also consume it, either by their respiration when it is aerobic (absorption of oxygen and release of carbon dioxide), or after death, when they decompose (the reduced carbon in the living organism will then oxidize). In fact, the atmospheric oxygen content can increase only if the carbon produced by living matter is isolated from any contact with atmosphere dioxygen. This means that this carbon must be buried, which is mainly the case today, in the shallow waters of the continental shelf. It is quite possible that the change in tectonic style that occurred after 2.5 Ga, with the appearance of plates whose size and thickness were similar to present day ones; would have favored the formation of large continental shelves on passive margins. Furthermore, at 2.1 Ga, the fragmentation of the Kernorland supercontinent was being completed. Thus the period was favorable to a general rise of sea levels, which increased the shoreline length and led to the formation of large continental shelves.

Iron Dissolved in the Ocean Must be Totally Oxidized

During the Archaean, the oceans were very rich in dissolved iron, which consequently, was reduced iron (Fe^{2+}). The oxygen released by cyanobacteria was initially consumed in the oxidation of the surface layer of the ocean. Subsequently, thanks to the oceanic circulation, the latter would progressively mix with the deeper layers, which were still reducing. This would lead to the iron oxidation and would cause its massive precipitation. This process, which consumed the oxygen dissolved in the oceans, may be able to account for the existence of the huge Banded Iron Formations that are characteristic of the Late Archaean and Early Proterozoic. In this scenario, the dioxygen would have begun to accumulate in the atmosphere only after that the whole iron dissolved in the oceans would have been oxidized.

Oxygen Trapped in the Mantle

Water injected into the mantle in the subduction zones returns to the atmosphere through volcanoes. In the case of the Archaean mantle, which was more reducing than today, this water could have been used to oxidize the mantle iron (the upper mantle contains about 6% iron), so volcanoes would have only released dihydrogen. This volcanic hydrogen reacted with atmospheric oxygen to give water. Some authors consider that the upper mantle oxidation was not completed before about 2.3 Ga, it was only later that the accumulation of atmospheric dioxygen would have started. However, no matter how seductive this theory may be, it is not consistent with the fact that lavas formed from the mantle and erupted since 3.8 Ga have the same state of oxidation as present-day lavas.

7

The Sun Needs to be "Hot Enough"

The amount of dioxygen produced is proportional to the abundance of eukaryotes and cyanobacteria. It is possible that cyanobacteria could not have proliferated to such an extent as to enrich the atmosphere in dioxygen before 2.4–2.0 Ga, and this for two reasons. At present, such bacteria develop in shallow, well-lit waters, which is on the continental shelf. Yet, the surface area of the continental shelves increased considerably after 2.5–2.3 Ga.

In addition, the luminosity of the Sun was much weaker in the Archaean and Proterozoic than it is today (☐ Box 6.2). The amount of solar energy that reached the Earth's surface was thus much less than that we receive today. Under such conditions, the Earth's surface would have been completely frozen, while the geological data demonstrate the opposite. We have, for example, seen in Chap. 6 that the temperature of the Archaean ocean was higher than today (> 50 °C, ☐ Fig. 6.22), and that climatic models estimate the temperature of the Earth's surface as being approximately 30 °C at 2.9 Ga (☐ Box 6.2). The only way of reconciling a cold Sun with a warm Earth surface is to invoke a significant greenhouse effect, which would have preserved the Earth from a global glaciation. Among the ideas put forward, it has been suggested that this greenhouse effect could have been caused by the presence of atmospheric methane (10^{-4} bars), possibly resulting from the activity of methanogenic organisms. But although methanogenic organisms proliferate in anoxic environments, dioxygen is highly toxic to them. So, the increase in the O_2 atmospheric content linked to the activity of cyanobacteria would be accompanied by a rarefaction of methanogenic organisms, and thus of a decrease in the atmospheric methane content. In other words, oxidation of the atmosphere would result in a decrease in the greenhouse effect and thus of the temperature. Cyanobacteria, which live in shallow waters, would have been more sensitive to this cooling such that their population would have strongly diminished. Their decreasing activity would cause a lowering of the atmosphere oxygenation; this latter becoming again favorable to the development of methanogenic organisms, etc.

If this scenario is correct, the Earth must have undergone an alternating cycle of glacial and warm periods. However, from about 2 Ga, the solar luminosity would have reached a critical threshold such that the presence of atmospheric methane would no longer be necessary to prevent the periodic freezing of the Earth's surface. Cyanobacteria would then have been able to proliferate and to irreversibly enrich the atmosphere in O_2.

Disruptive Events

Alongside the "revolutionary episode" that was the appearance of atmospheric oxygen, the Earth's history is full of events whose occurrence is unpredictable, but which could have had a major impact on the living beings, their survival and their development. These events could have been meteoritic falls or, to a lesser extent, global glaciations. We shall give here, a glimpse of the main disruptions of this kind that affected the Earth during the Proterozoic.

Glaciations and the "Snowball Earth"

During the Proterozoic, it seems that our planet underwent repeated episodes of glaciation, which even went so far as global glaciation (known as the "Snowball Earth"). The mechanism is simple. It always starts with a cooling, the causes of which may be multiple. This may be a temporary decrease in solar activity. It is equally possible that the aggregation of a supercontinent, at tropical latitudes where precipitation is abundant, favored the weathering of silicates and thus the trapping of CO_2 in carbonates (◘ Fig. 7.6a, *see also* ▶ Chap. 3). This pumping of atmospheric CO_2 lowered the greenhouse effect, which caused a drop in surface temperature. The presence of a supercontinent also favored the development of a continental climate with cold winters. The consequences were the formation of ice caps. However, icy surfaces, which are white, reflect a greater amount of the incoming solar energy (increasing the albedo), thus contributing to the lowering of the temperature and to the development of even larger ice caps, and so on.

Once initiated, the cooling would thus accelerate and the Earth would be completely frozen. However, a "Snowball Earth" is not a stable situation.

In fact, beneath the ice cap, the alteration of silicates cannot proceed and ceases and with it the sequestration of atmospheric CO_2 in carbonates (◘ Fig. 7.6b). Yet, the atmosphere continues to be fed by the CO_2 released by the volcanoes; CO_2 which, for lack of a mechanism to trap it, accumulates there. The greenhouse effect then increases and the planet warms up again, causing the ice caps to melt and leading ineluctably to the end of the glacial period. As soon as the ice sheets melt, the newly exposed continents are again subjected to silicate weathering, such that the huge amounts of CO_2 that have accumulated in the atmosphere are massively trapped in carbonates. This is the reason why, any significant glacial period is immediately followed by the deposition of thick carbonate formations.

Obviously, such major glaciation episodes would have significant repercussions on the development of living beings, and that they might even have caused mass extinctions. However, the very fragmentary nature of the Proterozoic fossil record makes it very difficult to know what were the exact consequences of these glaciations on life.

Major Glaciations of the Past

The oldest traces of glaciations recorded at middle latitudes, date from the Late Archaean, indeed, they are found in South-Africa in a sedimentary formation dated to 2.9 Ga and called the Pongola supergroup. The latter contains diamictites of glacial origin[1] as well as striated rocks or pebbles. Unfortunately, only little work has focused on this formation and, above all, it has not been possible to establish whether or not this glaciation affected the whole Earth.

---

1. The term diamictite describes sedimentary terrigenous rocks consisting of heterogeneous fragments that are poorly to non-sorted. This term is purely descriptive. However, its use is often restricted to glacial formations, then it becomes equivalent to the term tillite.

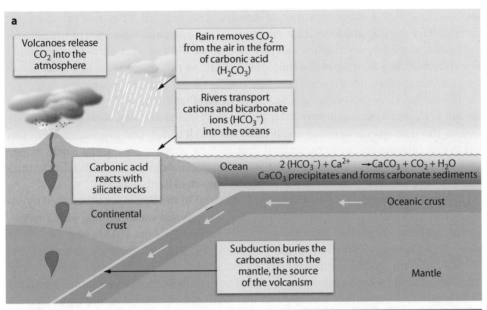

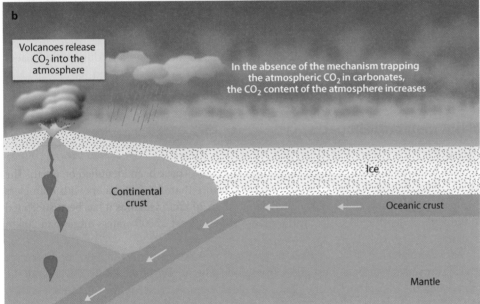

7

Fig. 7.6 Schematic section of a subduction zone during a warm period (a) and during a global glaciation (b). a. During a warm period, CO_2 is removed from the atmosphere by rain in the form of carbonic acid. The latter takes part in the weathering of silicate rocks and the rivers carry HCO_3^- ions to the ocean, where they are trapped in the form of carbonate rocks. In parallel with this trapping, volcanic activity injects CO_2 into the atmosphere, which results in some kind of equilibrium between the CO_2 amounts injected and extracted. If for any reason whatsoever (a temporary decrease in solar activity, etc.), the Earth's surface cools, the area covered by ice caps will increase. Glaciated surfaces, which are white, reflect a greater amount of the incoming solar radiative energy (with an increase in their albedo), thus contributing to the temperature drop. This mechanism can only accelerate and ends with the completely frozen Earth ("Snowball Earth"). **b.** However, in a glacial period, the ice cover prevents the alteration of silicate rocks and thus the trapping of CO_2. But as the latter continues to be released by the volcanoes, it accumulates in the atmosphere, increasing the greenhouse effect and leading to a progressive warming.

Between 2.45 and 2.2 Ga, three glacial episodes were recorded in the Huronian supergroup in Canada. Witnesses (diamictites, striated rocks, etc.) of this great Huronian glaciation have also been found in Australia, in North America and in South Africa. It is often considered to be the first episode of "Snowball Earth" actually proven. This period in the Earth's history is also associated with a major expansion of stromatolites, such that it has been proposed that the trigger for this glaciation was the production of huge amounts of dioxygen by cyanobacteria. Indeed, as we have seen earlier, dioxygen is toxic for methanogens, whose disappearance would have led to a decrease in the atmospheric methane content and thus to a reduction in the greenhouse effect, which would have lowered the Earth's surface temperature.

This scenario is fine; however, it is flawed by the fact that the age of the studied diamictites has not been established accurately: taking account of analytical errors, it ranges from 2.5 to 2.0 Ga. This time span is so broad that it could quite well correspond to a succession of several glacial episodes or, due to continental drift, to the passage of continental blocks at high latitudes. Finally, it does not seem that the Huronian glacial deposits are overlain by the thick carbonate deposits that might be expected after a "Snowball Earth" episode.

Between 0.9 and 0.58 Ga (the Cryogenian), the end of the Proterozoic is marked by another, very important glacial episode. Here again, it seems that we are dealing with a succession of three glacial episodes, each of which would have lasted about 100 Ma, and would have taken place around 715 Ma (the Sturtian glaciation), 635 Ma (the Marinoan glaciation), and 580 Ma (the Varangian glaciation or Gaskiers glaciation). Although several palaeomagnetic data indicate that these glaciations have affected low latitudes, many uncertainties remain as to their extent. Some authors consider that they were global, whereas according to others, a belt of unfrozen sea would have persisted at the equator. It would have constituted a refuge for the metazoans, which, once the glaciation was over, would have experienced a spectacular expansion and diversification, giving rise to the Ediacaran fauna (~600 Ma, *see later*).

Finally, we may note that the glacial episodes that occurred on Earth during the Palaeozoic (the Andean-Saharan between 450 and 420 Ma, and the Karoo between 360 and 260 Ma) did not achieve a global extent.

Major Meteoritic impacts

To date, only 168 meteoritic impact craters – of all sizes – have been identified on Earth. The Moon, although much smaller, is peppered with more than 500 000 craters with diameters greater than 1 km and about 1700 craters larger than 20 km. Insofar as it has been subjected to the same meteoritic flux, the Earth ought to show at least 22 000 craters more than 20 km in diameter.

As we explained in Chap. 5, plate tectonics as well as weathering and erosion processes have destroyed and deleted most crater marks from the surface of the Earth, and this, all the more when the impact is old. The age of the oldest oceanic crust known to date is less than 180 Ma, so, all impacts in oceanic regions older than 180 Ma have totally and irremediably disappeared in the subduction zones. Similarly, on continental regions, only very few impact structures of Proterozoic age are known, and none dating from the Archaean.

However, among the rare Proterozoic craters, two correspond to major events: the Vredefort giant crater (in South Africa), dated to 2.023 Ga and whose diameter is 300 km (◨ Fig. 7.7), and the Sudbury crater (in Canada), aged 1.85 Ga and with a diameter close to 250 km. The Acraman crater (in southern Australia), aged 580 Ma, has a diameter estimated at 85–90 km.

The absence of any impact crater in the Archaean record does not mean, as will be understood, that this period was spared the fall of meteorites, on the contrary. The Archaean craters having been destroyed, other traces and clues exist and testify to their existence. Indeed, craters are not the only records of meteoritic impacts: during the collision a large amount of material is ejected, sometimes to hundreds or even thousands of kilometers. These ejecta are then incorporated into sediments. Among the material ejected and preserved in the sedimen-

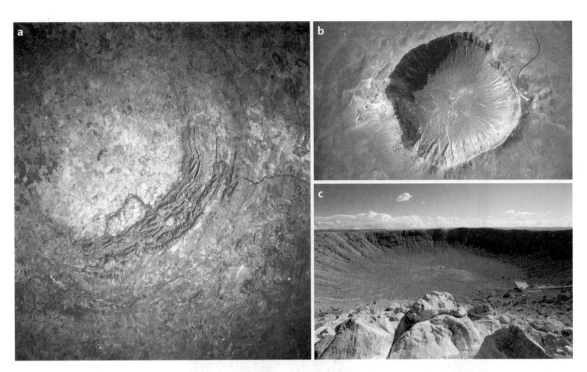

☐ **Fig. 7.7 Two terrestrial impact craters. a.** The giant Vredefort crater (South Africa); its diameter is 300 km and its age 2.023 Ga; **b.** and **c.** an aerial view and a view of the interior of Meteor Crater in Arizona (United States); this small crater (1200 meters in diameter and 180 meters deep) has been generated, about 50 000 years ago, by the impact of an iron meteorite, of about 45 meters in diameter and with a mass of 300 000 tonnes.

tary record we find, primarily, nickel-rich spinels, shocked minerals and impact spherules. Sediments may also record the geochemical signature of the extraterrestrial impactor, such as an abnormally high content of platinoids, such as iridium.

The study of recent impacts (such as that of Chicxulub in Mexico, aged about 65 Ma) has shown that, among all the ejecta, the spherules are the easiest to recognize. These are droplets of molten rock (or even vaporized rock) that were ejected during the impact and solidified in the atmosphere before falling down far from the crater. Generally, they are spherical in shape, but may also have "aerodynamic" shapes. The spherules may be exclusively glassy (microtektites) or crystalline (microkrystites).

The Archaean cratons contain some thick spherule-rich layers, which were emplaced during two distinct periods: between 3.47 and 3.24 Ga (in the Barberton greenstone belt in South Africa; ☐ Fig. 7.8) and between 2.65 and 2.50 Ga (Western Australia and South Africa). In South Africa the spherule layers are associated with breccias and iridium anomalies as well as with chromium isotopic anomalies. All these parameters bear unambiguous witness to the extraterrestrial nature of the event that produced them. The special feature of these Archaean layers is that they are thick (> 10 cm) and made up of large-size spherules, whose composition is mainly basic. These criteria, associated with the absence of any shocked quartz crystals, lead researchers to interpret them as the result of impacts that took place on oceanic crust rather than on continental crust.

In the Proterozoic terrains located on the west coast of Greenland, a thick (20 cm) spherule layer has recently been discovered, interbedded with dolomites. Its age is estimated as being between 2.13 and 1.85 Ga. Similarly, near the border between Ontario (Canada) and Minnesota (United States), a layer, 50 cm thick has been discovered, which is rich in spherules and in shocked quartz. Dated between 1.88 and 1.84 Ga, it is considered to have originated from the Sudbury impact crater. At the end of the Proterozoic (0.58 Ga), a layer of

7

□ **Fig. 7.8 Impact spherules, dated at 3.24 Ga and outcropping in the Barberton greenstone belt (South Africa).** This rock consists of an accumulation of small spherical items – the spherules – which are small droplets of rock melted during the meteorite impact and ejected, sometimes very far from the crater. (Photo: H. Martin.)

spherules (with a thickness of less than 40 cm) was deposited, following the Acraman impact, in southern Australia.

Of course, these few clues do not bear witness to all meteorite impacts that the Earth underwent during the Archaean and Proterozoic times – far from it. These are just some major impacts, sufficiently important for their ejecta to be scattered over large areas. Consequently, it is highly probable that they have widely and sustainably altered the environment of our planet's surface.

Indeed, apart from the devastating effect of the impact itself, the released energy hurls millions of tonnes of dust and aerosols into the upper atmosphere, where they may remain for months. The presence of these particles in the atmosphere reduces the amount of solar radiation that reaches the Earth's surface. There follows a drop in temperature of up to twenty degrees, giving rise to what is called an "impact winter". The dust and aerosols also darken the atmosphere, thus decreasing the brightness of the Sun, which causes a significant reduction of the photosynthetic activity. Finally, the great amounts of water that are vaporized react with sulfur, which thus, augment the concentration of sulfate aerosols in the atmosphere. These latter fall down to Earth as rains of sulfuric acid, which greatly affects living beings. The effects of major impacts on the terrestrial environment and on the biosphere may therefore be catastrophic...

In the case of impacts later than 0.45 Ga, after the first plants had colonized the emerged lands, it is possible that the energy released by the impact would have ignited any combustible material present on (or near to) the surface (vegetation, coal or hydrocarbon deposits, etc.). This general conflagration would not only have destroyed much of the flora and fauna, but would also have contributed to spreading ashes and soot into the atmosphere, making it even darker. So, the sediments deposited at the Cretaceous–Tertiary boundary (65 Ma) and formed during or immediately after the Chicxulub impact in the north of the Yucatán peninsula (in Mexico), contain soot and cenospheres, that is, small, microscopic spheres of carbonaceous material formed by hydrocarbon combustion.

The Volcanism of the Traps and Oceanic Plateaus

In many Archaean and Proterozoic terrains, geologists have found basalt formations that exhibit all the petrological and geochemical features of present-day traps and oceanic plateaus. Today, they consist of the stacking of multiple layers of solidified basalts that were emplaced at the surface of continents in the case of the traps, and under water for the oceanic plateaus. They have been erupted over a short period of time, in a hot-spot geodynamic environment. For example, 65 Ma ago, it is estimated that nearly 1.5 million km³ of basaltic lava were erupted in slightly less than 30 000 years, in the current region of the Deccan in India. Similarly, 250 Ma ago, in Siberia, between 1.5 and 4 million km³ of basalt (according to estimates) were erupted in less than one million years (◻ Figs. 7.9 and 7.10). Traps and oceanic plateaus form what geologists call "Large Igneous Provinces".

The initial effect of such eruptions is to eject thousands of tonnes of dust into the atmosphere. As for a meteorite impact, the amount of solar radiation that reaches the Earth's surface decreases, lowering temperature and causing a "volcanic winter". However, volcanoes also release very large volumes of carbon dioxide which, once the dust particles have settled out, causes a greenhouse effect that, in a few thousand years, increases the surface temperature. Such environmental changes definitely had consequences on the development of living organisms from the Archaean to the present day. By way of example, it has been calculated that the eruption of the Siberian traps results in a global warming of about 5 °C. However, although such warming could have significant environmental consequences, it is not, by itself, sufficient to account for the mass extinctions that took place at the end of the Palaeozoic (the Permian–Triassic event).

7

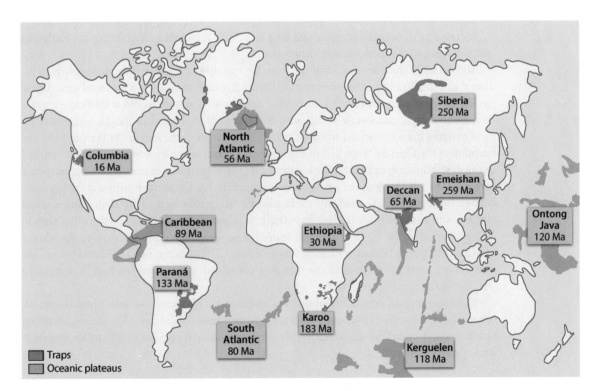

◻ **Fig. 7.9 The distribution and age of the main oceanic plateaus (in green) and traps (in red) that have been emplaced during the last 250 Ma.** Evidence for these recurrent volcanic episodes (linked to a hot-spot geodynamic environment) are known from the Archaean. The volumes of lava that have accumulated are enormous. They bear witness to significant volcanic activity, the repercussions of which on the environment could have been catastrophic, even going so far as to contribute to mass extinctions (extinction crises).

7

☐ **Fig. 7.10 The Columbia River basalts (United States).** The basaltic flows erupted about 16 Ma ago. The plateau (or trap) consists of a stack of several individual flows. Its thickness may reach 1800 meters. (Photo: H. Martin.)

It has recently been suggested that methane clathrates could have played a prominent role in certain climatic and biological crises. A clathrate consists in a lattice of water ice that encloses molecules of a trapped gas (CO_2, H_2S, CH_4, etc.). For instance, the melting of 1 m³ of solid methane clathrate, would release 168 m³ of methane (gas). At present, at the high pressures that prevail at the bottom of the oceans, methane reacts with cold water to form methane clathrates. Some authors have estimated that the mass of methane trapped at the bottom of the oceans in the form of clathrates is about 10 Pkg (Pkg = Petakilogram = 10^{15} kg)! An increase in the ocean's temperature by 4 to 5 °C is sufficient to destabilize the clathrates and to release all the methane that they contain.

Theoretically, the warming resulting from the trap-like volcanism is therefore likely to induce the destabilization of the methane clathrates. Because methane is a powerful greenhouse gas (23 times as efficient as CO_2), its release into the atmosphere would amplify the surface temperature augmentation. It is this sort of scenario that is often proposed to explain the most extreme mass extinction which occurred 250 Ma ago, at the end of the Permian: (1) The eruption of the Siberian traps heated the atmosphere and ocean by 4 to 5 °C; (2) this resulted in the destabilization of the methane clathrates; (3) the methane release into the atmosphere increased the efficiency of the greenhouse effect; such that the surface temperature again increased by about 5 °C. Overall, the surface temperature rose by 9 to 10 °C, causing a major climatic and biological crisis. This hypothesis is supported by measurement of the carbon isotopic variations ($\delta^{13}C$) in sediments. Indeed, a sudden increase in the relative abundance of ^{12}C at the Permo-Triassic transition was revealed. Clathrates are rich in ^{12}C and the release into the atmosphere of the methane that they contained would have resulted in decreasing the $^{13}C/^{12}C$ ratio and thus the $\delta^{13}C$.

The eruption and emplacement of traps and oceanic plateaus, also inject into the atmosphere, large volumes of hydrogen sulfide (H_2S) and sulfur dioxide (SO_2). These gases, first produce sulfuric acid, which subsequently condenses as acid rain, according to, for instance, a reaction as follows:

$$2\,SO_2 \;+\; O_2 \;\rightarrow\; 2\,SO_3$$

sulfur dioxide   dioxygen   sulfur trioxide

$$SO_3 \;+\; H_2O \;\rightarrow\; H_2SO_4$$

sulfur trioxide   water   sulfuric acid

Moreover, when they reach the stratosphere, these gases react with ozone. The destruction of the ozone layer also results in harmful consequences for the biosphere.

All these mechanisms, well-known on the present-day (Phanerozoic) Earth, undoubtedly played a significant role during the ancient history of our planet, from the Proterozoic – perhaps even from the Archaean – to the present.

The Evolution of the Prokaryotes

As we have seen in Chaps. 4 and 6, the prokaryotes appeared and diversified quite rapidly from an ancient ancestor well before the appearance of modern eukaryotes (possessing mitochondria). The initial diversification of the bacteria and the archaea was accompanied by the evolution of varied metabolic strategies. Most of these metabolisms probably emerged before 2.5 Ga. Indeed, because the bacteria were, as we will recall, the result of a radiation, that is an extreme diversification in a short space of time, if cyanobacteria – and thus oxygen photosynthesis – already existed by 2.7 Ga, as the fossil record and the subsequent oxygenation of the atmosphere from their activity indicate, we must conclude that most of the other lineages of bacteria – and thus also metabolisms – had equally appeared by that time. This includes aerobic respiration, which must have emerged as soon as a source of oxygen was available (and this was certainly the case, as we shall see later in this chapter). As for the archaea, they soon separated from the bacteria (probably before the latter diversified), but we do not know how to date this event exactly.

After this great initial diversification, the prokaryotes – and their metabolisms – continued to evolve until the appearance of eukaryotes (around 2–1.8 Ga, as we shall see soon), but probably at a lesser rate. This is explained by the fact that the major metabolisms had already appeared, and most of the virgin ecological niches had been colonized. That said, the massive oxygenation of the atmosphere must have led to a major ecological upheaval. Indeed, oxygen is a powerful and dangerous oxidizing agent. Its presence induces the formation of free radicals which can produce cellular damage. The organisms that could not adapt to oxygen by developing specific protection mechanisms and those whose metabolisms were strictly anaerobic were thus confined to more restricted habitats, such as anoxic sediments, hydrothermal vents, the terrestrial crust or even the interior of other unicellular or multicellular organisms (as symbionts).

However, the transfer of genes certainly remained active, which would have favored the adaptation of certain clades to new environments. In addition, and very rapidly, mutual symbiotic relationships were undoubtedly established, based on the exchange of metabolites. These symbioses are common in nature today, notably in sediments and other anoxic systems where the wastes produced by one type of organism are used by its symbiotic partner as a resource, and vice versa. Moreover, as we have already mentioned in Chapter 4, certain symbioses are quite obviously the origin of the eukaryotic cell itself and have conferred on it the capacity to respire oxygen, via the ancestor of the mitochondrion.

The appearance of eukaryotes constituted a key event. Indeed, eukaryotes, more complex and possessing new capabilities – such as phagocytosis, which allowed them to carry intracellular digestion (heterotrophic prokaryotes secrete their hydrolytic enzymes outside the cell) – had the possibility of colonizing a whole range of equally new ecological niches. They thus underwent an evolutionary radiation almost as significant as that undergone by bacteria when they diverged from their last common ancestor.

At the same time, the appearance of the eukaryotes favored a new diversification of the prokaryotes and more particularly of the bacteria. In fact, eukaryotes (notably the multicellular ones, such as animals) constituted so many new ecological niches for bacteria, and to a lesser extent for the archaea. Parasitism therefore developed extensively and pathogenic bacteria appeared. With the eukaryotes, a whole variety of prokaryote lines also emerged that were commensal or beneficial to their hosts. Today they form most of the microbial flora of the intestines, the skin or the mucous membranes. In humans, for example, the intestinal microbiota consists of several thousand species of bacteria and, to a much lesser extent also archaea (notably methanogens), of which only about one thousand may be cultivated in the laboratory.

All these species of prokaryotes living in association with eukaryotes have evolved and become specialized by following the evolution of their hosts, but, in most cases, they belong to major groups (phyla) that already existed. It seems as if practically no large group of bacteria has emerged since the first great radiation. However that may be, the evolutionary process that bacteria and archaea have undergone since around 2 Ga was continuous and punctuated by periods when it proceeded faster (for example during evolutionary radiations) as well as longer or shorter episodes when the speed of evolution slowed down. Prokaryotes, just like the eukaryotes never stopped evolving!

The Origin and Diversification of the Eukaryotes

The terms "prokaryote" and "eukaryote" were introduced with their present-day significance by the microbiologists R. Stanier and C.B. van Niel in 1962. They described the two major types of cellular architecture that exist on our planet. Eukaryotes, whether they are unicellular (microalgae, dinoflagellates, amoeba, etc.) or multicellular (plants, animals, fungi, etc.) are characterized by three major properties (▢ Fig. 7.11):
- a more or less complex system of internal membranes that define a region known as the nucleus, surrounded by a double membrane, which contains the genetic material
- organelles bounded by membranes: the mitochondria, where respiration takes place and, in photosynthetic eukaryotes, chloroplasts, the site of photosynthesis;
- a highly developed cytoskeleton consisting of filaments of actin and microtubules of tubulin.

For a long time, prokaryotes were defined negatively by the absence of these features: the absence of a true nucleus and organelles (some prokaryotes also possess a cytoskeleton and systems of internal membranes, but less developed than those in the eukaryotes). Today, we know that bacteria and archaea are linked by a fundamental property: translation and transcription are coupled together (in eukaryotes the two processes are separate: transcription takes place inside the nucleus and translation in the cytoplasm).

In as far as it has now been shown that mitochondria and chloroplasts derive from ancient endosymbiotic bacteria (from alphaproteobacteria and cyanobacteria, respectively), eukaryotes must have appeared after the prokaryotes diversified (*see* ▶ Chap. 4). Certain eukaryotic lineages, though, lack easily identifiable mitochondria and, for a long time, it was believed that they were well and truly lacking them. It was then thought that they formed a primitive lineage (the "Archezoa"), which predated the mitochondrial endosymbiosis. It was subsequently discovered, however, that these apparently amitochondriate lineages possessed mitochondrial genes and greatly reduced mitochondria, which disproved the "Archezoa" hypothesis. The last common ancestor of extant eukaryotes possessed mitochondria.

How did the evolutionary history of the eukaryotes start? Most authors think that they derive from prokaryotic ancestors. This hypothesis is mostly based on the simpler organization of prokaryotes. Nevertheless, the existence of two independent lines among prokaryotes – the bacteria and the archaea – complicates the question. What is more, comparative analysis of the complete genomes of organisms in the three domains of life has revealed a paradox: the eukaryote genes involved in the replication of DNA, transcription and translation resemble their counterpart archaeal genes, while the genes involved in energy and carbon metabolism

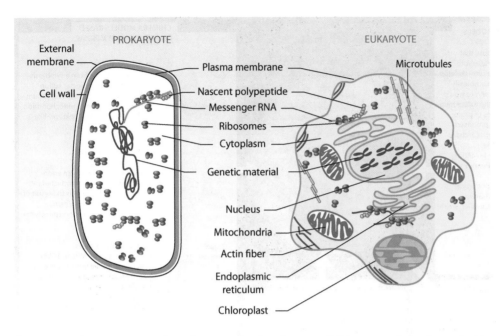

7

◘ **Fig. 7.11 Schematic structure of a prokaryote cell and a eukaryote cell.** Eukaryoteic cells are characterized by three essential features: a more or less complex system of internal membranes that define a nucleus which contains the genetic material; organelles bounded by membranes: mitochondria, where respiration takes place and, in the case of photosynthetic eukaryotes, chloroplasts; and a highly developed cytoskeleton consisting of filaments of actin and microtubules of tubulin which give them, among other things, the ability to carry out phagocytosis. Prokaryoteic cells (bacteria and archaea) are linked by a fundamental property: lacking a nuclear membrane, transcription and translation are coupled together.

resemble the corresponding genes in bacteria. How to explain this "mixed ancestry" in eukaryotes? Let us return to this question that we have already raised in Chapter 4.

Hypotheses About the Origin of Eukaryotes

We can distinguish two major groups of models that attempt to explain the origin of the eukaryotes (◘ Fig. 7.12).

Autogenous Models

The first family, the more classical of the two, is still widely accepted by a large section of the scientific community. It groups together models said to be autogenous. Here, the prokaryote-to-eukaryote transition would be explained by the increase in complexity of ancestral structures in a prokaryotic lineage independent of the archaea and bacteria, which would have developed a system of endomembranes defining a nucleus and a complex cytoskeleton favoring the emergence of phagocytosis (that is to say the capacity of including, in vesicles, microorganisms or particles within the cytoplasm). This property would have been the basis for an essential event: the incorporation by this proto-eukaryote of the ancestor of the mitochondrion. The latter was an alpha-proteobacterium which would have been engulfed by phagocytosis, but not digested, and which would have become an endosymbiont.

In this group of hypotheses, eukaryotic bacterial-like metabolic genes would have come through lateral transfer from the mitochondrial ancestor, and the rest of the genes of the proto-eukaryote host would resemble archaeal genes by vertical inheritance. Indeed, in the most widely accepted model, archaeal-like genes would have been present in an independent proto-eukaryotic line that had separated from that of the archaea, after an initial bifurcation

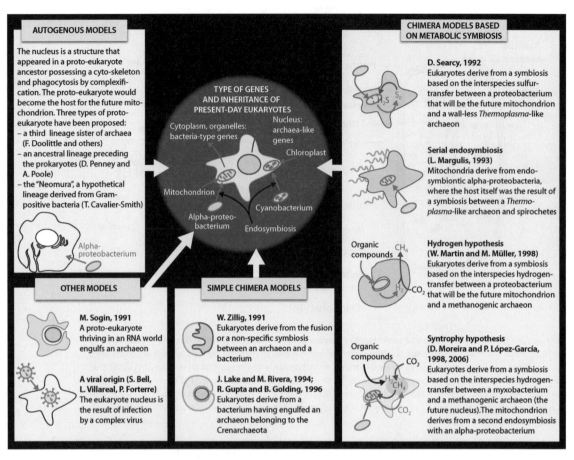

Fig. 7.12 Some hypotheses for the origin of the eukaryotes. As one can see, this question is by no means resolved. Models for the origin of eukaryotes may be classed in three major families: autogenous models, chimera models, and the others. Certain models are discussed on the previous page and below.

from the cenancestor between the bacteria on one side and a line that would lead to archaea and eukaryotes on the other side (☐ Fig. 4.27a in Chap. 4).

In other models, far more controversial, archaeal-like genes are also present in eukaryotes by vertical inheritance. Accordingly, the model proposed by T. Cavalier-Smith, a researcher at Oxford working in eukaryotic systematics and evolution, is very close to the most widely accepted model, namely the existence of a lineage of proto-eukaryotes, sister to that of the archaea. However, to him the ancestral line leading to these two sister branches (that he calls the Neomura) would not derive from the last common ancestor, but from Gram-positive bacteria. Accordingly, the evolution of eukaryotes would have been quite a recent event (about 800 million years ago). That would explain, in his eyes, the presence of a single plasma membrane in the archaea, the eukaryotes and the Gram-positive bacteria, whereas the Gram-negative bacteria have two. However, this hypothesis comes up against numerous criticisms, among which is the fossil record, which proves quite definitely that the existence of eukaryotes precedes by about one billion years the date that the researcher proposes (*see later*).

Chimera Hypotheses

The second major family of models explaining the origin of the eukaryotes has gained a lot of ground since comparative genomics has been practiced: chimeric models. These postulate

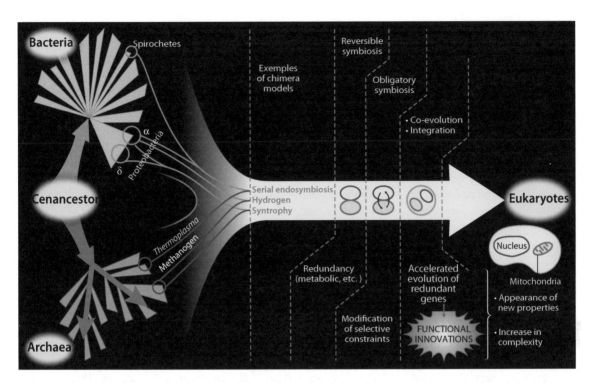

□ **Fig. 7.13 The origin of eukaryotes: chimeric hypotheses.** The diagram illustrates the fundamental concepts behind this family of models where the eukaryotes are the result of one (or two) symbiotic events between a bacterium and an archaeon (□ Fig. 7.12 and the text for the various hypotheses). A symbiotic association is far more than the sum of its parts: the genetic redundancy allows an increase in the evolutionary rate of the "duplicated" genes, thus producing conditions that are particularly favorable for the emergence of new properties. It was perhaps in just this type of nursery for evolutionary innovation that the eukaryotes were born.

that the eukaryotes are the result of a symbiotic event involving an archaeon and one or more bacteria (□ Fig. 4.27b in Chap. 4).

These archaea-bacteria symbioses would explain the mixed nature of the eukaryote genome without the need to invoke the existence of a hypothetical third line in which the principal traits specific to eukaryotes would have evolved *before* the incorporation of the mitochondrion. These traits would, in effect, be an intrinsic consequence of the symbiosis between bacteria and archaea. We may imagine that initially this symbiosis was reversible; then once a certain number of genes had been transferred from one partner to the other and lost in the donor – the endosymbiotic transfer of genes is a common phenomenon – the symbiosis became obligatory. Under these conditions, certain of the redundant genes for the same function in the symbiotic consortium (the association) were freed from selective constraints, which allowed them to undergo an accelerated rate of evolution and thus eventually produce new functions (□ Fig. 7.13).

The fact that symbioses are efficient nurseries for evolutionary innovations is thus one of the fundamentals of these chimera models. A symbiotic consortium is not the simple sum of the elements of which it consists. As a consequence of the genetic redundancy and the increase in the rate of evolution with duplicated genes, this consortium can, in effect, gain new properties. The fact that mitochondria and chloroplasts have evolved in the setting of endosymbioses is a reliable witness of the evolutionary significance of this phenomenon.

One of the pioneering chimeric hypotheses was the "serial endosymbiosis" model, devised by Lynn Margulis, who worked at Amherst College in Boston in the United States until she recently passed away. L. Margulis first brought back to scientific debate the idea – stated by the Russian scientist C. Mereschkovsky in 1905, but subsequently largely forgotten – that chloroplasts derived from cyanobacteria. (Mereschkovsky also suggested that the nucleus derived from bacteria). She additionally proposed that mitochondria also derived from endosymbiotic bacteria After having been harshly criticized in the 1970s, these ideas were confirmed thanks to molecular phylogeny at the end of the same decade. Subsequently, L. Margulis suggested that the host that had incorporated the ancestor of the mitochondrion was itself the result of a symbiotic event between wall-less archaea, similar to present-day wall-less *Thermoplasma*, and spirochetes (bacteria), which would have provided the consortium with motility. This part of the of serial endosymbiosis model remains speculative: it has not received any confirmation from the comparison of genomic sequences carried out to date.

A second group of chimeric hypotheses considers that the eukaryotes are the result of the establishment of a metabolic symbiosis between an archaeon and a bacterium that would evolve into the mitochondrion. Among these models, we may mention:

– the hypothesis proposed by Dennis Searcy, an American researcher at Amherst College who first proposed that *Thermoplasma*-like archaea were involved in the symbiosis that led to the origin of eukaryotes. According to him, the metabolic symbiosis was based on the interspecies sulfur transfer.

– the "hydrogen hypothesis" by William Martin, a researcher at Düsseldorf, and Miklos Müller, emeritus professor at Rockefeller University of New York, who make interspecies hydrogen transfer the key to the symbiosis between a fermentation bacterium that will become the mitochondrion and an archaeon, probably a methanogen, that consumes it.

All the chimeric models that we have mentioned so far struggle, however, to explain the appearance of a fundamental characteristic of the eukaryotes: the nucleus. They are constrained to propose an emergence *de novo*, but the selective forces in action are not clearly formulated, or are far from being convincing. To get around this difficulty, a third family of chimeric models suggests that the nucleus itself is the result of another endosymbiosis. Thus the "syntrophy hypothesis" by David Moreira and Purificación López-García, researchers at the University of Paris-Sud, postulates that eukaryotes are the result of two successive endosymbioses: first the endosymbiosis of a methanogenic archaeon, which would become the future nucleus, and a fermentation bacterium belonging to the myxobacteria, very complex bacteria that form multicellular structures and which have life cycles resembling those of certain social amoeba. As in the case of the "hydrogen hypothesis", this symbiosis would be based on interspecies hydrogen transfer. The archaeon would consume the hydrogen produced by the fermenting myxobacterium. The mitochondrion would be the result of a second symbiosis with an alpha-proteobacterium endowed with a versatile metabolism: it was, like certain bacteria in this group today, capable of using methane for its metabolism and to recycle it within the symbiotic consortium, and, depending on the environmental conditions, it could equally respire oxygen.

Yet Other Hypotheses

To finish, we will mention other models, less widespread and less supported. Among them, some suggest that the eukaryotes are ancestral (this is the view of David Penny and Patrick Forterre, for example) and that prokaryotes derived from them by reduction (❏ Fig. 4.27c in Chap. 4). Other models suggest that certain viruses have "produced" the nucleus by infecting archaea (P. Forterre) or even primitive cells in a hypothetical RNA world (M. Sogin).

The Diversification of Eukaryotes and Ancient Trace Fossils

Once the first eukaryotes had appeared, they could, as we have said, colonize new ecological niches. From then on, they probably diversified very rapidly, that is, they underwent an evolutionary radiation. This partly explains the difficulty that biologists find in resolving the phylogenetic relationships of the different major eukaryotic groups, that is, in establishing the relative order in which they appeared.

Eukaryote Diversity

Currently, scientists recognize seven large "supergroups" of eukaryotes (or "kingdoms"), which include very varied phyla (■ Fig. 7.14). It is possible that the root of the eukaryote tree, that is to say the most ancestral split, is located between two major initial lines proposed by T. Cavalier-Smith: that of the unikonts (consisting of the opisthokonts [which include the metazoans and fungi] and amoebozoans), whose ancestor would have possessed a single flagellum located at the rear of the cell, and that of the bikonts, consisting of the remaining groups, whose ancestor would have had two flagella. However, after that, the relative order of appearance of these major supergroups remains very uncertain. It is the same for the various phyla within the major supergroups.

It is only in the case of the opisthokonts that reconstructing the order of emergence of the different lineages has been possible with relative certainty. Fungi (including the basal lineage LKM11 or Cryptomycota) and their sister-lineage the nuclariid amoebae were the first to diverge, followed by the choanoflagellates and their related group, the metazoans (or animals) (■ Fig. 7.14).

Most of the eukaryotic lineages consist exclusively of unicellular organisms, which are referred to as "protists", but in certain groups we find multicellular species, with more or less organized tissues. Such is the case with the animals (metazoans) and the fungi in the kingdom of the opisthokonts, with the terrestrial plants and the green algae as well as certain red algae in the kingdom of the archaeplastids, and with the brown (e.g. kelps) algae in the kingdom of the stramenopiles or heterokonts (■ Fig. 7.14). Multicellularity has therefore evolved independently several times in the eukaryote domain, and the first eukaryotes were undoubtedly unicellular.

The Oldest Eukaryote Fossils

Can we date the appearance of the earliest eukaryotes? Two types of approach allow us to attempt to answer this question: use of the fossil record and molecular dating.

As regard the most ancient traces of eukaryote fossils, scientists are not in general agreement. In 1999, steranes were extracted from kerogens in sedimentary rocks dating to 2.7 Ga, and which were interpreted as being derivative fossils of sterols from eukaryote membranes (molecular fossils; ■ Box 6.4). However, these results were retracted in 2008: these biomarkers were in fact the result of contamination more recent than 2.7 Ga.

What about morphological fossils? It has been suggested that the oldest of these correspond to *Grypania spiralis*, aged 2.1 Ga. It is the large size of this microfossil that has led some scientists to make it a eukaryote. However, size is not an absolute criterion to differentiate between prokaryotes and eukaryotes, because certain prokaryotes are as large as eukaryotes (for example certain sulfur-oxidizing bacteria such as *Thiomargarita*) and, conversely, certain eukaryotes are as small as prokaryotes (the picoeukaryotes, for example, have cell diameters less than 2 μm). In addition, from its morphology, *Grypania* could simply correspond to large filamentous cyanobacteria.

A more reliable morphological feature to differentiate eukaryotes and prokaryotes seems to be the decoration of the cell's surface because, very often, the unicellular eukaryotes possess scales or other structures on the surface of their cells. The oldest, decorated, unicellular fossils – and about whose eukaryote nature consensus has been reached – are the acritarchs, aged about 1.5 Ga. They have been identified in the north of Australia in the Roper Group, in terranes dating from the Mesoproterozoic (■ Fig. 7.15). Acritarchs do not resemble any group of present-day eukaryotes. The first morphological fossils that could possibly be assigned to a modern group of eukaryotes correspond to *Bangiomorpha*

7

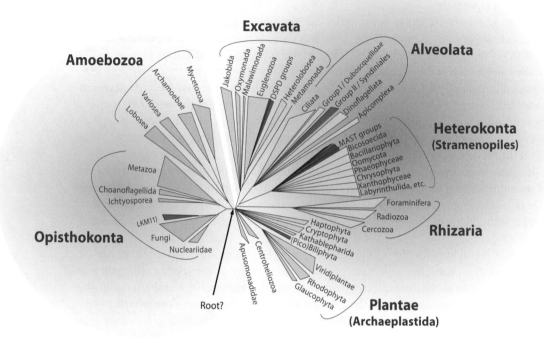

UNIKONTS

BIKONTS

Amoebozoa

Excavata

Alveolata

Heterokonta
(Stramenopiles)

Rhizaria

Opisthokonta

Plantae
(Archaeplastida)

Root?

◻ **Fig. 7.14 The diversity of eukaryotes.** The most ancient divergence among eukaryotes is perhaps that which saw the emergence of the unikonts on the one hand and the bikonts on the other. The eukaryote diversity includes seven large super-groups, or kingdoms, but the relative order in which these were established is very uncertain. They are: the Opisthokonta, the Amoebozoa, the Excavata, the Alveolata, the Heterokonta (or Stramenopila), the Rhizaria, and the Archaeplastida. Only certain groups include multicellular species (they are indicated by a star). The animals (the metazoans) correspond to only a very tiny section of the whole diverse range of the eukaryotes. The organisms represented are: Opisthokonta – *Boletus edulis* (ceps); Amoebozoa – *Echinostelium minutim* (myxomycete or slime mold); Excavata – *Euglena spirogira* (Euglena); Alveolata – *Paramecium tetraurelia*; Heterokonta – *Fucus vesiculosus* (bladder wrack); Rhizaria – *Globigerina* sp. (foraminifera); Archaeplastida – *Abies nordmannia* (the Nordmann pine). In red: taxons known only from their genetic signature. Diagram: Dominique Visset.

pubescens, between 723 Ma and 1.2 Ga, which several authors interpret as being red algae. Finally, we may note that for a minority of researchers, in particular T. Cavalier-Smith, the oldest eukaryotes would only date from 850 Ma, which corresponds to the time when the fossil record becomes plentiful.

The environment in which the oldest eukaryotes that left those fossils lived was probably aquatic, but we are far from knowing that with certainty. Some might have been associated with the surface of sediments or damp soils, for example.

To date, the current shortage of eukaryote microfossils does not allow us to establish the date of the appearance of this group with any robustness. Some researchers have therefore tried to date this event by using the evolutionary information contained within gene sequences (◻ Fig. 7.16). Molecular dating rests on the hypothesis of a molecular clock, that is to say a rate of evolution (the number of mutations per time unit) that is more-or-less constant over time. The more the sequences of homologous genes differ, the older the divergence of the taxa to which they belong. Under such conditions, if one knows the average rate of evolution, one can, in theory, estimate the date of divergence.

In reality, numerous problems introduce bias in these estimates, the most important of which being that the hypothesis of a constant molecular clock is unrealistic: in different lineages, the genes evolve at different rates. This is why the molecular dating of the oldest diver-

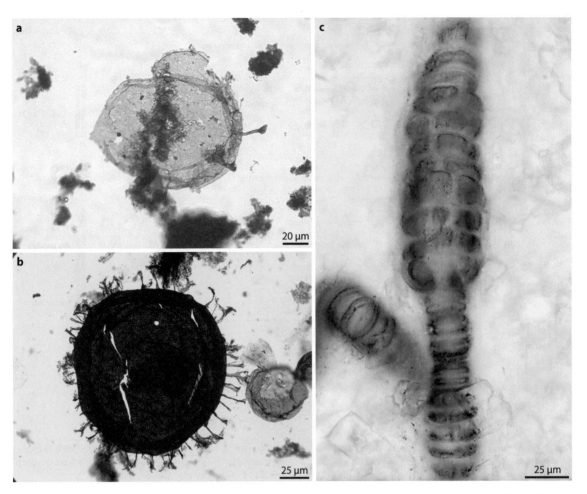

7

Fig. 7.15 Some ancient eukaryote microfossils: a. *Tappania plana*, Roper Group (Australia), 1.5 Ga; **b.** *Shuiyousphaeridium macroreticulatum*, Ruyang Group (China), about 1.3 Ga; **c.** *Bangiomorpha*, which some interpret as a red alga aged 1.2 Ga. (Photos: E. Javaux and N. Butterfield.)

gence among the eukaryotes – that which separates the unikonts from the bikonts – varies enormously depending on the authors and the method of reconstruction that is used (between 1.085 and 2.309 Ga). Nevertheless, improvements to this type of approach, with the incorporation of "relaxed clocks", that is to say models that allow different rates of evolution to be taken into account as a function of the different lineages, seem promising.

The Cambrian Explosion

The end of the Proterozoic (a period called the Neoproterozoic, which extends from about 1000 to 542 Ma) coincides with a period of global change, probably initiated by the break-up of the Rodinia supercontinent about 900 Ma ago (*see earlier*).

The Neoproterozoic is marked by three severe glaciations, which have left isotopic signatures and large quantities of glacial sediments distributed across the whole planet: the Sturtian glaciation (~710 to 725 Ma), the Marinoan (~636 to 600 Ma) and the Gaskiers glaciation (~580 Ma). The first two are perhaps the most extreme glaciations that the Earth has experienced. Some authors maintain that during these periods, our planet was entirely covered with ice (the "Snowball Earth" hypothesis, *see earlier*). Under such conditions, the primary productivity through photosynthesis would have been almost annihilated, with radical conse-

7

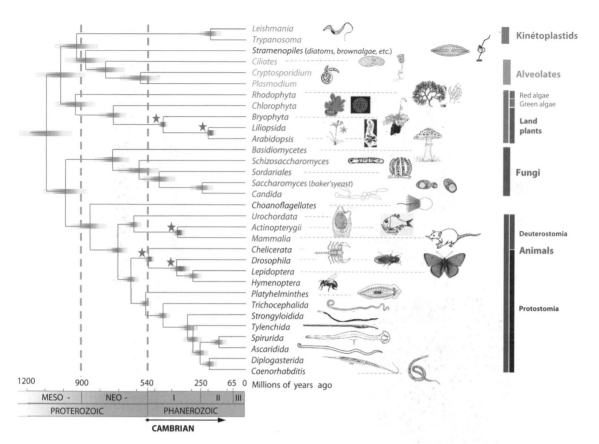

Fig. 7.16 Approximate age of the emergence of several major eukaryote lines obtained recently by molecular dating using a "relaxed clock" and, as calibration points, known fossils (stars). The phylogenetic tree was rooted using *Dictyostelium*, an early-diverging amoebozoan as outgroup. The yellow rectangles indicate the confidence intervals for the various node's dates.

quences for a large portion of terrestrial ecosystems. Other scientists tend, however, to favor scenarios that are less drastic and assume that, during the glaciations, more-or-less extensive oceanic regions would have remained free of ice, thus constituting refuges ("refugia") for a major portion of the biological species. The end of these glaciations would be explained by an accumulation of CO_2 in the atmosphere, leading to a sufficiently strong greenhouse effect to counteract the effect of the high albedo.

Just after these glaciations in the Neoproterozoic, large quantities of fossil imprints of multicellular organism with soft tissues start to appear. That is, of organisms without a shell or any other form of external or internal skeleton. They probably represent the first fossils of multicellular eukaryotes belonging to the metazoan phylum. These fossils, with a very varied morphology are collectively known as the "Ediacaran fauna", after the name of the Australian hills where they were first described (Fig. 7.17). The presence of these imprints is so characteristic that a period known as the Ediacaran has recently been created to define this geological period, which extends from the end of the Marinoan glaciation to the appearance of organisms with shells, which characterize the Cambrian (540 Ma).

The Ediacaran fauna was widely distributed across the planet. It has been recorded at more than 30 sites on the 5 continents. The fossil imprints are very diverse in terms of their morphology, both in size (from a few centimeters to more than a meter) and as to their structural plan (there are several forms of symmetry), which shows the presence of sophisticated organisms. Some authors have assumed a possible relationship between this fauna and the jellyfish, whereas others have created for it a group of organisms that is completely extinct

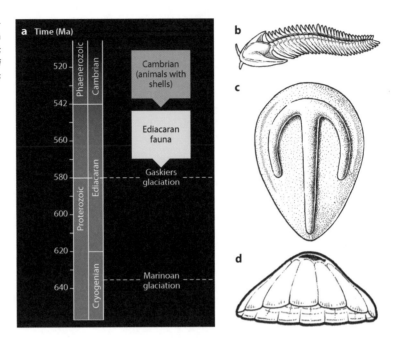

□ Fig. 7.17 The Ediacaran fauna: stratigraphic location (**a**) and reconstruction of a few representatives: *Spriggina* (an annelid; length: 3 cm) (**b**); *Parvancorina minchami* (an arthropod; length: 1 cm) (**c**); *Mawsonites* (a medusa; diameter: 40 cm) (**d**).

nowadays (the *Vendobionta* group). There is currently a consensus in favor of the following idea. A large part of the Ediacaran fauna could correspond to ancestral representatives of present-day phyla: sponges and cnidarians (jellyfish, corals) for organisms with radial symmetry, *Arthropoda, Mollusca* (in these last two cases, they would be possible "ancestral representatives" because of their body plan and not because of the presence of a skeleton or a shell; they should not be thought to resemble modern arthropods or mollusks), annelids and echinoderms for organisms with bilateral symmetry. Another portion of the Ediacaran fauna would represent extinct lines than had no evolutionary successors.

The Ediacaran fauna disappears quite abruptly from the fossil record just at the boundary of the Cambrian (540 Ma). This transition is not well understood. It has been suggested that it may have been caused by a major anoxic event. The beginning of the Cambrian is marked by a new massive radiation in the fossil record of eukaryotic organisms, notably of animals, this time possessing skeletons or shells: this is the Cambrian Explosion. Among the best-known fossiliferous formations dating from this period, we may mention the Burgess Shale in Canada. Fossils belonging to all the major present-day phyla of metazoans have been recorded from this period. No other major phylum of multicellular organisms seems to have appeared in the 500 Ma that have followed. As for terrestrial plants, they did not appear in the fossil record until the Silurian, 425 Ma ago.

Other Planets, Other Living Worlds ?

Chapter 8

Suns, Earths ... Life?

In the universe, there is one particular galaxy, among billions of others ... In that galaxy, there is one Sun, among 200 billion others ... And around that Sun, a small, blue, "Goldilocks" planet, which is neither too hot, nor too cold, that today shelters an incalculable number of living beings. This living planet, the Earth, seems unique in the universe because it is ours.

What about the other places in our Solar System, such as planets and their satellites? Do they show any sign of life? And, farther away, what about planetary systems around other stars? Do they exist, and if so, do they harbor planets that could be living worlds, at least based on what we have learned about the Earth in the preceding chapters of this book, in other words "life as we know it"? Astronomers, geologists, planetologists, and biologists, interested in the search for extraterrestrial life, have made tremendous progress over the last decade or so, giving many answers to these questions, but at the same time, raising other equally tantalizing interrogations and yet unsolved problems. In the present chapter we attempt to summarize this very rapidly expanding, multidisciplinary field of science, from the point of view of an astronomer.

◻ **Artist's view of the Alpha Centauri A+B system (α CenA+B), showing in the foreground an Earth-mass planet discovered in summer 2012 with the ESO 3.6m telescope in Chile.** The Alpha Cen A+B system (and more distant third member, Alpha Cen C or "Proxima Centauri"), contains the stars closest to the Sun, only about 4 light-years away. The planet, called Alpha Cen Bb, orbit the stra Alpha Cen B with a period of 3.236 days. It is therefore much closer than Mercury is to the Sun (0.04 AU), so must be very hot and cannot be inhabited. The tentalizing speculation is that Alpha Cen Bb is just the closest member of a planetary system possibly containing another Earth-mass planet, but this time far enough from the star that there may be liquid water at its surface, and thus, as discussed in this chapter, be "habitable". (Credit: ESO)

M. Gargaud et al., *Young Sun, Early Earth and the Origins of Life*,
DOI 10.1007/978-3-642-22552-9_8, © Springer-Verlag Berlin Heidelberg 2012

Life Elsewhere in the Solar System?

The Solar System contains several thousand bodies (planets, satellites, asteroids, comets, etc.) exhibiting a remarkable diversity of physical and chemical environments. Space missions, that have obtained observations *in situ* or in close proximity to these bodies, have allowed us, for some fifty years, to become aware of the breadth and complexity of the problem. *A priori,* because of their distance from the Sun, three terrestrial (rocky) planets are "habitable," that is to say that, in principle, they could possess liquid water at their surfaces: the Earth and, more marginally, Venus and Mars. Yet only the Earth does contain liquid water and is inhabited. A crucial constraint appears here: the presence of an atmosphere and the greenhouse effect that accompanies it, in which water, with its triple ice/liquid-water/vapor cycle plays an essential role in the energy exchanges in a planetary atmosphere, and in the light that it receives from its parent star.

We know that the current temperature of Venus is much too hot (about 500 °C) – forbidding the presence of liquid water – and too acid to harbor any form of life. Consisting of 95% carbon dioxide, the atmosphere is very dense and has probably been the site of a runaway greenhouse effect, such that it has been incapable of maintaining a sufficiently moderate temperature, because of its proximity to the Sun.

As far as Mars is concerned, it is penalized by its low gravity, because that has enabled the lighter gases to escape. This was particularly the case with water, which, evaporating into the upper atmosphere, was progressively dissociated in an irreversible manner by the Sun's ultraviolet radiation. Its atmosphere is thus very tenuous today and consists mainly of carbon dioxide, although, as we shall see, in quantities too small to generate a significant greenhouse effect. However, the most recent exploration carried out by orbiting space probes (as, for example, by ESA's *Mars Express* or NASA's *Mars Reconnaissance Orbiter, MRO*) have revealed evidence of sedimentary layers on its surface, such as furrowed geology, suggesting that the "red planet" had liquid water at its surface for at least a billion years. This length of time is *a priori* sufficient for life to have appeared and started to develop. If such was the case, was it able to resist the Late Heavy Bombardment that peaked around 600 million years after the beginning of the formation of the Solar System (*see* ▶ Chap. 5), or the inexorable disappearance of liquid water from the surface? Luckily for the scientists searching for the first traces of life, the surface of Mars is not continuously remodeled by tectonic movements, as is the case on Earth. As a result, not all hope of finding fossil traces of any possible ancient life has been entirely lost. This is why it has been actively sought in the soil, from the *Viking* missions in 1975, to the small mobile robots *Spirit* and *Opportunity* and the *Phoenix* lander, which found water ice just beneath the dust on the north polar cap, confirming the observations made from orbit by *Mars Odyssey*. And although from the evidence no "Martian" exists at the surface, it is not impossible, if life has existed at some given time on Mars, that it still persists in some specific niche in the subsurface, at a level where there might be liquid water. Recent observations suggest that even today there are still seasonal salty water flows in some craters (◘ Fig. 8.1). In addition, Mars does retain some geological activity: the last volcanic eruption dates back to only 2 million years. So we cannot completely rule out the presence, on the red planet, of a mixture of oxidized and reduced compounds that allow any potential forms of life a way of obtaining energy.

The consideration of plate tectonics has other implications. A specific feature of the Earth, when compared with Mars and the other terrestrial planets, is precisely that it is the only planet having been the site of plate tectonic activity since its formation. And it is quite possible that this activity is an important factor in the appearance and development of life. On the Earth, because it allows the coexistence of both reduced and oxidized compounds (for example at hydrothermal vents, which are abundant along the mid-ocean ridges), this tectonic activity at the boundary between the interior and its surface can provide a source of chemical energy, an alternative to solar energy and one that is stable over time.

◨ **Fig. 8.1 Oblique view of warm season flows in Newton Crater on Mars.** This image, obtained on 30 May 2011, combines orbital imagery from MRO (Mars Reconnaissance Orbiter) with 3-D modeling. It reveals evidence for flows that appear in spring and summer on a slope inside the crater. Sequences of observations recording the seasonal changes at this site and a few others with similar flows might be evidence of salty liquid water active on Mars today (McEwen et al. 2011). The image has been re-projected to show a view of a slope as it would be seen from a helicopter inside the crater, with a synthetic Mars-like sky. (Credit: NASA/JPL-Caltech/Univ. of Arizona.)

8

However, the principal role that plate tectonics plays on Earth at present is its action in regulating the climate. Indeed, on the geological time scale, plate tectonics builds chains of mountains and causes the continents to emerge, whose erosion traps atmospheric carbon dioxide in the form of carbonate sediments, those sediments subsequently returning to the mantle at the subduction zones. In contrast, volcanism, which also results from plate tectonics, reintroduces carbon dioxide into the atmosphere. In other words, plate tectonics provides, over a very long time scale, a system of regulation (a cycle) of the atmospheric CO_2 content and therefore maintains a relatively constant greenhouse effect. This property ensures the maintenance of a surface temperature that is compatible with the presence of liquid water. Such a mechanism probably plays an important part in whether a planet is "habitable" or not (*see below*).

Returning to water, does it exist *in liquid form* elsewhere in the Solar System? The answer is yes, probably, but in very special environments. For instance, it is thought that foglike water droplets may exist at some depth in the warmer, inner layers of the atmospheres of Jupiter and Uranus, and that Neptune may have a hot, highly compressed ocean below its thick atmosphere. It is probably hard to imagine how life may have appeared in such conditions, but given the resilience of bacteria in extreme environments on Earth (*see* ▸ Chap. 4), if some were brought into these atmospheres (by meteorites, for instance), it is not impossible that they could have survived.

By contrast, the most frequent occurrence of water at the surface of the Solar System objects is in the form of *ice.* Apart from the polar caps of Mars, the only other "planet" to be covered with ice is Pluto – but Pluto is now considered more like a "dwarf planet" (or large asteroid) belonging to the "Kuiper Belt," a distant system of numerous, low-mass planetary objects dating from the very early stages of the formation of the Solar System. On the other

8

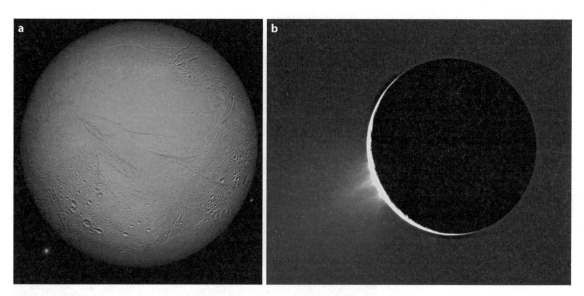

■ **Fig. 8.2 (a) Enceladus, one of Saturn's satellites, as seen by the Cassini spacecraft, front lit by the Sun, and covered by ice showing evidence for cracks. (b)** When backlit, the view reveals the presence of "geysers," springing out of the frozen surface of the satellite. These geysers consist mainly of water, but a small amount of carbon dioxide and methane has also been found. The main implication is the presence of an underground liquid ocean, below the frozen surface.

hand, small deposits of ice have been found in some deep craters of Mercury by NASA's *Messenger* probe.

Most of the other large objects of the Solar System are frozen satellites that orbit the giant planets, but both observations and planetary structure theory suggest that *liquid water may exist below their surfaces*. The most spectacular proof of this was the recent discovery by the *Cassini-Huygens* probe[1] of genuine geysers on Enceladus (■ Fig. 8.2), one of the small satellites of Saturn (its diameter is no more than 500 km). The existence of these geysers has a double implication: on the one hand it shows that liquid water does exist or is being formed beneath the layer of ice that covers Enceladus, and, on the other, it proves that this body produces energy (as a result of tidal effects from Saturn's proximity). In addition, carbon dioxide and organic molecules have been detected there. That implies the existence of internal chemical reactions of an unknown nature, which could form potential sources of energy for a metabolism in the absence of sunlight.

Europa, the icy satellite of Jupiter that is almost as large as the Moon, is also a "candidate for supporting life". The striated morphology of the surface suggests a form of pack ice in slow movement and the models of its internal structure predict the existence of a liquid ocean at depth. Even if it is difficult to see how life could have started under the conditions that prevail there today, Europa has become the target of several projects for space missions aiming to search for possible traces.[2]

This section wouldn't be complete without a few words about Titan, the biggest of Saturn's satellites, having a diameter of 5150 km, slightly larger than Mercury. This is the only planetary satellite in the Solar System to hold an atmosphere. Even from the ground, it is clear that it is very cold and cannot contain liquid water at its surface, but only ice. Yet, the fact that it possessed an atmosphere made it sufficiently unique to be the target of the *Huygens*

---

1. The *Cassini-Huygens* interplanetary mission is a joint ESA/NASA/ASI (Italian Space Agency) project. It was launched on 15 Oct. 1997. It consists in the *Cassini* spacecraft, put into orbit around Saturn on July 1st, 2004, and the *Huygens* descent vehicle, which landed on Titan on 14 Jan. 2005.
2. Movie fans may recall that Europa is also the enigmatic, sub-ice life-bearing target of the "Discovery" expedition in the film "2010," a sequel to Stanley Kubrik's "2001, A Space Odyssey," released in 1984.

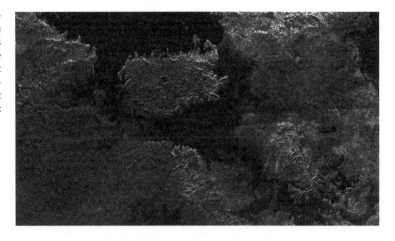

Fig. 8.3 This radar image of Titan, obtained by Cassini's radar instrument during a near-polar flyby on 22 Feb. 2007, shows a big ice island (called Mayde Insula) in the middle of one of the larger methane lakes at the highest latitudes (Kraken Mare). This island is about 90 km by 150 km across, about the size of the Big Island of Hawaii. (Credit: NASA/JPL-Caltech/ASI.)

8

probe, jettisoned in flight from the *Cassini* spacecraft to land on Titan after an interplanetary voyage of more than eight years. The atmosphere, made almost exclusively of nitrogen (N_2), with small amounts (1.6 %) of methane and ethane, was thought to be a good approximation to the Earth's primitive atmosphere (though much colder: ~ −200 °C). The mission was extremely successful, the probe parachuting safely down to the satellite's surface. *Huygens,* as well as several flybys by *Cassini,* revealed that Titan was in fact covered with ammonia and CO_2 ice mixed with water ice, and vast oceans of methane and ethane (**□** Fig. 8.3). The atmosphere can be usefully compared with the Earth, featuring winds and methane rains, as well as seasons. Truly not a very hospitable environment, yet rich in complex organic, perhaps prebiotic molecules. The inner structure of Titan may include an underground liquid water ocean. While the possibility of some form of microbial life has been advocated, no evidence (necessarily indirect, given the purely chemical nature of the on-board experiments) has been found so far.

Unexpected Worlds Beyond the Solar System: Exoplanets

Well beyond the Solar System, the discovery of exoplanets (or "extrasolar planets") – in other words, planets orbiting stars other than the Sun – will remain one the major results in the overall history of astronomy. Admittedly, the idea of other inhabited worlds goes back to Antiquity (Metrodoros of Chios, 400 BC; Epicurus 341–270 BC); and has persisted throughout the ages. To mention but a few: Giordano Bruno, *De l'infinito, universo e Mondi,* 1584 – he was eventually burnt at the stake for heresy by the Inquisition at Rome in 1600; Cyrano de Bergerac and his *Histoire comique des États et Empires de la Lune,* 1650; Fontenelle and his *Entretien sur la pluralité des mondes,* 1686 (**□** Fig. 8.3); Voltaire and his *Micromégas;* Jules Verne and his *Voyage dans la Lune,* 1865 and made into a film (so to speak!) by Georges Méliès in 1902; Camille Flammarion and his whole series of works on Earth and the heavens, such as *Les étoiles et les curiosités du ciel* [*Stars and Curiosities of the Heavens*], 1881, among others; and, of course, a whole host of works in the 20th century.

All these fictional accounts assumed, more or less explicitly, the existence of worlds similar to the Earth, or to the planets in the Solar System as envisaged in the imagination or as seen through instruments of inadequate optical quality. (One thinks of the Giovanni Schiaparelli's famous martian "canals" of 1877, "confirmed" by Percival Lowell in 1906.) To astronomers the question was then: "Do other Solar Systems exist?," with the implicit assumption that systems might differ in detail, but which would include "Earths" near their central stars, and more distant "Jupiters".

This extremely anthropomorphic view of other worlds in space came into conflict with the very first discovery of extrasolar planets. The news was announced in 1992 by Alex Wo-

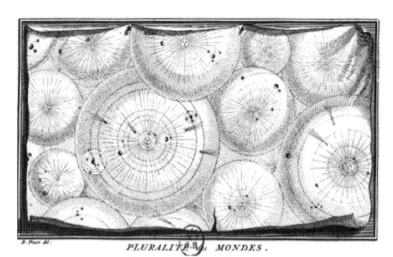

PLURALITÉ DES MONDES.

8

szczan and Dale Frail, who were studying pulsars using the large Arecibo radio telescope on Puerto Rico: these neutron stars, with masses of about 1.4 solar masses, very dense and strongly magnetic, emit a periodic radio signal, which, in this particular case had a period of about one millisecond. Extremely accurate timing of the signal from the pulsar PSR B1257+12 showed variations in the period that were themselves periodic, amounting to some tens of days. Analysis showed that these perturbations could be caused, not by one, but by three bodies orbiting the pulsar, with masses of 0.02, 4.3 and 3.9 Earth masses, and thus planets, with periods of 25, 66 and 98 days – which would have placed them (with the exception of the third) well within the orbit of Mercury (period 88 days), if they were part of the Solar System. However, it was obvious that, given their proximity and the extremely energetic radiation produced by the pulsar, there was no question of being able to envisage any form of life whatsoever on their surface.[3]

The decisive discovery, however, was made by Michel Mayor and Didier Queloz, who, using the 193-cm telescope at the Observatoire de Haute-Provence (in France), discovered the first planet orbiting a star similar to the Sun – 51 Pegasi, lying at a distance of 15 pc (48 light-years) away from us – by using the radial-velocity method (or velocimetry, ■ Box 8.1). The surprise came from the nature of the planet that was detected (known as 51 Peg b)[4]; its minimum mass was 0.468 times the mass of Jupiter, and its period was just 4.231 days, placing it at an orbital radius of just 0.05 AU from the star. In other words, this radius was one twentieth of the Earth's distance from the Sun, and one eighth of Mercury's! As a result, the surface of the planet reaches a temperature of 1000 °C, which is why the expression "hot Jupiter" was quickly applied to this class of planets. From the very beginning, this discovery raised a whole series of fundamental, and completely unexpected, questions: how could so massive a planet lie so close to its parent star? 51 Peg being even older than the Sun (7.5 Ga as against 4.5 Ga), how could this planet have survived despite the evaporation it undergoes as a result of such a high temperature? More generally, given that other, less massive and more distant planets might well exist, even if they were not detectable (■ Box 8.1), how could such a planetary system, so fundamentally different from the Solar System, have been formed?

It is difficult to know the answers at this stage, with planets of such different nature and location. It is not appropriate for us to explain in detail here the formation of planetary sys-

---

3. Just two other pulsars are known to have planets, and they are very different systems, each with a single planet, with periods of 2.18 hours and 10 years, respectively.

4. The designations of exoplanets are far from being homogeneous. The Geneva school of thought (led by Michel Mayor and his colleagues) designates them by the name of the parent star, followed by lower-case letters of the alphabet, in increasing order of their distance from the star. So the name of the planet in orbit around the star 51 Peg is 51 Peg b.

tems, which we have discussed in Chapter 1. Let us simply say that (even before the discovery of 51 Peg b), some experts in dynamics had already foreseen that within a planetary system, some planets might *migrate*, as a result of mutual gravitation interactions, either towards the interior, or towards the exterior of the system. Nevertheless, the question remains wide open at the present time. That being said, it is now thought that, even in the Solar System, planetary migration has been an important phenomenon: the Late Heavy Bombardment (*see earlier*, and ▶ Chaps 1 and 2) is the probable result of the mutual migration of Jupiter and of the three other giant planets. Calculations show, in fact, that through resonance effects, the smaller bodies in the early Solar System were destabilized, and crashed onto the Earth and the Moon, in particular.

From "Hot Jupiters" to "Super-Earths"

Since 1995, observations and discoveries have progressed in leaps and bounds, both from the ground and from space. The number of exoplanets known today is estimated to be around 2000, more than 700 having been confirmed around about 600 stars. The others await confirmation. When all the methods of detection (▢ Box 8.1) are taken together, the distances range from some ten parsecs (i.e., some thirty light-years) to the galactic center, 8500 pc away (28 000 light-years).[5] Given the observational bias (described in the box), it is probable that – at least as far as "solar-type" stars are concerned (between 0.3 and 1.5 solar masses) – most, if not all, are accompanied by at least one planet, even though the effective detection rate is no greater than 10 per cent.

How was this spectacular result achieved? Among exoplanet-detection methods, two clearly stand out by the harvest that they have allowed us to glean. They are the "radial-velocity" method (also known as "velocimetry"), and the "transit" method, both of which may, in certain specific cases, be combined to obtain a better understanding of the planet (its orbit, possible atmosphere, etc.). The "radial-velocity" method, when applied to a new sample of stars, or to an existing sample that has been observed year by year – which allows the detection of planets of lower and lower mass, and at greater and greater distances from their parent star (or both) – enable us to obtain precise, individual values, such as their orbital parameters (period, eccentricity, semi-major axis), and their minimum mass. If a transit is also involved, it is possible to determine the orbital inclination and the radius. We thus have access to an increasing number of planets for which we can measure the density, and thus determine their state: rocky (terrestrial) planets or gaseous (like the giant planets).

Using these methods we have also discovered that about 10 per cent of planets are members of planetary systems that include up to 6 planets (HD 10180 and Kepler 11). Of all these systems, *not one* resembles the Solar System! The planetary masses range from 6 Earth masses to 13 Jupiter masses,[6] the orbits range from 0.01 AU (closer than 51 Peg) to 11.6 AU (equivalent to somewhere between Saturn and Uranus in the solar system), and some are very eccentric (▢ Fig. 8.5) Statistically, the position of massive planets relative to "small" planets appears to be independent of their distance from the star. (There is a continuous sequence between the "hot Jupiters" that were first discovered, the "warm Jupiters" and the "cold Jupiters," such as those in the Solar System.) One could say that the 84 systems with two or more planets could be the result of a completely random process governing their formation. One signifi-

---

5. For detailed observational data on exoplanets, see the online *Encyclopaedia of extrasolar planets*: http://exoplanet.eu/ .

6. The limit for the fusion of deuterium is 13 Jupiter masses. Beyond this the object is considered to be a star, because it generates its own energy. (It is a "brown dwarf," up to the hydrogen fusion threshold at 0.08 solar masses or 80 Jupiter masses).

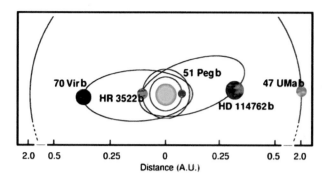

8

Fig. 8.5 Orbits of a few "hot Jupiters" (named after their parent star), including the first discovered exoplanet orbiting a solar-type star, 51 Peg. Distances are expressed in astronomical units (AU). Note that these hot Jupiters all have orbits closer to their stars than Mercury's orbit lies from the Sun (orbital radius 0.39 AU). In contrast, the figure also shows the distant inner planet around 47 UMa, another solar-type star. It is now known that 47 UMa is in fact surrounded by a planetary system consisting of 3 planets: the inner one (47 UMa b, shown), orbiting at 2.13 AU, and 47 UMa c and d orbiting farther out, at 3.79 and 11.6 AU respectively. For comparison, Jupiter is situated at 5.20 AU from the Sun, and Saturn at 9.54 AU.

cant correlation is, however, quite clear: the higher a star's metal content – the abundance of iron relative to hydrogen, measured from its spectrum –, the more likely it is to harbor at least one planet, which brings us back to the problem of the mechanism, or mechanisms, that govern planetary formation.

The effectiveness of the second method, the "transit" method, results from a fundamentally different observational approach. Whereas with the velocimetric method any orbital inclination whatsoever (except one that it exactly in the plane of the sky) is, *a priori*, able to produce a reaction in the motion of the star, the transit method is limited to the cases where the orbit of a planet is sufficiently close to lying at right angles to the plane of the sky, so that the planet should reduce the light from the star to a measurable extent when it passes in front of it (◖ Box 8.1). For a star taken in isolation, the probability of success is extremely low, of just a few per cent (depending on the type of planet being considered). Naturally, this method is extremely useful for stars where velocimetry has already determined the existence of planets (in establishing radii, the possible existence of an atmosphere, etc.). But in the general case where one is working "blind," the aim is to compensate for the low probability of any individual detection by observing a large number of stars simultaneously, and thus over a wide field of view, using digital detectors that are able to measure tiny fluctuations in the intensity of illumination on each pixel.

Because of the essential stability required by the instrumentation, and the necessity of pointing in exactly the same direction for months on end, the transit method cannot be used effectively except from space. The first satellite to use it on a large scale is an international project, led by the French space agency (CNES), and known as CoRoT (for "Convection, Rotation and planetary Transits"). Launched in 2006 and expected to function until 2013, this satellite carries a relatively modest telescope (27 cm in diameter, with an overall field of view of $2.7° \times 3°$, covered by 4 CCDs, of which two are reserved for the detection of transits). It combines asteroseismology and the search for planets by means of the same diagnostic methodology (but on different time scales), namely by detecting extremely faint, periodic variations in stellar luminosity, down to 10^{-4} parts. To date, some twenty exoplanets have been discovered from space for the first time, in uninterrupted observing sessions ranging from 20 days to five months. The complete program will involve about 150 000 stars. The American *Kepler* satellite, launched by NASA in 2009, represents a quantitative jump over CoRoT. With its primary mirror having a usable diameter of 95 cm, and its field of view of 105 square degrees, *Kepler* allows about 160 000 stars to be observed simultaneously. To date, more than 1200 candidate objects, observed in front of about 1000 stars during the first four-and-a-half months of the mission, have been analyzed. It is estimated that parasitic effects, such as magnetic activity (◖ Box 8.1), should not, eventually, lead to more than 10 per cent "false detections". While confirmation is awaited from ground-based observation (mainly through velocimetry), these transits correspond to stars that are candidates for harboring planets or planetary systems. For the time being, apart from special cases, the results that have been obtained primarily possess a *statistical* value, and do not give information about

■ **Fig.8.6 The "transiting" systems discovered by Kepler, that have the greatest number of planets, with from 6 (top) to 4 (bottom) members, in order of orbital period (Lissauer et al. 2011).** The size of the circles is proportional to the actual size of the planet; for comparison between the systems, the colors are indicative of the planets ranked by size. Note that all periods are less than ~100 days. The spectral types of the central stars are not yet known, so their masses are also unknown. It is remarkable, however, to note that if the parent stars were all solar-type, all these planets (up to 6: KOI-157 = Kepler 11) would orbit well inside the orbit of the Earth (compared to 2 for the solar system)! Note that, because of its transit methodology, Kepler (and CoRoT) primarily detect planets orbiting close to their parent stars. It is entirely possible that other, undetectable planets orbit at greater distances, just as in the case of the solar system.

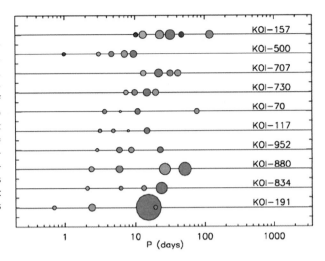

all the individual stars. (It may be noted in passing that the success rate, around 0.6 per cent is, at present, slightly lower than expected on the basis of the respective apparent sizes for a planet and its parent star.)

Remarkable results have emerged from these observations. (i) A high proportion (about 1/3) of the transiting "candidate planets" are part of planetary systems (as against about 1/10 found by velocimetry, which illustrates the significance of observational bias), including between two (115 systems) and six planets (1 system). (ii) Because of the method employed, most of these planets are close to their parent star: for the great majority, their periods lie between about 2 and 90 days (■ Fig. 8.6), which, in the Solar System, would mean that they would lie *inside* Mercury's orbit! (iii) The radii of these candidate planets lie between 1 and 10 Earth radii, and thus between that of Earth and Jupiter, most with 2 to 3 Earth radii: these are *super-Earths* (■ Fig. 8.7).[7] In three instances so far, a planet has even been found to orbit a close binary system, where the stars' orbital periods are 20–40 days!

This adventure in the field of astronomy, this "hunt for exoplanets," is far from having come to an end. It has already revealed completely unexpected planetary systems. After opening with the discovery of "hot Jupiters," the tally of achievements now has added compact planetary systems, consisting of almost as many planets (closer to their star than Mercury is to the Sun), as in the whole Solar System itself, which is, however, nearly 80 times as extensive. That is without, of course, counting planets that are smaller or more distant (or both), which are currently undetectable.

So we now know more than 700 individual planets, to which may be added around 1300 "candidate" planets that have been studied statistically. In other words, about 2000 planets in total. There can hardly be any doubt that hundreds, or indeed thousands of others remain to be discovered. However, among these, some will be the particular subjects of our attention, namely *habitable exoplanets*.

"Habitable" Exoplanets: Other Earths?

The "Holy Grail" in the search for exoplanets remains the discovery of planets comparable with the Earth, following the postulate that these planets could harbor a form of life. Or, to be more precise, the discovery of planets that are not necessarily similar to the Earth, but

---

7. A very recent announcement (December 2011) revealed that Kepler had discovered two planets of the size of the Earth, as well as three the size of Mars, but very close to their parent stars. These are not, therefore, "Earths" comparable with ours.

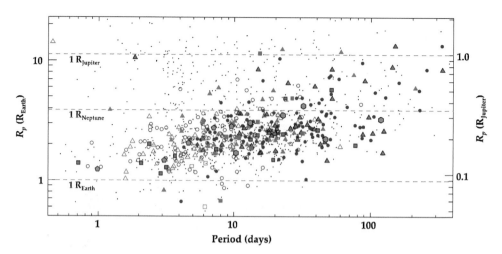

☐ **Fig. 8.7 The planetary period (in days) vs. radius (in Earth radii), for a sample of 1199 planetary candidates discov-ered by Kepler (Lissauer et al. 2011).**   Small dots: single planets. Planetary systems with 2 planets are shown by circles; 3 planets: triangle; 4 planets: squares; 5: pentagon, and 6 (Kepler 11): a hexagon. Colors indicate the location of the planets within their respective systems. Within this period range, we may note the relative paucity of Jupiter-sized planets, espe-cially in multiple systems. Also, smaller planets tend to have shorter orbital periods (they orbit closer to their parent star). These recent "statistical" findings will put new constraints on models of planetary formation.

which are likely to be "habitable" – that is, fulfilling certain conditions considered necessary (although not necessarily sufficient) for life to have developed or where life could develop one day. The question of these conditions is an extremely subtle one, and the answer, or answers, may be considered somewhat arbitrary, because to date we know just one single place in the universe where life has appeared: the Earth. The latter harbors a vast range of conditions, including extreme ones, that are favorable for life – or its survival. In addition, we must guard against an anthropocentric view, and not consider, *a priori,* that the search for life is synony-mous with the search for "intelligent" life, or even "technological" life. In this context, the only field where astronomers have some legitimacy (because they have the appropriate tools: observations, theories, numerical simulations, etc.), in addition to physics, of course, is that of "astrochemistry," in the light both of our knowledge of the origin and abundance of the ele-ments on the one hand, and of terrestrial biology on the other.

Astrochemistry may be defined as the science of molecules in space. For some forty years, radioastronomical techniques, most notably at millimeter wavelengths, have enabled the de-tection of numerous molecules, whether in nearby celestial objects such as comets or plan-etary atmospheres, or in our galaxy's cold interstellar medium, or even in other galaxies. A major result has emerged: among the 160 known interstellar molecules, the great majority are dominated by carbonaceous structures, most often in chains, but also in rings. In particular, numerous common organic molecules have been discovered, such as ethanol or acetic acid. The most complex gaseous molecule identified to date consists of a chain of 11 carbon atoms ($HC_{10}CN$). Study of interstellar dust in the infrared domain has also shown the general pres-ence of polycyclic aromatic hydrocarbons (or "PAHs") which are analogous to soot from com-bustion, or derivatives of oil (*see* ▶ Chap. 1). The most spectacular is undoubtedly "Buckmin-sterfullerene" (also dubbed "footballene"), a spherical fullerene consisting of 60 carbon atoms.

It is not surprising that carbon should dominate astrochemistry. In addition to its phys-ical properties that allow it to link readily with numerous other abundant atoms, such as hydrogen, oxygen or nitrogen, carbon itself is the most abundant element in the universe, after hydrogen and helium (an inert gas that does not react chemically), and oxygen (1/1000 relative to hydrogen), compared to which it is almost as abundant. And astrophysics explains extremely well why this is so. Unlike helium, which was primarily produced by "primordial nucleosynthesis" (three minutes after the Big Bang, as was deuterium), the "heavy" atoms are produced by stars during the course of their evolution. If life exists elsewhere in the universe other than on Earth, there is every chance that it will be based, in one way or another, on carbon. Some authors have raised the possibility of life base on silicon (some physical and chemical properties of which resemble those of carbon). That appears very unlikely, because observation shows (and stellar nucleosynthesis clearly explains) that the abundance of silicon is one tenth of that of carbon. In addition, the Si-H chemical bond is much weaker that the C-H bond, which means that silane (SiH_4), for example, reacts extremely violently in the presence of even a very low concentration of an oxidizing compound, unlike methane (CH_4), whereas, by contrast, silica (SiO_2) is very stable and not gaseous, unlike CO_2. This is the reason why molecules based on silicon generally exist in an oxidized, rather than in a hydrogenated form, as with carbon.

The other, even more significant factor that astronomy has revealed, and which has recently been confirmed by the ESA *Herschel* satellite, is the almost universal presence of water, whether in a liquid, solid (ices), or gaseous (molecular) form. Water has indeed been observed in extremely different environments: planets, the interstellar medium, disks and envelopes around stars that are forming, distant galaxies, etc. Moreover, we know that water is a unique solvent, promoting numerous reactions between organic molecules. It is, therefore, perfectly natural, if one is searching for the conditions necessary for the appearance or the development (or both) of life on other planets to start by assuming that these conditions are similar to those that we observe on Earth, whether present-day ones or former ones – "life as we know it". This point of view is reinforced by the universal nature of the abundances of the chemical elements, and also of chemistry based on carbon. This is thus the common thread running through the concept of the "habitability" of exoplanets.

The "Habitable Zone"

Introduced by Jim Kasting and his colleagues in 1993, this term came to be widely accepted. A planet is said to be "habitable" if it possesses an atmosphere and when the physical conditions at the surface as such that water, if it is present, can exist in the liquid form. We must emphasize from the outset that *"habitable" does not mean "inhabited"*. Here, we restrict ourselves to the concept of the minimal essential conditions. For an atmospheric pressure of 1 bar, as on Earth, the presence of liquid water at the surface naturally corresponds to a temperature between 0 °C and 100 °C.[8] These values change when the pressure is different, which means that the *greenhouse effect* caused by the trapping of the radiation from the parent star in the atmosphere should be taken into account. This effect raises the surface of the planet to a higher temperature than it would have in the absence of an atmosphere. For example, the average temperature of the Earth would be -18 °C if there were no atmosphere. (The planet would be entirely glaciated and would resemble Jupiter´s satellite Europa, for example.) As against that, the Earth´s average temperature is actually +15 °C.

Simple physical considerations provide us with some interesting orders of magnitude. Accepting a minimum temperature T_{min} and a maximum temperature T_{max}, provides the definition of a "habitable zone" like a ring, with an inner radius $R_{min}(T_{max})$ and an outer radius $R_{max}(T_{min})$ around the parent star. The simplest hypothesis is that the planet absorbs the stellar radiation L_* and reradiates it as a "black body" at temperature T. This temperature at a given distance will be a function of L_*, of the radius of the planet, and also of its "albedo" A, that is, the fraction of the energy that is actually reradiated relative to the energy received. ($A = 0$ for

8

8. This pressure regime does not apply to all living organisms. On Earth, certain "extremophile" bacteria can survive at great depth and at temperatures above 100 °C, or both (◧ Table 4.4, Chap. 4). Expressed simply, however, the regime mentioned applies to the vast majority of organisms that we know about.

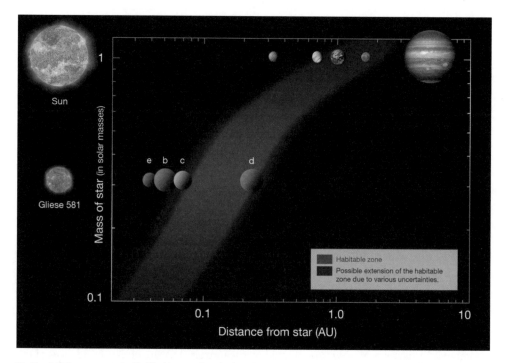

■ **Fig. 8.8 The concept of a "habitable zone" applied to the Solar System (top), and to the planetary system orbiting the star Gliese 581, a red dwarf of 0.3 solar masses and temperature of 3500 K (bottom).** The diagonal band corresponds to the habitable zone in the astronomical sense: the planetary surface temperature is such that water, if it exists, is in liquid form. This band is drawn as a function of the mass of the central star (Sun = 1) and of the distance of the planet from that star (in astronomical units). This zone may be extended somewhat if one takes into account other effects, such as the greenhouse effect caused by the presence of an atmosphere (dark blue area on the diagram). The two exoplanets Gliese 581c and Gliese 581d, discovered in April 2007, lie at the limits of the central star's habitable zone. (Based on a diagram by Franck Selsis, Univ. of Bordeaux. Credit: ESO)

a completely reflective surface, and $A = 1$ for a matt black surface, for example, that is completely absorbent.) For the Earth, $A = 0.29$ if its cloudy atmosphere is taken into account; its value would be $A = 0.13$ without it. If the mass (and thus the luminosity) of the parent star is varied, we obtain a ring that is closer or more distant from that star (■ Fig. 8.8). It is reassuring to find that, even with this simplified, purely astronomical, calculation, the Earth lies well within the Sun's habitable zone! We also need to consider the case of habitable planets orbiting stars that are different from the Sun, in particular, cooler stars – and thus redder ones, which may alter many factors in our perception of the possibility of life on other planets. This consideration is significant in practice, because in terms of astronomical observation, the contrast is, in this case, more favorable to the detection and the determining the character of planets (■ Box 8.1).

That being so, because there is no possibility of habitability (at least at the surface) without an atmosphere, more precise calculations need to be carried out taking into account model atmospheres and the associated greenhouse effect. It is even possible to wonder about the long-term persistence of habitability. For example, on the primordial Earth (*see* ▶ Chap. 2) the atmosphere was reducing and dominated by CO_2, and the Sun itself was less luminous. As a result, we need to adopt a more realistic definition of the habitable zone. The form that is generally adopted is as follows. Assuming a "rocky" planet, of terrestrial type (such that there is a solid surface), and incorporating sufficient water (in other words, with the presence of oceans), the inner, "hot" boundary (R_{min}), corresponds to a maximum greenhouse effect, such that the water is completely vaporized, and then dissociated by stellar UV radiation (which is present even in stars cooler than the Sun as a result of their magnetic activity: with flares, etc.), creating free hydrogen, which irreversibly escapes into space. The external, "cold" boundary (R_{max}), also correspond to a maximum greenhouse effect, but of a completely different nature. In this case, the CO_2 (produced, by example, by tectonic activity as on Earth) eventually can no longer be absorbed by the water to form carbonates. In a complex effect, once a threshold has been passed at which the increase in CO_2 causes heating, an increase in the albedo A occurs, caused by the appearance of clouds consisting of CO_2 ice. The planet then reflects the stellar radiation more efficiently, and causes the opposite effect, stopping heat absorption. The radius R_{max} first increases, moving habitability farther from the parent star (◘ Fig. 8.9), but past a certain point eternal glaciation prevails …

However, it is clear that other theories are possible, notably if one takes the effects of long-term stellar and planetary evolution into account. We therefore find that there are several possible definitions of the zone of habitability, as a function of the hypotheses adopted concerning the composition of the planetary atmospheres at any given time. The first consequence is a *broadening* of the "astronomical" habitable zone, at the hot boundary as well as at the cold one. However, this zone (without an atmosphere), as described earlier, remains an acceptable first approximation. For the various models including a CO_2 atmosphere, the inner radius R_{min} is decreased by about 10 per cent; the outer radius R_{max} may, however, be increased by 50 per cent (for a very high concentration of CO_2). Obviously, this distinction becomes crucial for exoplanets whose orbits are just within the intermediate zone, even if, for example, giant planets are involved. In such a case, we cannot exclude envisaging *habitable satellites* (considered in the same sense as the "habitable planets" themselves).

Here again, the comparison with the Solar System is instructive. Basing our view on the Earth's atmosphere, we see that not only the Earth, but also Mars, is within the Sun's habitable zone, Venus is excluded. We now know that Mars must have had water on its surface for about 1.5 billion years – but we suspect that cessation of its tectonic activity, possibly the

8

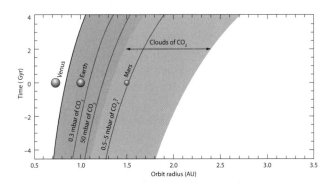

◘ **Fig. 8.9 Evolution of the location and width of the habitable zone (HZ) in the solar system (in AU; shaded in blue)**, as a function of two factors: (i) the time dependence of the Sun's luminosity since its arrival on the main sequence (**4.5 Ga ago**, *see* ▶ Chap. 1): ordinate, in Ga; (ii) the strength of the greenhouse effect, here calculated as a function of the partial pressure of CO_2 (width of the HZ) (from Selsis 2007). As it evolves, the Sun becomes hotter and hotter, so the HZ recedes farther away; as the CO_2 content increases, the HZ becomes wider. For reference, the partial pressure of CO_2 today is 0.3 mb for the Earth, and 6 mb for Mars. We see that the Earth is particularly "resilient," staying within the HZ at all times and CO_2 concentrations. Indeed, we know for example that the concentration of CO_2 on the Earth 500 Ma ago was about 7.5 mb, i.e., comparable with Mars now and perhaps 25 times higher than today (the exact value **at the time** is still debated). For very high concentrations of CO_2, the result becomes uncertain (light blue zone), because CO_2 ice starts to form in the atmosphere, and it is not clear whether this would make the greenhouse effect still stronger (continuing warming), or weaker (higher reflectivity of the atmosphere to solar light). In the best case, to be habitable, Mars would need an atmosphere containing almost 100 times more CO_2 than at present. (Diagram by F. Selsis, credit Springer))

result of its too low a mass, led to irreversible evaporation, as described earlier. In addition, despite the large amount of CO_2 in the atmosphere of Mars (with a partial pressure of 6 mb, or 20 times that in the Earth's atmosphere), the greenhouse effect is insufficient to heat it. Its average temperature is –50 °C, and it is only at the equator, in summer, that the temperature reaches +15 °C.

So we can simultaneously assess the "operational" side of the notion of the habitable zone – it may be estimated relatively accurately by astronomers for a given type of star – and its limitations. (For a given habitable zone, certain planetary factors, such as an inadequate greenhouse effect, or else the existence or absence, of tectonic activity, may nullify the habitability.) As some authors have emphasized, *sensu stricto* we should speak of a *"potentially habitable zone,"* solely in the astronomical sense – and not speak of *"potentially habitable planets,"* because, for the time being, we have no information about the composition of the atmosphere of planets said to the "habitable," if any such exist.

"Biomarkers"

In spite of these reservations, let us assume that the conditions for "habitability," in the sense just described, are combined for a given planet. How might we recognize that life actually exists on its surface? There again, we are forced to draw conclusions from the only planet on which we know life to have developed: the Earth, in trying to answer the question: If we ourselves were living on an exoplanet, what proofs would we have of life on Earth? Here, it is not images that count, but the chemical information provided by spectroscopy, from the visible region into the infrared. In 1993, Carl Sagan, the pioneer in the search for life on Mars, analyzed a spectrum of the Earth from space to search for signatures of life. He employed three simultaneous criteria: large quantities of oxygen O_2, and traces of methane CH_4, as well as strong absorption in the red, attributable to chlorophyll. Another method consists of taking a spectrum of the light from Earth that is reflected by the Moon ("Earthshine"). In it one can clearly see absorption lines not just of oxygen and methane, but also of ozone O_3, alongside lines from the greenhouse gases, water H_2O and carbon dioxide CO_2. The presence of methane, oxygen and ozone in sufficiently significant quantities to be detected by spectroscopy (◘ Fig. 8.10), as well as N_2O, is considered as a good indication of the existence of life – at least of life similar to life on Earth at present, where photosynthesis plays a vital role. It may be noted that most of these gases may also be produced by abiotic reactions, but it is the *simultaneous* presence of CO_2, H_2O, and O_3 that is nowadays considered to be the most significant signature, because these molecules cannot remain in chemical equilibrium on the scale of a whole atmosphere if they are not permanently produced or consumed by biological processes.

This search for biomarkers from spectral signatures in the atmosphere of exoplanets is the fundamental issue in the search for life on their surface. Whereas *astronomical* studies nowadays allow us to determine the structural characteristics of an increasing number of exoplanets (their mass, radius, and density, and thus their rocky or gaseous nature, or even that their atmosphere is evaporating in the case of certain – uninhabitable – "Hot Jupiters"), the *physical and chemical* study that would allow any potential atmosphere to be defined by spectroscopy (whether those exoplanets are, or are not, located within the habitability zone), remains an observational and technological challenge that is very difficult to meet. Nevertheless, the era of "exoplanet characterization," i.e., the determination of their structure, atmospheric composition, etc., has already begun. An initial stage has already been achieved with the acquisition of medium-resolution spectra ($\lambda/\Delta\lambda = 100$ to 1000) of several hot Jupiters, cold Neptunes, and even super-Earths – but, of course, none of these are yet habitable planets.

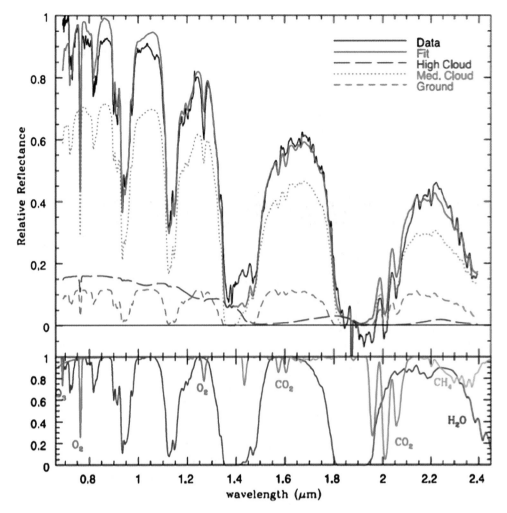

Fig. 8.10 The Earth as a living planet: Near-IR reflectance spectrum of the integrated Earth (in arbitrary linear units), obtained by observing the Moon ("Earthshine," after Turnbull et al. 2006). **Top panel:** observed spectrum. The fit to the data (red line) is obtained by summing several components of the Earth's atmosphere at the time of the observation (see key): high-altitude clouds at 10 km (cirrus ice particles) and above; mid-altitude clouds at 4 km (cumulus water droplets) and above; ground grey reflectance and the overlying air mass. **Bottom panel:** identification of molecules for a clear-air spectrum. Prominent throughout the spectrum are the broad absorption lines of water (H_2O), then several weaker narrow lines of CO_2, CH_4, O_2, O_3, and very weak N_2O (not shown).

To go any farther, given the extremely high contrast that exists between the parent star and its one or more planets (■ Box 8.1), it is nowadays feasible to block the light from the star as far as possible, either by blocking the star's light with a coronagraph, by destructive interferometry ("nulling"), or by a combination of the two. Nevertheless, the luminosity of the planet still remains very low. It is one thing to "see" an exoplanet, as explained in ■ Box 8.1, but it is quite another, far more difficult, to disperse its light to obtain a spectrum. First-generation proposals exist for space-borne spectroscopy of giant planets at a contrast level of 10^{-5} relative to their parent stars.

To achieve the decisive step, in other words to obtain contrasts of around 10^{-9} (a terrestrial atmosphere in the habitable zone of a G star, like the Sun), it will be essential to use far more powerful, future-generation instruments. This means the "ELT" (Extremely Large Telescopes, about 40 m in diameter), currently projected (by ESO at Cerro Armazones in Chile) and which might enter service in 2025 at the earliest, or else, a fleet of interferometric satellites of the *Darwin* type. Such satellites are envisaged on an even longer

8

time scale, because they require technology that has not yet been mastered, both as regards the detectors as well as "formation flying". This as yet unproven technique requires knowing to a high degree of accuracy the relative positions of several satellites that are in orbit close to one another.

By that time in the future, the odds are that thousands of exoplanets will have been discovered. Perhaps one of them will prove to be more favorable for spectral study, without having to wait for the next generation of giants ... But then, what are ten more years for us to be patient, compared with the 24 centuries that have elapsed since the time of Metrodoros of Chios?

■ Box 8.1   How can we find exoplanets?

Detecting a planet orbiting another star than the Sun (an "exoplanet") is a technological challenge that has been within our capabilities only for the last twenty years, even though the basic principles are, in fact, extremely simple. *Seeing* an exoplanet similar to the Earth is, by contrast, far more difficult, and is one of the major issues in the whole field of astronomy in the coming decade, or even the following one, with current projects for telescopes with diameters of around 40 m.

Some history and the principles involved

Let us begin with a short recap: Kepler's First Law (1609), reformulated within the framework of universal gravitation introduced by Newton in his *Principia* (1687), states that two bodies in orbit around one another follow elliptical paths, with a common focus at their center of gravity (their "barycenter"). For example, if we have a binary system consisting of two identical stars, these will periodically approach and recede from one another in a symmetrical manner. In the opposite case with two bodies of very different mass, such as a star and a planet, to a first approximation one may say that the planet "revolves round" the star, but in reality the center of the star will "revolve" around the system's barycenter. If, for example, we were to reduce the Solar System to the Sun + Jupiter system (Jupiter being the most massive of the planets in the Solar System, even though it is one thousandth of the mass of the Sun), that system would, in fact, revolve around a point that is not situated at the center of the Sun, but slightly outside, exactly 1.07 solar radii from the center.

This example of a Sun-Jupiter system in itself gives an idea of the observational constraints. Let us assume that we want to detect such a system at a distance. Jupiter's orbital period (the jovian "year") is 11.861 terrestrial years. To be able to detect the resulting motion of the Sun, it would be necessary to observe for several years, under conditions that were as close as possible from one observation to another (requiring extremely

stable instrumentation). The proper motion would be about 0.5 mas (milliarcsecond – thousandths of a second of arc), which is currently unobservable[1]. The amplitude of the variation in radial velocity is about 12 m/s (and only 9 cm/s for the Sun-Earth system). Without attaining this accuracy initially, it was nevertheless in this manner that the first exoplanet orbiting a normal star[2], 51 Peg, was discovered in 1995. Since that date, more than 700 exoplanets, as well as more than 1300 exoplanet "candidates" (which remain to be confirmed) have been discovered by various methods, which we will now briefly describe.[3]

Method 1: radial velocities

The first method, successfully used to discover the first exoplanets orbiting solar-type stars, is known as the "radial-velocity method" (or "velocimetry"). Its principle derives directly from the factors mentioned earlier: it involves measuring the motion of the star along the line of sight by means of the periodic oscillations in its radial velocity synchronized with the orbital period of the planet that one is attempting to detect. This is the so-called "reflex motion" of the star. The measurements are made thanks to the Doppler effect on the star's spectral lines, which have the same oscillations.

Although simple in principle, the method is very difficult to implement for four main reasons: (i) small size of the effects to be measured (a target value of about 0.1 m/s for a system comparable with the Sun-Earth

---

1. However, a ten-fold improvement will soon be obtained by the use of the interferometric method on the VLT (VLTI), and one hundred times better with the European *GAIA* astrometric satellite, which will be launched in 2013.

2. In fact the first exoplanets were discovered in 1992, but these were planets orbiting pulsars (neutron stars that are rotating extremely rapidly), and were found by measurements of variations in the pulsations (*see* text). Because this is a very specialized environment, the term "exoplanet" will be used here exclusively for a planet orbiting a normal star.

3. Reference to the Encyclopaedia of Extrasolar Planets (http://exoplanet.eu/) will be found to be of great benefit.

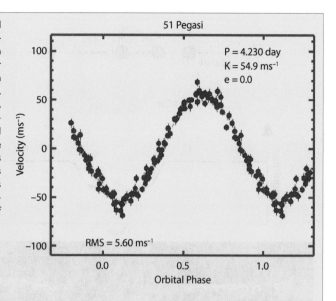

Fig. 8.11 Radial velocity curve as a function of orbital phase, for the first-discovered extrasolar planet, orbiting 51 Peg, a solar-type star located 48 light-years from the Earth. The discovery was first published by M. Mayor and D. Queloz in 1995, from data taken with the 193-cm telescope at the Haute-Provence Observatory in France. The figure shows the sinusoidal "reflex motion" of the star, exerted by the planet, i.e., the variations in its radial component, as seen by the observer. The figure was obtained at Lick Observatory by G. Marcy et al. (the "California & Carnegie Planet Search Team") in 2003, after eight years of observations. This planet was the first of a new class of planets, unknown in the Solar System: a Jupiter-mass body orbiting at 0.05 AU from the star, i.e. about one-tenth of Mercury's distance from the Sun! This class of planets is now known as "hot Jupiters."

system, as indicated earlier); (ii) the duration of the orbital periods concerned, which requires an almost perfect stability of the spectrograph over periods that may extend to years; (iii) parasitic "noise" caused by motions at the surface of the star, like that of the granulation produced by convection at the surface of the Sun, or else by atmospheric turbulence; (iv) periodic oscillations of the whole star caused by large-scale convection (as studied by stellar seismology). The first two constraints depend, respectively, on the masses of the parent star and the planet on the one hand, and on the orbital characteristics of the planet on the other.[4] The other two factors are related to the structure of the parent star itself, which should, therefore, be understood as fully as possible.

The relative significance of these constraints will depend on the specific star and planet involved. In particular, we can understand intuitively that the perturbation that the planet causes on the star will be greater, the greater its mass, and the closer it is to the star. There will be major variations in the radial velocity and a short orbital period. In this context, we can understand the surprise created by the detection of the first exoplanets. The initial aim was to search for pairs comparable with the Sun-Jupiter system, but the observations led to the discovery of unexpected systems: exoplanets that were both very massive and very close to their parent star (0.46 times the mass of Jupiter, and an orbital period of just 4.231 days in the case of

51 Peg, and thus closer than Mercury is to the Sun). For this reason, the planets were soon dubbed "hot Jupiters" (Fig. 8.11).

Since 1995, the radial-velocity method has progressed greatly with respect to the velocities measured (which are less than 1 m/s), to the accumulation of data over long periods (some stars having been observed now for more than 15 years), and to the understanding of stellar physics (through asteroseismology). To date, more than 700 exoplanets have been discovered by this method. The sensitivity is now such that it is possible to derive "variations of the variations" (or "periodograms"), which reveal the presence of different periods, i.e., of several planets around the same parent star – in other words, true *planetary systems*.

By combining these results with those obtained by the transit method (described later), it has been found that, among the planets known to date, there are more than 80 multiple systems, including as many as 6 planets (of which 2 systems are known, *see* text), and with masses ranging from "super-Jupiters" (13 M_J) to "super-Earths" (a few Earth masses), and where the detectable limit gets lower and lower, day by day. The detection of planets with the mass of the Earth – at least those relatively close to their parent star – is imminent (*see* text: "Habitable planets").

It should, however, be borne in mind that the radial-velocity method is marred by several observational biases. One is the intrinsic difficulty of detecting distant or low-mass planets (the motion induced in the parent star is very small, and is even more difficult to detect if there are other planets); another is because the orbit has too high an inclination. (The radial velocity observed is the projection of the true induced motion of the parent star: at the limit, a planet, even a massive, close-in one, that orbits in the plane of the sky is undetectable by this method.)

---

4. The three principal characteristics of a planetary orbit are: (i) the distance of the planet from the star (more precisely, the extent of the semi-major axis of the orbit); (ii) its eccentricity; (ii) its inclination relative to the star's equator. To simplify things here, and without going into particular cases, we shall assume that the planets describe circular orbits in the stars' equatorial planes. But apart from the inclination, which is more difficult to determine, these values may equally be measured in most cases.

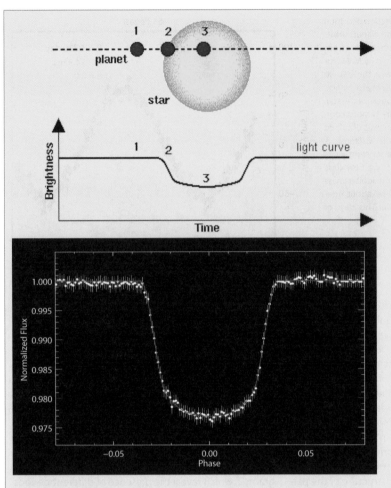

8

□ **Fig. 8.12 (Top) Light-curve of a distant star, the disk of which is partially occulted by a planet (a "transit").** The stellar light-curve suffers a dip by a fraction of a magnitude, then recovers its original value. The typical duration of such a transit, where the planet is close to the star (say less than 1 AU) is of the order of a few hours. *(Bottom)* An actual transit of the CoRoT-Exo-1b planet, discovered by the CoRoT satellite, showing a dip of 0.25 per cent in the stellar flux (Credit: CoRoT exo-team).

Method 2: transits

Retaining the same concept of a planet orbiting a star, there is another effect that may be sought. This is the decrease in the light from the star when the planet passes in front of the star's disk. This is known as a "planetary transit." Such transits, in front of the Sun, are well known from Earth, with transits of Venus (the last being in 2004, followed by one in 2012), and of Mercury (the last in 2006 and the next in 2016). Naturally, at stellar distances, it is impossible to obtain a direct image of the phenomenon, but the decrease in illumination, although faint, is measurable (□ Fig. 8.12). The luminosity of the star, and the obscuration by the planet, are proportional to their apparent areas, that is, the square of their respective radii. The attenuation then depends on the distance between the star and the planet. Another aspect is that the closer the planet is to the star, and the longer the duration of the transit, the greater the probability of detection. In the final analysis, this method favors the detection of planets that are *close* to their parent stars.

The great advantage of this method is that the measurement of luminosity is, in principle, simpler than ob-

taining a spectrum. Current, space-borne instruments are able to attain an accuracy of about 10^{-4} in terms of luminosity. There are, nevertheless, two restrictions on the efficiency of this method. The orbits of the planets must be almost exactly at right angles to the plane of the sky, and also sufficiently close to the parent star. (It is, obviously, highly unlikely that one would accidentally observe a transit if the planet is so far distant that its orbital period amounts to centuries ...) It should also be noted that the presence of spots on a rotating star, linked to magnetic activity (as on the Sun, *see* ► Chap. 1), may result in a decrease in the luminosity, and may thus simulate a transit, resulting in a so-called "false positive".

All that being so, another significant advantage accrues from the factors just mentioned. Thanks to present-day digital detectors, it is possible to measure *simultaneously* the luminosities of a very large number of stars that are present in a given field of view. Typically these may number tens to hundreds of thousands, which tends to compensate, to a certain extent, for the limitations mentioned above. To increase the chances of success (and to determine the periods), it

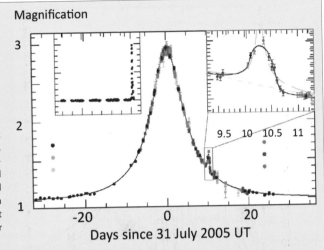

Fig. 8.13 Detection of an extrasolar planet by "microlensing." The method relies on the fact that, in agreement with general relativity, light rays are bent by gravitational fields. When a background star passes exactly behind a foreground star, its light is beamed, leading to a temporary "magnification" (lensing). The best conditions for detection are met when the background contains millions of stars, like the bulge around the center of our galaxy. The figure shows a prominent, Gaussian-shaped light-curve, corresponding to lensing by a single star, but here another object has also intercepted the same beam, as may be seen from the other, small Gaussian superimposed on the main one. Modeling of the amplitude and width of the small feature proves that the signal is caused by a planet orbiting the star (OGLE-2005-BLG-390Lb). In this case, the planet's mass is ~5.5 Earth masses, and it orbits a 0.2 solar-mass star at a distance of 2.6 AU. After Beaulieu et al. (2006).

8

is then necessary to consider monitoring over long, continuous spans of time, which implies making observations from space. It may also be noted that in this way a large number of planets is detected (with the reservation that they must be confirmed, *see* text), and this enables us to approach the detection of exoplanets from a different point of view, namely that of the *statistics* obtained from a large number of stars taken all together.

Method 3: gravitational microlensing

Here, the basic idea is a generalization of the transit method: instead of considering the transits of a planet in orbit around a given star, why not take the case of transits in front of a background star that is not associated with the planet? Obviously, to obtain a certain likelihood of detection, such a method cannot work unless there is a sufficiently dense "stellar background," as is the case in the direction of the galactic center. But that is not sufficient, because such a background is, by definition, highly luminous. To observe the transit of the planet in front of a star, it thus becomes necessary to find some other criterion than a decrease in luminosity. This is provided by general relativity, thanks to the phenomenon known as "gravitational lensing."

What is this? Putting it simply, general relativity, introduced by Albert Einstein between 1907 and 1915, states that gravitation is just a manifestation of the curvature of space-time through the effects of mass. In particular, the path of a ray of light is not straight, as in classical mechanics, but curved in the vicinity of a massive body. This was first confirmed during the total

solar eclipse of 1919[5]. The apparent positions of several stars close to the Sun's disk, once it was occulted by the Moon, were found to have moved closer to it, as predicted by theory. The presence of a mass therefore creates a "focusing" effect on background light. This is why the effect is known as a "gravitational lens."

This phenomenon has now been widely verified and documented to a high degree of accuracy, most notably in extragalactic astronomy, where the largest masses in the universe, clusters of galaxies, are capable of deviating the light from background galaxies to form "gravitational images" in the form of flattened arcs. As far as planetary transits are concerned, we can only be dealing with bright dots. Under these circumstances, gravitational focusing is revealed, paradoxically, by an *increase* in the luminous flux during a transit, but which is relatively short, and rather like a flash from a lighthouse. Under these circumstances, the term "gravitational microlensing" is used (Fig. 8.13). As with the previous method, a wide field of view is required. For the region of the galactic center, the density of stars on the sky is such that it is possible to follow several million stars simultaneously.

This method has been successful as far as the principle is concerned, but until now its achievements have been relatively few when compared with the other methods: just 13 exoplanets (including 1 system with 2 planets). The result is more interesting from a qualitative point of view, rather than quantitively, because although 65 per cent of exoplanets discovered by velocimetry lie less than 100 pc (about 300 light-years) from the Sun, the exoplanets discovered by gravitational

5. The observations were conducted by A. Eddington, who took measurements during an expedition to the small island of Sao-Tomé-and-Principe, off the coast of Gabon, in equatorial Africa. The eclipse was also observed at Sobral, on the Brazilian coast.

8

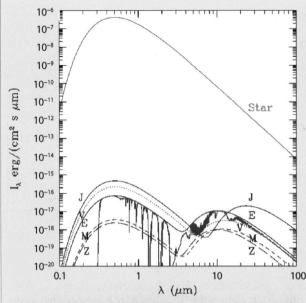

Fig. 8.14 Comparison between the absolute fluxes, as a function of wavelength, in the visible-to-far IR range, of the Sun and Solar-System planets (from top to bottom: Jupiter, Venus, the Earth, Mars, and Mercury) (Credit: SAO mode; from L. Kaltenegger et al. 2010). This figure illustrates the "contrast" problem affecting the direct imaging of planets in the vicinity of a solar-type star: seen from the Earth, a Jupiter-like planet is ~10^9 times fainter than the star in the visible (0.5 μm), and ~10^5 times fainter in the mid-IR (20 μm).

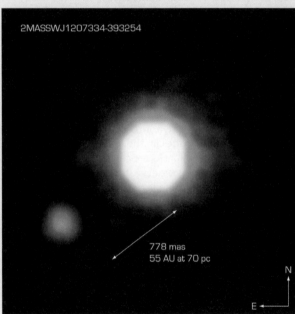

Fig. 8.15 Direct imaging of a planet around a brown dwarf star (Chauvin et al. 2005). This image was obtained by the ESO VLT equipped with the NACO adaptive-optics system. In this case, the imaging was made possible by two favorable factors: (i) the star is a "brown dwarf," i.e., a low-mass, very red star, where the contrast problem illustrated in Fig. 8.14 is much alleviated in the near-IR; (ii) the planet is sufficiently far from the star (~0.8 arcsec) that its image is not blended with that of the star. (Note that, contrary to what the image might suggest, in reality the star and the planet are point-like! They look extended only because of the non-zero diffraction limit of the mirror.)

microlensing lie at great distances, up to 6.5 kpc (that is, 20 000 light-years), and thus in an environment completely different from our own. An important limitation, however, is the non-repeatability of the detection, because there is only a random, short-lived link between the star and the foreground planet.

Method 4: direct imaging

The method that is, *a priori*, the most basic one for detecting an exoplanet is to see it directly. But here again, the difficulties are considerable. Basically, it is all a matter of contrast. By definition, a planet does not shine with its own light, but radiates a fraction of the light that it receives from the star. In the visible region the planet basically reflects light from the star: this is how we can see the planets in the Solar System. But the contrast is enormous (Fig. 8.14). For example, there is roughly a factor of 10^9 (one billion) between the Sun and Jupiter in the visible region (0.5 μm). On the other hand, in the infrared, planets re-emit some of the stellar radiation that they absorb (and this is even one of the key factors in the greenhouse effect). The contrast is lower, but still remains high: a factor of 10^5 at 20 μm, again between the Sun and Jupiter. At present, such values are not accessible to observation.

But if a cooler, and thus redder, star than the Sun is chosen, then the contrast decreases and, provided the apparent distance between the planet and the star is greater than the resolution of the telescope, it becomes possible to separate the star and the planet. Just recently, we have started to be able, in very favorable cases, to see exoplanets directly (◼ Fig. 8.15). Confirmation of a discovery, however, may take time, because it is necessary to follow the planet for weeks, or even months, to establish that it is really a planet (rather than a very low-mass brown dwarf, for instance) and to determine the parameters of its orbit.

8

Epilogue

The Earth: So Far a Unique Planet

At the close of this book, we may attempt to reflect about the implications of the existence of life on Earth, on the one hand, and the existence of hundreds, thousands, and perhaps millions (at the scale of our Galaxy) of (exo)planets located in the "habitable zone" (HZ) of their host stars, on the other. As discussed in the preceding chapter, the definition of the HZ is based on the necessary (but not sufficient) presence of liquid water at planet surface. We have stressed that two conditions must be met simultaneously: the planet must be in the astronomical HZ (ensuring minimal warming by the host star), *and* an atmosphere must exist, at some point in its evolution, such that its greenhouse effect (as measured by the CO_2 concentration, for instance) is strong enough to keep large quantities of water in liquid form (in other words, allow the presence of oceans).

Recall also that, for biologists, a "habitable" planet should present the physical and chemical conditions compatible with life, more precisely life "as we know it". These conditions are determined from studying the limits of life on Earth, which implies the presence of liquid water (but not necessarily at the surface, as we have seen), as well as temperatures compatible with life (between a few degrees below 0 °C and about 120 °C), as well as a source of energy, so that a metabolism can operate. The source may consist either of light, or of chemical oxidation-reduction (redox) reactions. Other sources of energy, which are not known to be used by terrestrial organisms, might perhaps also be envisaged as some researchers have suggested (radioactivity, tidal energy, or other exergonic processes). The long-term chemical stability of an atmosphere as well as, most importantly, the long-term availability of essential nutrients through biogeochemical cycling, most likely including the co-occurrence of reduced and oxidized molecular species necessary for (at least terrestrial) life's metabolism implies a third condition: that the planet is, one way or the other, geologically active.

For the time being, out of the hundreds to thousands of planets being progressively discovered around other stars, a handful fall into their astronomical HZ, but we don't have yet the spectroscopic technology to know whether they have a suitable atmosphere, based on our current best choice of "biosignatures". Still, there is a good hope that the relevant spectroscopic data will be obtained some time in the future. In addition, in the best cases (and increasingly so as astronomical observations become more complete), we are able to determine whether these planets are "rocky" (as opposed to gaseous), and even obtain an estimate of their masses, radii, etc. However, this is not sufficient to assess their possible tectonic activity. Perhaps planetary geology, drawn from studies of the Earth and of the diversity of large bodies in the solar system, will one day allow enough understanding about the internal structure of planets to attribute to some well-studied candidate habitable planets a probability of being tectonically active. On this basis, we may be reasonably optimistic in finding some day a sample of potentially habitable planets for which a "search for life" may start in earnest.

Yet, other considerations may come into play. Consider, not the age of the Earth (and of the solar system), i.e., about 4.5 Ga, but the age of the universe itself, i.e., about 13.5 Ga. Modern cosmology tells us that the first stars were formed about 400 Ma after the so-called "Big

M. Gargaud et al., *Young Sun, Early Earth and the Origins of Life*,
DOI 10.1007/978-3-642-22552-9, © Springer-Verlag Berlin Heidelberg 2012

Bang", during which, in the course of a few minutes only, the first atoms were formed (helium, deuterium, and a very small quantity of heavier elements up to beryllium). Thereafter, successive generations of stars synthesized the other elements, including perhaps the most important of all in our context: carbon. But as the universe expanded, the abundances of the elements quickly reached a quasi-steady state: stars that formed 12-13 Ma ago (and we know some, in the halo of our galaxy) already had, after only a few stellar generations, a composition not too different from that of the Sun and of younger stars today. From this we can rather safely infer that many (perhaps all?) stars of our galaxy in its infancy were already surrounded by planets, comparable to the planets we see today around stars in the solar neighborhood. In particular, we observe planets around stars that have a mass smaller than the Sun: these stars have a lifetime comparable or larger than that of the universe (the mass limit turns out to be just slightly smaller than a solar mass, ~0.9 $M_\odot$), and therefore *a very large number of planets in our galaxy must be much older than the Earth,* i.e., be nearly as old as the universe! Remember, for instance, that the first discovered planet orbiting a solar-type star, 51 Peg b, is 7.5 Ga old, i.e., about 3 Ga older than the Earth. In other words, if life developed in such planets on a more or less similar timescale than on the Earth (possibly adapted to other stars, like red dwarf stars), it would be extremely interesting to see how it evolved (if it did) and diversified as well as whether it has persisted over such long timescales. Among many other questions (*see below*), may the "biosignatures" we know now be detectable after such an enormously long time?

All arguments being considered, the truth is that, at least for the moment, we have at hand no planet or celestial body harboring life other than the Earth, this "terrestrial life" having appeared after the long series of steps, some of them still highly uncertain or even unknown, that we have described in the preceding chapters.

Life in the Universe: Chance or "Cosmic Necessity"?

Yet, as this book comes to its epilogue, we may want to broaden our approach. Although it may be for the biologist to ponder over the question of the origin of life and its evolution on Earth, for philosophers, chemists, geologists, astronomers, or even for the interested public, one fundamental question remains unanswered so far: *is life unique* and, consequently, life on Earth its only example?, or is it a *"cosmic necessity"* and life on Earth only a handy example? Before planetary systems were discovered around other suns, the former point of view, i.e., that life emerged only on Earth, and only out of chance and contingency, was strongly defended by Jacques Monod[25], Nobel Prize winner for Physiology and Medicine (1965). The latter opinion, i.e., that life is just the result of some form of determinism, when appropriate physical and chemical conditions are met, was maintained by Christian de Duve[26], another Nobel Prize winner for Physiology and Medicine (1974). The position of Monod must be understood in a historical context when molecular biology and its amazing complexity were being revealed for the first time. Many of the processes and molecules underlying those early discoveries may not seem as unique and rootless as they seemed at that time. The debate Monod – de Duve should be rather formulated in modern times as: "is life rare in the universe or is it frequent?"

25. "The universe is not pregnant with life, nor with the human biosphere. Our number came up from a game of roulette. It is astonishment that, like someone who has won a billion, we should feel about the strangeness of our condition." (J. Monod, *Le Hasard et la Nécessité*, Point-Seuil, 1970, p. 185, French edition)
26. "Life belongs to the very fabric of the universe. If it were not an inevitable consequence of the combinatorial properties of matter, it would have been absolutely impossible that it should have been born naturally." (C. de Duve, *Construire une cellule*, De Boeck-Wesmael, 1990, p. 291, French edition)

⬛ **Fig. 9.1.** Life around a binary system? (Credit: Roberto Retagna)

If indeed life on Earth was the result of a series of unlikely events and chance dominated during the process, reconstructing how it emerged will be extremely difficult. If life was, following Christian de Duve, a cosmic imperative, it would be rather easy to reproduce it and find the physico-chemical conditions where it emerged. Regardless of the relative weight of chance versus determinism, most scientists in this field now agree that, under determined –albeit unknown– conditions, some kinds of self-organization processes ruled only by the laws of physics and chemistry (some perhaps yet unknown) have a non-negligible probability to emerge, with features that resemble some of the characteristics of life. This may be for instance the ability of chemical replicators to obey selection rules based on *dynamic kinetic stability*, as proposed by Addy Pross[27] to define a kinetic state of matter. These properties may also be found in the peculiarities of *hypercycles,* as networks of reaction systems theoretically studied since 1970 by Manfred Eigen[28], or more generally to autocatalysis and network autocatalysis that may give rise to processes that are preferentially selected because they reproduce themselves at a faster rate. The main requirement for the emergence of life is the persistence of far-from-equilibrium conditions allowing this kind of fast-reaction selection process, i.e., a selection governed by *dynamic kinetic stability*. In other words, a necessary condition for features of life to emerge is an environment that is rich in chemical energy, and the universe has plenty to offer.

If, one day, we discover life in another planetary system, or if the developments of synthetic biology and systems chemistry lead to the spontaneous emergence of a chemical system whose behavior mimics the evolutionary capabilities of living beings in an appropriate

27. *Pross A: What is Life? How chemistry becomes biology. Oxford University Press, Oxford, 2012*
28. *Eigen M: Self-organization of matter and the evolution of biological macromolecules. Naturwissenschaften 1971, 58:465–523.*

chemical environment, we shall have the beginning of an answer. However, even if, in such a situation, the radical position of Jacques Monod has to be abandoned, we shall still be far from being able to decide definitively in favor of the kind of "necessity" advocated by Christian de Duve, because many other questions remain open. Judge for yourselves ...

If the *emergence* of life proves not to have been a phenomenon strictly exclusive to the Earth (and thus unique), the inevitable next question is: given the undoubtedly large number of "habitable" planets, some nearly as old as the universe itself, is life rare, both in space and in time, or is it widespread or even universal, after perhaps some gestation period that may differ from place to place? And, for all that, are the conditions for the emergence – not to say the survival – of life precisely definable anyway? In strictly physical and chemical terms (the presence of liquid water, of organic molecules, of energy in a form that can interact spontaneously with these building blocks, etc.), this is perhaps so, because such conditions involve compounds that are abundant and fairly uniformly distributed throughout the universe. But as long as the processes that led to life on Earth remain undeciphered, it will be impossible to arrive at a convincing answer.

And even if we were able to answer the preceding questions, what would be the *evolutionary path* followed, over periods of up to 13 billion years, by any extraterrestrial life? We have already seen the wide variety of planets that could harbor liquid water, and it is a good bet that any life around a star like Gliese 581 – a relatively cold "brown dwarf" radiating a lot in the infrared (with a surface temperature of just 3500 K rather than 5900 K for the Sun) at the same time as in the ultraviolet (because of its intense chromosphere) – would follow a very different evolution from that followed on Earth, driven by the selection of "solutions" adapted to that particular environment. How would energy exchanges take place, what would be their nature and their efficiency? What possibilities would there be for photosynthesis? How fast would that evolution take place and how far would it go? As already noted, stars with a mass only slightly smaller than the Sun have a lifetime larger than that of the universe! And we have to keep in mind that about half of the stars are in binary systems, and that planets have already been found to orbit such systems: how would life evolve in the light of two Suns?

These are only a few of the many puzzling questions to which nobody has an answer today. If, in the future, reliable indications of extraterrestrial life do come to be discovered, then perhaps we shall be in a better position to provide parts of the answer. It is even possible that we shall be able to come, through this extraordinary, but indirect, means, to understand the origin of life on our planet, that is, *our own origins*. Which would not be the least of paradoxes...

The Main Principles for Rock Classification

Classically, geologists distinguish between two major groups of rocks: (1) the igneous or **magmatic** rocks, formed through crystallization of a magma generated at depth, either within the crust or the mantle; (2) exogenous or **sedimentary** rocks, which form at the surface of the planet. In addition, a third group, called **metamorphic** rocks, consists of rocks that were initially magmatic or sedimentary, and which were transformed and re-equilibrated under new conditions of pressure and temperature.

Magmatic Rocks

These are the result of magma crystallization and their classification depends primarily on two criteria: (1) the place where the magma crystallized; (2) the chemical composition of the magma.

Place of Magma Crystallization

Due to their density, their viscosity, their temperature, etc., some magmas cannot reach the surface, so they accumulate at depth where they form large pockets known as plutons. There, they crystallize slowly, giving rise to large, stubby crystals forming a "granular texture"; such rocks are called "plutonic rocks".

On the other hand, if the magma reaches the surface, it cools rapidly, giving rise to numerous, small acicular crystals, forming a texture called "microlitic texture". In some cases, the cooling is so sudden that crystals do not have time to form. The magma then freezes as a glass and the texture is said to be "vitreous or hyaline". The magmas that erupted at the surface of the Earth, give rise to what is called effusive or volcanic rocks.

Composition of the Magma

The chemical composition of the magma controls the nature of the crystallizing minerals. Put simply, geologists recognize several families of magmas depending on both their SiO_2 and MgO contents. The magmas that are very rich in MgO (~30%) and poor in SiO_2 (~42%) are said to be *ultrabasic* whereas those that are poor in MgO ($\leq$ 1%) and rich in SiO_2 ($\geq$ 70%) are said to be acid. Between these two poles, there is a whole range of compositions that geologists gather, for the sake of simplicity, into two groups: one of basic rocks and the other of intermediate rocks.

The diversity of igneous rocks arises from the combination of these two parameters: place of crystallization and magma composition (◼ Fig. R1). Of course, this classification is extremely simplistic. This is why specialists have refined it and have proposed several additional subdivisions.

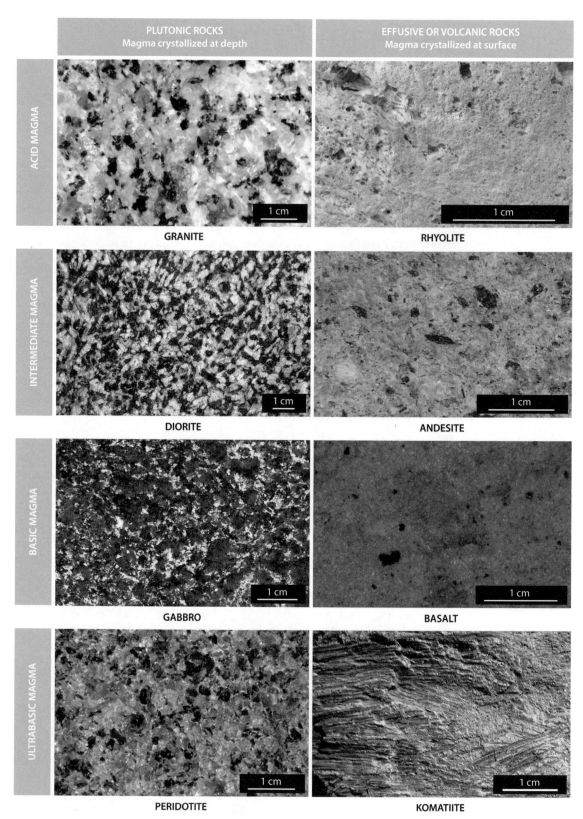

PLUTONIC ROCKS Magma crystallized at depth	EFFUSIVE OR VOLCANIC ROCKS Magma crystallized at surface
GRANITE	RHYOLITE
DIORITE	ANDESITE
GABBRO	BASALT
PERIDOTITE	KOMATIITE

ACID MAGMA · INTERMEDIATE MAGMA · BASIC MAGMA · ULTRABASIC MAGMA

▣ **Fig. R1 The classification of magmatic rocks relies mainly on the combination of two factors:** the place where magma crystallization took place; it controls both the shape and the size of the crystals (small, elongated crystals in volcanic rocks and large stubby crystals in plutonic rocks; and the composition of the magma, which determines the nature and the abundance of the crystallizing minerals. (Photos: H. Martin.)

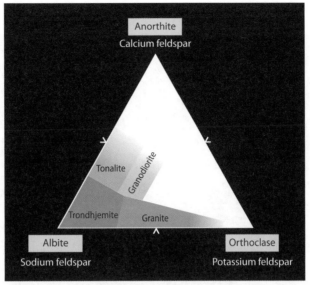

Fig. R2 Classification diagram for acid plutonic rocks containing more than 10% quartz. This classification is based on the relative abundance of potassium feldspar (Orthoclase), sodium feldspar (Albite) and calcium feldspar (Anorthite).

For example, in ▢ Fig. R1, the "granite" box corresponds in fact, to a range of silica-rich (quartz-bearing) igneous rocks that also contain plenty of feldspars. Classically the mineralogists separate the plagioclase feldspars, which form a continuous series between albite [$NaAlSi_3O_8$] and anorthite [$CaAl_2Si_2O_8$] and the potassium feldspar = orthoclase [$KAlSi_3O_8$]. Depending on the nature of the feldspars, the "granite" box in ▢ Fig. R1, can be divided into several rock types (▢ Fig. R2):

- **tonalite:** the feldspar is exclusively a calcic plagioclase;
- **trondhjemite**: the feldspar is a sodic plagioclase;
- **granodiorite:** there are two feldspars, but the plagioclase is far more abundant than the potassium feldspar;
- **granite:** there are two feldspars, but the potassium feldspar is as abundant or more abundant than the plagioclase.

The Sedimentary Rocks

Sedimentary rocks originate in the weathering and erosion of pre-existing rocks. Most often, the products of this alteration are transported, either in the form of solid particles (by rivers, glaciers, wind, etc.), or in the form of ions dissolved in water. This transport generally ends in a sedimentary basin (lake, sea, ocean, etc.), where unconsolidated sedimentary rocks are deposited (sand, for example). There, they become progressively compacted and cemented, the induration of sand giving a sandstone.

There are thus (▢ Fig. R3):

- **detrital rocks**, which are formed through the accumulation of debris transported in the form of solid particles;
- **chemical rocks**, formed by precipitation of the ions dissolved in water.

Derital Rocks

The classification of these sedimentary rocks is based on the size of the particles. If the latter are small (< 63 μm), the loose rock is called a **silt**, if larger (a size between 63 μm and 2 mm), it is a **sand**, and if larger in size than 2 mm, **gravels** or **pebbles**. Once indurated, these loose rocks become **lutites, arenites** or **sandstones**, and **conglomerates**, respectively.

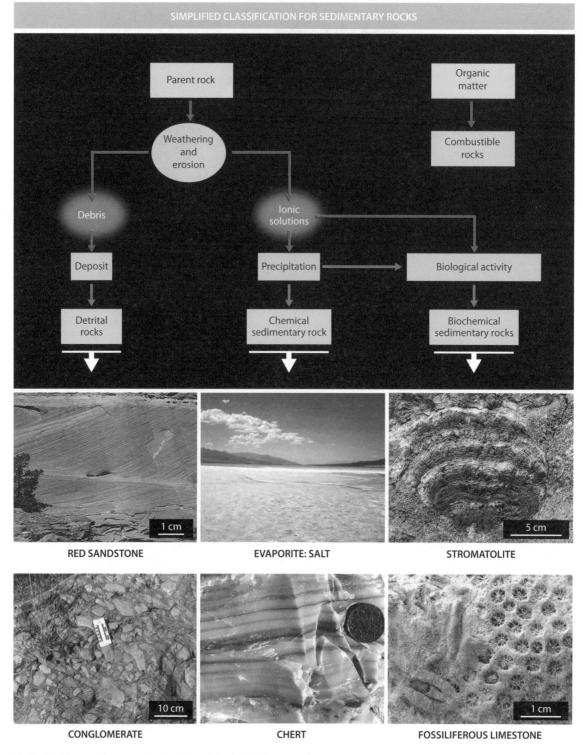

SIMPLIFIED CLASSIFICATION FOR SEDIMENTARY ROCKS

RED SANDSTONE

EVAPORITE: SALT

STROMATOLITE

CONGLOMERATE

CHERT

FOSSILIFEROUS LIMESTONE

Fig. R3 Diagram illustrating the broad principles for the classification of sedimentary rocks. The photographs represent examples of detrital rocks (left-hand column), rocks of chemical origin (center column), and rocks of biological origin (right-hand column). (Photos: H. Martin.)

Chemical Rocks

The precipitation of ions dissolved in water may occur in two ways.

The precipitation may be direct as, for example, by evaporation of the depositional environment. The ions are then deposited in the form of salts, such as sodium chloride, gypsum, etc. Due to their origin, these rocks are called **evaporites**. This precipitation may also occur due to change in the physicochemical conditions of the environment, as is the case when hydrothermal fluids enter the ocean. The siliceous rocks that are formed in this way are known as **cherts**.

The precipitation may be achieved through the intervention of living organisms. This is particularly the case with organisms that remove ions from their environment to construct their shells or skeleton. After their death, the accumulation of shells may form a rock called rock of biochemical origin.

Combustible Rocks

There exists a third type of sedimentary rock. These are combustible rocks, also linked to living beings. They form directly through the accumulation of organic matter: they are the lignites, coals, and petroleum.

Metamorphic Rocks

A metamorphic rock is a magmatic or sedimentary rock that, generally, due to plate tectonics, has been transported to an environment where the conditions of pressure and temperature differ from those of its place of origin. Let us take as an example, the basalts of the oceanic crust. When, at a subduction zone, they sink into the mantle, the deeper they go, the higher the pressure and temperature become. The minerals of which they consist, which were in equilibrium under the low-pressure and low-temperature conditions at the bottom of the oceans, become unstable and are replaced by new minerals, that are stable under the new conditions of pressure and temperature.

All these reequilibration reactions take place in the solid state. They also usually give rise to a modification of the rock texture. For example, when flakes (micas) or acicular (amphibole and pyroxenes) minerals grow during a metamorphic event, they align themselves perpendicular to the main pressure, thus defining a plane of preferential orientation in the rock, which is called the foliation plane (◘ Fig. R.4).

◘ **Fig. R4 A metamorphic rock (gneiss) showing an almost horizontal foliation plane.** The crystals of mica (black) have all recrystallized perpendicular to the principal pressure, thus defining a preferential plane of orientation (the foliation plane). The large feldspar crystals (white) have also been deformed and they are aligned within the foliation plane. (Photo: H. Martin.)

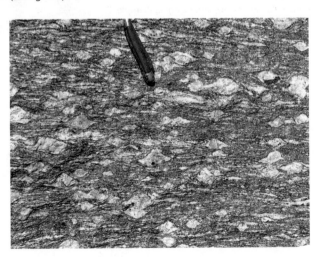

SOME OF THE MAIN METAMORPHIC ROCKS

Micaschist
A rock that mainly consists of an association of mica and quartz, and that does not contain feldspar (or very little). Because of the abundance of mica crystals, the foliation plane is particularly well marked. The micaschists thus often display planar cleavage.

Gneiss
Metamorphic rock containing quartz and feldspars as well as, most often, micas. Depending on the origin of the rock that has undergone metamorphism, it is called paragneiss (former sediments) or orthogneiss (former granites).

Paragneiss

Orthogneiss

Amphibolite
A rock mainly consisting of amphibole and plagioclase feldspar, which may also occasionally contain a small amount of quartz. The amphibolites are often metamorphosed basalts.

Eclogite
A rock that consists in an association of garnet (red) and pyroxene (green). The eclogites are produced by high-pressure metamorphism of basalts (at great depths). For example, when an oceanic basalt is subducted, it transforms into an amphibolite, then into garnet-bearing amphibolite, and finally, into eclogite. (Sample: F. Cariou.)

Migmatite
At high temperature, a gneiss may start melting, giving rise to small veins of granitic magma. Such a partially molten rock, is called migmatite. In the photograph, the white veins correspond to the granitic liquid produced by melting of the surrounding gneiss, whereas the dark areas are unmelted remnants.

◨ **Fig. R5 Some of the main metamorphic rocks to which reference is made in this book.** (Photos: H. Martin.)

The 14 Chronological Stages in the Origin of the Earth and Life

1   The Solar System's gestation

2   The Solar System's birth

3   The Solar System's infancy

4   Bringing water to Earth

5   The beginning of the differentiation of the Earth

6   The first continents, the first oceans

7   The Late Heavy Bombardment

8   The transition from non-life to life

9   Biological evolution begins

10   Oxygen begins to accumulate in the ocean and the atmosphere

11   The appearance of eukaryotes

12   The appearance of the first multicellular organisms

13   The Cambrian Explosion

14   The explosion of macroscopic life

▪ 1 Around about 4750 years ago

The Solar System's gestation
From an interstellar cloud to a circumstellar disk

4.57 billion years ago, somewhere in our galaxy, the Milky Way, a condensation of gas and dust within an interstellar cloud, collapsed under the influence of gravitation.

At its center, an embryonic core formed, the origin of our future star, the Sun. The rest of the condensation became a relatively isolated envelope, which fell more slowly onto that core via a circumstellar disk (which surrounded the star).

After one hundred thousand years, a protostar was born.

This star-to-be subsequently underwent a true metamorphosis: in a million years its envelope became progressively more tenuous, the central star grew at the expense of the circumstellar disk, along with ejection of material in the form of bipolar jets.

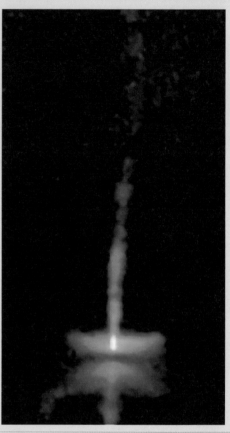

☐ **Fig. C.1 The young star HH30 (in the constellation Taurus).** Less than a million years after its birth, our Sun must have resembled this star. We can see the circumstellar disk, with a flared shape, and the bipolar jets, symmetrically leaving the star perpendicular to the disk. (The radius of the disk is 4 times that of the present-day Solar System.)

▪ 2 Between 4570 and 4560 million years ago

The Solar System's birth
The formation of the giant planets

The circumstellar disk underwent significant transformations. Through mechanisms that are still very poorly understood, the grains of dust that it contained, grew rapidly, until they formed "clumps", a kilometer in radius, which are known as planetesimals. The latter, in turn, collided with one another and, in a few million years, gave birth to planetary embryos. Subsequently, these embryos very rapidly "accreted" the gas from the disk in their vicinity, until they reached several tens of Earth masses. This is how the giant planets were born. At the end of ten million years, the disk had disappeared, so in the Solar System there remained:

– the giant planets, formed in the outer regions of the disk;

– those planetesimals that had escaped the planetary-formation process;

– meteorites, which are the debris resulting from collisions between planetesimals.

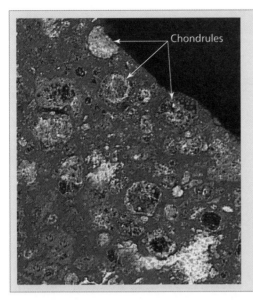

Fig. C.2 A fragment of the Allende meteorite. This meteorite, which fell in Mexico in 1969, contained, at the time when the Solar System was formed, radioactive elements, the traces of which may be found today. Certain of these traces have allowed us to date, very accurately, the age of the Solar System: 4.5685 billion years.

■ 3 Between 4560 and 4500 million years ago

The Solar System's infancy
The formation of the rocky planets

From the multitude of planetesimals that still existed and influenced by the gravitational perturbations caused by the giant planets, a second category of bodies formed, but much slower this time. These are the rocky or terrestrial planets, Mercury, Venus, Earth and Mars.

Collisions that became rarer and rarer, but which were also violent, occurred over some tens of millions of years. It was the last of these collisions, between the young Earth and a body the size of Mars, that gave birth to the Moon.

Fig. C.3 The formation of the Moon. For several tens of millions of years after its birth, numerous collisions took place in the Solar System. Some debris could have remained in orbit around one of the objects involved and then gather together again, forming a new satellite body. This is how the birth of the Moon is envisaged, by a gigantic collision between the forming Earth and a Mars-sized impactor, about 70 million years after the beginning of the Solar System.

▪ 4 Between 4560 and 4500 million years ago

Bringing water to Earth

The significance of extraterrestrial sources

Water was brought to the Earth at the end of the formation of the rocky planets – or in the period immediately following it – when our planet had cooled sufficiently.

Comets, consisting of ice and dust, could be one of the sources. However, study of their hydrogen-isotope composition (the hydrogen/deuterium ratio) indicates that 20% at the most of the Earth's water could be of cometary origin.

It seems, therefore, that the water in the oceans must have been gained either during the final phases of accretion, by the planetesimals originating in the outer asteroid belt, or somewhat later, by meteorites, through a meteoritic bombardment after the Earth's formation. Thereafter the water provided by micrometeorites represented a non negligeable supplementary supply to the hydrosphere.

▫ Fig. C.4 The asteroid 433 Eros as seen by NASA's NEAR-Shoemaker spaceprobe. This small body (33 km long) now belongs to our Solar System's asteroid belt. Between 4.56 and 4.50 Ga, such bodies, which were then lying sufficiently far from the Sun to contain a lot of ice, probably contributed a significant amount to the water on Earth.

▪ 5 Round about 4500 million years ago

The beginning of the differentiation of the Earth

The separation of the core and the mantle, the formation of a magma ocean

During the Solar System's first 70 million years, planetary accretion led to the formation of a homogeneous Earth. Around 4500 million years ago, the iron and silicates separated. Due to its greater density, iron concentrated at the center of the planet and formed the core. The silicates, being lighter, remained outside and formed the mantle. The rotation of the solid inner core within the liquid outer core is the source of the Earth's magnetic field which, even today, still protects the surface of the planet from the solar wind.

At the same time, the gravitational energy released during Earth's accretion together with that due to the disintegration of radioactive elements (which were very abundant), caused the melting of the whole outermost portion of the mantle, which gave rise to a magma ocean.

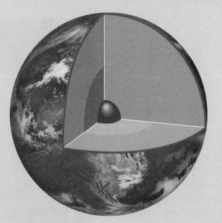

▫ Fig. C.5 A schematic section of the present-day Earth, showing its structure in concentric layers. A solid inner core (turquoise), a liquid outer core (grey), a mantle (green), oceanic and continental crusts (brown).

■ 6 About 4400 million years ago

The first continents, the first oceans
Towards an Earth that was potentially habitable?

Because of the high temperature that prevailed on Earth, water was vaporized first into the atmosphere. Subsequently, it condensed and formed oceans.

The analysis of oxygen isotopes within zircon crystals discovered in Australia, and with ages between 4400 and 4300 million years, indicate the presence of liquid water (and thus perhaps of oceans) on the surface of the planet 4400 million years ago.

Those same zircon crystals also show that a stable, granitic, continental crust existed at that time, less than 200 million years after the beginning of the Earth's formation. With a continental crust and oceans, the Earth was thus potentially habitable from 4400 million years ago … which does not mean that it was inhabited.

Fig. C.6 The oldest terrestrial material known: a zircon discovered at Jack Hills (in Australia), aged 4.4 billion years. Zircons are very resistant to alteration and they contain radioactive elements such as thorium and uranium, which allows them to be dated easily. They are thus excellent "memories" of Earth's history.

■ 7 About 3900 million years ago

The Late Heavy Bombardment
An temporarily uninhabitable Earth?

Over the first 500 million years of the Earth's history, the frequency of collisions between planetesimals greatly decreased.

However, the 1700 lunar craters, dated to around 3900 million years ago, bear witness to an episode of meteoritic bombardment, as unique as it was intense, which has been attributed to the late reorganization of the orbits of the gas-giant planets.

Extrapolating, it is estimated that more than 22 000 craters (of which 200 would have had a diameter greater than 1000 km) would have formed on Earth at that period. If life already existed, either it would have completely disappeared and the process would have had to start again from scratch (undoubtedly in a different form), or there was a mass extinction, but microorganisms living at the greatest depths of the oceans or in subsurface rocks were protected, and they might have subsequently repopulated the planet.

Fig. C.7 The Schrödinger crater, formed on the Moon during the intense bombardment 3900 million year ago. With a diameter of 320 km, it is not, by far, one of the largest lunar craters.

■ 8 Between 4300 and 2700 million years ago

The transition from non-life to life
From prebiotic chemistry to the first cells

Life appeared on Earth at some uncertain date, between 4300 and 2700 million years ago and, probably, between 3800 million (after the Late Heavy Bombardment) and 3500 million years ago. (The latter date being the age of mineral microstructures attributed – without any absolute certainty – to being microfossils.)

During this interval of time, thermal or photochemical processes in the atmosphere (which primarily consisted of N_2, CO_2, and H_2O), the action of reducing minerals in hydrothermal systems at the bottom of the oceans, as well as the fall of specific meteorites (the carbonaceous chondrites) were the source of organic material.

Networks of chemical reactions developed, and the products became associated with one another in supramolecular systems which, after stages that have not yet been fully elucidated, gave birth to the first cells possessing three fundamental properties: a membrane, a metabolism, and a reproducible genetic system that served as a basis of evolution by natural selection.

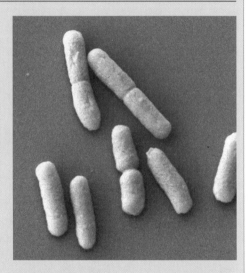

■ **Fig. 6.8** Bacteria under a scanning electron microscope. Living organisms undoubtedly quite rapidly assumed the form of the most simple prokaryote cells known today. (Like the bacterial cell shown above.)

■ 9 Between 3500 and 2700 million years ago

Biological evolution has started
The first traces of life and the diversification of prokaryotes

The first cells diversified and adapted to various ecological niches. They gave rise to prokaryotes, with a simple structure, which subsequently separated into bacteria and archaea. The appearance of new types of metabolisms (photosynthesis, methanogenesis, and various forms of respiration) accompanied that separation. Photosynthetic bacteria synthesized organic matter from CO_2, using the energy of light and an electron donor: H_2S, Fe^{2+}, H_2, or perhaps even then, H_2O (in oxygenic photosynthesis).

Stromatolites, laminated organo-sedimentary structures formed by complex communities of prokaryotes that induce carbonate precipitation, were widespread. The oldest of these date back to 3450 million year ago, but their biological origin is questioned, as are those of other very ancient fossil traces. The oldest non-controversial stromatolites have an age of 2700 million years.

■ **Fig. C.9** The oldest fully-recognized stromatolites known (Tumbiana, in Australia). These structures, 2700 million years in age, were formed by the precipitation of carbonates associated with complex microbial communities.

10 About 2400 million years ago

Oxygen begins to accumulate in the ocean and the atmosphere
The impact of life on the environment

The oxygenation of our planet is the consequence of the presence of life and, in particular, of that of cyanobacteria.

This group of bacteria developed oxygenic photosynthesis which breaks apart H_2O molecules, releasing oxygen as waste. In parallel, oxygen respiration spread in different microbial lines. Those organisms that did not succeed in respiring oxygen or tolerating it, found themselves confined to anoxic ecological niches, such as sediments.

Oxygen that had not been entirely consumed by aerobic respiration, first oxidized minerals in the surface rocks of the planet, and then accumulated in the ocean, the atmosphere and in sediments (forming the so-called "banded-iron formations").

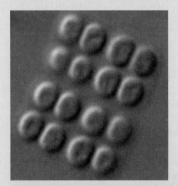

◼ **Fig. C.10** Different types of cyanobacteria (coccoidal, top, filamentous, bottom). These bacteria produce oxygen during photosynthesis. They contributed to the oxygenation of the Earth's atmosphere.

11 About 2000 million year ago

The appearance of eukaryotes
The first cells with nuclei and their fossil traces

The eukaryotes consist of cells that include a nucleus (which contains the genetic material) and organelles: mitochondria, where oxygen respiration takes place, and in the case of plants, chloroplasts, where photosynthesis occurs.

Mitochondria and chloroplasts derive from ancient bacteria incorporated by a eukaryote cell during endosymbiosis. The eukaryotes are thus partly the product of symbioses that involve prokaryotes.

The first eukaryotes were unicellular. They did not have a mineral skeleton, but an organic wall. However, in the fossil record they can be distinguished from the prokaryotes by their size, generally (but not always) larger and, above all, by the ornamented structure of their walls. The most ancient eukaryote micro-

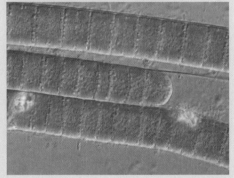

◼ **Fig. C.11** Tappania plana, one of the most ancient eukaryote fossils (from the Roper Group, Australia). It is 1500 million years old.

fossils (acritarchs) are about 1500 to 1800 million years old.

■ 12 Between 1200 and 540 million years ago

The appearance of the first multicellular organisms
The first traces of algae, animals and fungi

In the fossil record, the first fossils of multicellular eukaryotes date back to 1200 million years ago, then others appeared between 1000 and 750 million years ago (algae, fungi, and unidentified organisms).

The most ancient animal fossils date back to about 550–600 million years ago. These are the famous Ediacara fauna, consisting of microscopic fossils, and then macroscopic forms. They were soft-bodied organisms (without a skeleton or shell), which exhibited an enormously rich range of body plans. The ability of the animals to precipitate minerals and to form skeletons appeared just before the Cambrian (around 540 million years ago). Plants emerged later, as revealed by fossil moss spores that are about 440 million years old.

Differentiated cellular structures, specialized for carrying out specific functions (cellular tissues) appeared in plants and animals.

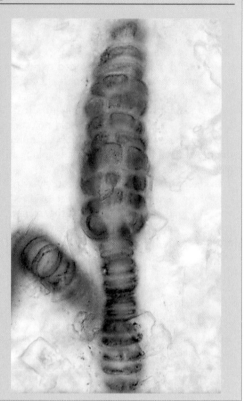

▣ **Fig. C.12** Bangiomorpha pubescens, which is, perhaps, a red multicellular alga, 1200 million years old.

■ 13 540 million years ago

The Cambrian Explosion
Animal life diversifies, shells and carapaces develop

Most of the representatives of the Ediacara fauna disappeared mysteriously at the boundary of the Cambrian.

We then once again see the appearance of a great diversity of fossil animals, but this time, with a lesser range of body plans, but with the presence of shells, carapaces, spines and various appendages. The evolutionary radiation (great diversification in a short period of time) was caused by the appearance of biological innovations (structures for protection and predation, and new lifestyles) that allow new ecological niches to be populated.

In parallel with this, prokaryotes and unicellular eukaryotes continued to diversify and evolve.

▣ **Fig. C.13** Evidence of the Cambrian Explosion. Anomalocaris, a predator that swam in the ocean 505 million years ago. (It was 45 cm long.)

■ 14 Between 540 million year ago and the present

The explosion of macroscopic life
Biological evolution continues ...

The last 540 million years are distinguished by the explosion of macroscopic life, which first spread in the oceans, and then on dry land.

This period was accompanied by sudden environmental changes – linked to glaciations, to intense volcanism and to the fall of one or more meteorites – that led to massive extinctions of species, particularly macroscopic ones.

The impact of the Chicxulub meteorite, 65 million years ago, eliminated most of the dinosaurs and favored the expansion of present-day lines of mammals, of which one, that of the primates, would evolve into the species Homo sapiens, about 200 000 years ago.

In parallel with this, other animals, plants, unicellular eukaryotes, and microscopic prokaryotes have continued to evolve. All present-day organisms, from bacteria to humans, have undergone the same degree of evolu-

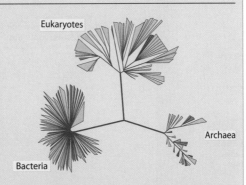

Fig. C.14 The diversity of life today. Two domains of prokaryote organisms are recognized (the bacteria and the archaea), and one domain of eukaryote organisms, among which there are several multicellular lines.

tion. They have all travelled along the same long evolutionary path. And they continue to evolve...

THE RIVER OF TIME

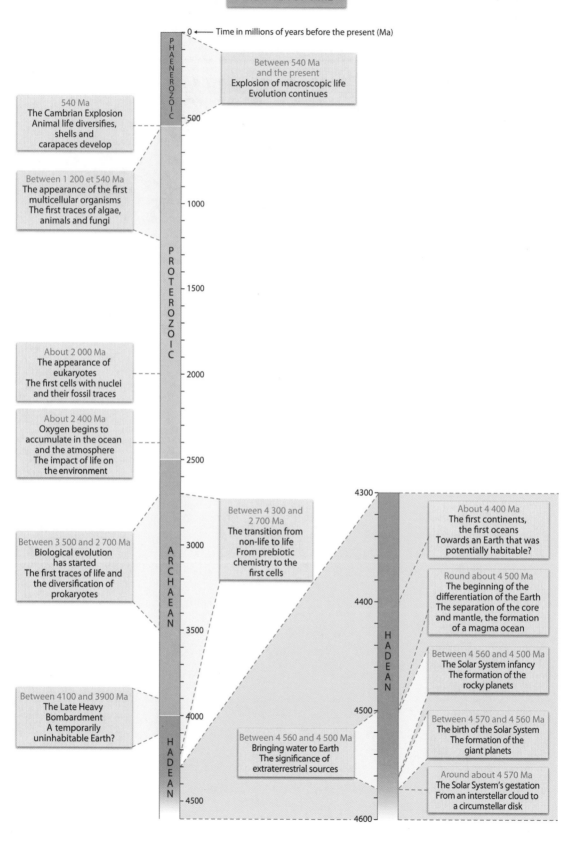

Time in millions of years before the present (Ma)

Between 540 Ma and the present
Explosion of macroscopic life
Evolution continues

540 Ma
The Cambrian Explosion
Animal life diversifies,
shells and
carapaces develop

Between 1 200 et 540 Ma
The appearance of the first
multicellular organisms
The first traces of algae,
animals and fungi

About 2 000 Ma
The appearance of
eukaryotes
The first cells with nuclei
and their fossil traces

About 2 400 Ma
Oxygen begins to
accumulate in the ocean
and the atmosphere
The impact of life on
the environment

Between 3 500 and 2 700 Ma
Biological evolution
has started
The first traces of life and
the diversification of
prokaryotes

Between 4 300 and 2 700 Ma
The transition from
non-life to life
From prebiotic
chemistry to the
first cells

Between 4100 and 3900 Ma
The Late Heavy
Bombardment
A temporarily
uninhabitable Earth?

Between 4 560 and 4 500 Ma
Bringing water to Earth
The significance of
extraterrestrial sources

About 4 400 Ma
The first continents,
the first oceans
Towards an Earth that was
potentially habitable?

Round about 4 500 Ma
The beginning of the
differentiation of the Earth
The separation of the core
and mantle, the formation
of a magma ocean

Between 4 560 and 4 500 Ma
The Solar System infancy
The formation of the
rocky planets

Between 4 570 and 4 560 Ma
The birth of the Solar System
The formation of the
giant planets

Around about 4 570 Ma
The Solar System's gestation
From an interstellar cloud to
a circumstellar disk

PHAENEROZOIC

PROTEROZOIC

ARCHAEAN

HADEAN

HADEAN

Glossary

* Another glossary entry
** See the classification of rocks, p. 267

Abiotic Organic Chemistry

Organic chemistry that is not carried out by living organisms, whether before or after the appearance of life.

Accretion, Magnetospheric

The process by which solar-type stars acquire their mass. The material (gas and dust) is accreted from the circumstellar disk* and falls onto the star following the lines of force of the stellar magnetosphere.

Adsorption

The accumulation of a gas, a liquid or a solute on the surface of a solid.

Alteration

Modification of the physical and chemical properties of rocks and minerals by atmospheric, biological, chemical and other agents.

Amphibole

A family of minerals. Inosilicates (water-bearing double-chain silicates). This family includes (among others) the calcium amphiboles, such as actinolite $Ca_2(Fe,Mg)_5Si_8O_{22}(OH,F)_2$ and hornblende $Na_{0-1}Ca_2(Fe^{2+},Mg)_{3-5}(Al,Fe^{3+})_{0-2}Si_{8-6}Al_{0-2}O_{22}(OH,F)_2$.

Amphiphile

Describes a molecule that has one polar, hydrophilic region and one hydrophobic region.

Anabolism

A set of chemical reactions of the cellular *metabolism that lead to the synthesis of organic molecules, including *macromolecules.

Anorthosite

A plutonic** magmatic rock consisting solely of plagioclase feldspar. Most frequently it results from the accumulation (through flotation) of crystals of plagioclase. Relatively rare on Earth, these rocks are abundant on the Moon, where they form the primitive "crust".

Anoxic

Describes an environment without oxygen.

Anticodon

A nucleotide triplet carried by tRNA* possessing a complementary sequence to that of the codon*, and that is specific for a particular amino acid.

Apatite

A mineral. A phosphate with the formula $Ca_5(PO_4)_3(F,Cl,OH)$.

Archaeon (Pl. Archaea)

A prokaryotic organism belonging to the domain Archaea, one of the three domains of life.

Asteroid Belt

The belt in the form of a ring between Mars and Jupiter, containing most of the asteroids known in the inner Solar System.

Astronomical Unit (AU)

By definition, the distance between the Sun and the Earth, namely 149.6 million km. The Solar System has a radius of 30 AU, out to Neptune but not including Pluto (now defined as a trans-Neptunian dwarf planet belonging to the Kuiper Belt*).

ATP (Adenosine Triphosphate)

A molecule used by all living organisms to store chemical energy in the form of two high-energy bonds between phosphate groups; it is also the precursor of certain co-enzymes, such as coenzyme A.

Autocatalysis

Describes a chemical process whose product catalyzes its own formation.

Autotrophy

The capacity of certain living organisms to synthesize organic matter from inorganic sources. The energy required for this process may be from light (photosynthesis) or originate from inorganic redox reactions (chemo-autotrophy).

Bacterium

A prokaryotic organism belonging to the domain Bacteria, one of the three domains of life.

BIF (Banded Iron Formation)

A sedimentary rock abundant in Archaean terranes (4.0–2.5 Ga) and that is no longer formed nowadays. It consists of an alternating set of centimetre-thick layers rich in iron (magnetite) and of white silicaceous beds (chert). The abundance of banded iron formations before 2 Ga reveals the absence of oxidizing conditions at the surface of the continents and thus also the low abundance of atmospheric dioxygen.

Biogenic

Of biological origin.

Biomarker

An organic molecule of undisputable biological origin.

Bipolar Jet

The phenomenon where matter is ejected to a considerable distance on both sides of a circumstellar disk* around a young star, and directed along the polar axis. It is linked to the presence, around the star, of an extensive stellar magnetosphere*.

CAI (Ca-Al-rich Inclusion)

A millimetre- or centimetre-sized assemblage of refractory minerals (especially the oxides and silicates of calcium and aluminium) often present in chondrites. This type of refractory inclusion is formed at a high temperature by condensation, melting and crystallization (or both) in the early gaseous nebula*. The CAI are considered to be the most ancient solid particles in the Solar System.

Carbonate

A family of minerals containing the CO_3^{2-} ion. Examples are: calcite $CaCO_3$, dolomite $MgCa(CO_3)_2$, siderite $FeCO_3$.

Catabolism

A set of chemical reactions of the cellular *metabolism that break down organic molecules within the cell.

Catalysis

The action of a chemical species capable of increasing the speed of a reaction and to be unchanged itself at the end of the process. It is the result either of a temporary interaction involving a reactant and a catalyst, or of a sequence of reactions in a loop in which at least one species is recycled.

Cenancestor

The last common ancestor of all living organisms, also known as LUCA (*Last Universal Common Ancestor*).

Chloroplast

The organelle of photosynthetic eukaryotes (algae, plants) where photosynthesis takes place. It derives from an ancient cyanobacterial endosymbiont (*see* Symbiosis).

Chondrule

A small spherical aggregate, millimetre-sized, consisting primarily of olivine or pyroxene (or both), arranged in a radial manner and enclosed in a silicate glass or an interstitial matrix known as mesostasis. Chondrules are abundant in undifferentiated stony meteorites (primitive meteorites).

Chondrite

A primitive, undifferentiated, stony meteorite that has never melted, frequently rich in chondrules*. The chondrites are classified into several groups, based on the degree of oxidation of the iron that they contain and on their metamorphic degree. The C1 chondrites are considered to be the most primitive and their overall chemical composition is very similar to that of the Sun, except for volatile elements.

Chromosome

An organized structure of DNA, associated with proteins, that carries genetic information in cells.

Clathrate

A form of water ice whose crystalline structure traps large quantities of gas (CO_2, H_2S, CH_4, etc.).

Coacervate

A colloidal (suspended and unstructured) assemblage of various hydrophobic organic molecules which occur as stable dispersion in an aqueous solvent.

Codon

A triplet of nucleotides in DNA or RNA that usually encodes one amino acid. Stop codons indicate the end of a coding gene.

Coenzyme

A small non-proteinic organic molecule which provides to an enzyme chemical functions not available from the protein amino acids, facilitating its action.

Cofactor

A chemical compound or a mineral ion the association of which with an enzyme is necessary for the catalytic action.

Core, Terrestrial

The central zone of the Earth (16% by volume, and 1/3 by mass), primarily consisting of iron, with a small amount of nickel and some traces of sulfur. The core is divided into a solid, inner core (between −5155 to −6378 km) and a liquid, outer core (between −2891 and −5155 km).

Craton

See Shield.

Crust

The solid outermost envelope of the Earth. Its lower portion is in contact with the mantle. The boundary between the crust and the mantle is called the Mohorovicic discontinuity (or Moho). With the rigid upper part of the mantle it forms the lithosphere. Two types of crust are recognized: (1) oceanic crust, of basaltic composition with an average thickness of 7 km, which forms the bottom of the oceans; (2) continental crust, of granitic composition, whose thickness is 30 to 80 km, which forms the continents.

Diapirism

The ascent of material (magma or rock) governed by the force of gravity. In general, it relates to material of low density that rises through denser rocks.

Differentiation (in Geology)

The separation of an initially homogeneous compound into several physically and chemically distinct phases. In the case of the Earth, for example, these phases are, the core, the mantle, the continental crust, the oceanic crust, the oceans and the atmosphere.

Disks, Circumstellar

In general terms, disks of dust that are very extensive (which may attain a radius of several hundred AU*) surrounding young stars, such as T Tauri* stars. Depending on the stars being considered and their stage of evolution, it is possible to recognize accretion disks (stellar formation), protoplanetary disks (the start of planetary formation) and debris disks, an advanced stage in the formation of planets where collisions between "small bodies" (planetesimals) still plays an important part.

DNA (Deoxyribonucleic Acid)

Nucleic* acid consisting of deoxyribonucleotides (*see Nucleotide*).

Dyke

A sheet of magmatic rock that is injected (in the form of magma*) into a fissure or a fault within pre-existing rocks.

Enzyme

A protein with catalytic properties (able to increase the rate of a chemical reaction) which may require one (or more) cofactor(s)* or one (or more) coenzyme(s)* (or both).

EPS (Exopolymeric Substance)

Polysaccharide macromolecules of greater or lesser complexity synthesized by numerous microorganisms (notably the cyanobacteria). They are secreted outside the cells and provide protection, cohesion, or adherence to a substrate.

Equilibrium State

In thermodynamics, an *isolated* system evolves towards a state of equilibrium, that is to say in which no change is possible and entropy is a maximum.

Erosion

The process of degradation and transformation of relief or rocks, produced by an external agent (rain, ice, wind, etc.).

Eruption, Stellar Magnetic

A large-scale phenomenon taking place at the surface of young stars, analogous to solar eruptions, but far more intense. Particularly visible in X-rays, the eruptions are the result of magnetic reconnections, a sort of short-circuit resulting from the small-scale random motions of the star's magnetic loops of opposite polarity. This magnetic field arises, as with the Sun, in convective motions in the outermost layers of the star (the dynamo effect).

Escape Velocity

The minimum velocity that a body needs to attain, theoretically, to indefinitely escape from a body and thus to overcome the gravitational attraction exerted by the latter.

Eukaryote

From the Greek *eu-*, true, and *karyon*, nucleus. An organism that belongs to the domain Eucarya, one of the three domains of life. The eukaryotes are characterized by the presence of a nucleus, surrounded by a membrane, containing the genetic material (chromosomes), and by the presence of organelles called mitochondria.

Exoplanet

A planet orbiting a star other than the Sun. Also known as an extrasolar planet. The first exoplanet was discovered in 1995. By October 2012, more than 840 were known, as well as over 1300 candidates to be confirmed.

Extremophile

An organism that carries out its life cycle in an environment where the physical and chemical conditions approach the limits that are compatible with life and which are lethal for most organisms.

Feldspar

A family of minerals. Tectosilicates (three-dimensional silicates). From the chemical point of view, they consist of two major families: the alkali feldspars (sodium endmember = $NaAlSi_3O_8$ = albite, and potassium endmember = $KAlSi_3O_8$ = orthoclase); the plagioclase feldspars (sodium endmember = $NaAlSi_3O_8$ = albite and calcium endmember = $CaAl_2Si_2O_8$ = anorthite). Albite is an endmember that simultaneously belongs to both families. The feldspars are the most abundant minerals in the continental and oceanic crusts (52%).

Fermentation

A chemical, energy-conversion reaction whereby electrons released from organic molecules are ultimately transferred to molecules obtained from the breakdown of those same molecules without involving the consumption of external redox agents.

Fossil, Molecular

A biogenic macromolecule, more or less transformed, associated with ancient sedimentary rocks, and which bears witness to the past existence of organisms. Generally, the term is applied to modified lipids (hopanes, steranes) present in the fossil record.

Garnet

A mineral. A nesosilicate (a silicate formed of individual SiO_4 tetrahedra). Its composition generally ranges between three endmembers: pyrope $[Mg_3Al_2(SiO_4)_3]$, almandine $[Fe_3Al_2(SiO_4)_3]$, and grossular $[Ca_3Al_2(SiO_4)_3]$.

Gene

A unit of biological information. In present-day cells, the information corresponding to a gene is carried by DNA and it codes for the synthesis of a functional polymer (RNA or a protein).

Genome

The overall set of genetic material carried by an organism. It consists of all the genes, but also other regions that may be non-coding (regulatory regions, for example). The genome may consist of one or several chromosomes, as well as extra-chromosomal elements.

Genotype

An individual's genetic inheritance. In the case of a specific gene, the term describes a type (a variant) with a particular sequence (an allele).

Geodynamo

A theory describing the mechanisms by which, in the terrestrial core, rotation and convection of electrically conducting fluids create and maintain a magnetic field.

Geothermal Gradient

The thermal gradient that results from the increase in temperature as a function of depth toward the interior of a planet. In the Earth's continental crust, the average geothermal gradient is $30°\,C.km^{-1}$.

Glaciation

A cold period in the Earth's history, characterized by the development of a vast cryosphere. Frozen water accumulates on the continents forming ice sheets. Water is thus unable to return to the ocean, so sea level drops (marine regression). The main glaciations known took place in the Precambrian, at the beginning of the Cambrian, at the end of the Tertiary, and in the Quaternary.

Greenhouse Effect

Heating of the surface of a planet arising when the atmosphere traps the infrared energy radiated by the ground.

Greenstone Belt

A geological formation primarily consisting of volcanic rocks (basalts and komatiites**), and sometimes of volcano-sedimentary rocks. They have an elongated shape (about 100 km long and only a few tens of km wide), whence the name of 'belt'. Their green color is due to the metamorphism of the basalts and komatiites. Such formations are very common in Archaean terranes (4.0–2.5 Ga).

Gypsum

A mineral. A sulphate with the formula $CaSO_4,2\,H_2O$.

Halite

A mineral (salt). A halogenic compound of formula NaCl.

Hertzsprung-Russell Diagram

A diagram relating the luminosity of a star to its surface temperature. Depending on their evolutionary state, stars occupy very specific locations on this diagram. The majority of stars are found along a line known as the Main Sequence*.

Heterotrophy

Metabolism* of living organisms that obtain nourishment from pre-existing organic compounds.

Hopanoids

Polycyclic molecular compounds found in the plasma membrane of many bacteria, providing strength and rigidity.

Hot Spot, Geology

An area of volcanic activity that is the result of the ascent of a column of hot mantle material, whose origin lies at the core-mantle boundary or the lower mantle-upper mantle boundary. When it arrives near the surface, the ascending mantle may melt through adiabatic decompression and give rise to oceanic-island magmatism (La Réunion, Hawaii, etc.) to submarine plateaus (Ontong-Java, Kerguelen, etc.) or to traps (Deccan, Columbia River, etc.).

Hydrolysis

Breaking a molecule by the addition of a molecule of water.

Hydrophobic

A property of molecules (hydrocarbons, for example) to be immiscible with water. It can be also applied to parts of molecules (chemical groups).

Hydrothermal

Describes the circulation of water through rocks of the crust, which occurs, in particular, near sources of heat.

Hyperthermophiles

Organisms whose optimal temperature for growth is above 80 °C and which are able to live up to temperatures of 110–120 °C.

Ice Line

In a protoplanetary disk* surrounding a star of any given type, the line within which the grains of circumstellar dust are sufficiently heated by the star for the ice that covers them to be evaporated.

Inclusion (geology)

A small-size (1 to 100 μm) cavity within a mineral. It frequently contains fluid phases (gas or liquids) trapped at the time the host mineral crystallized.

Interstellar Dust

Small grains, sub-micron in size, present in interstellar clouds and nebulae, consisting mainly of silicates*, carbon compounds and water ice.

Interstellar Molecules

To date, about 160 molecules have been identified in the interstellar medium. The greatest number are organic molecules, which contain up to 11 atoms of carbon. The discovery of glycine, the simplest of the amino acids, has been announced and awaits confirmation. Other simple molecules, not containing any atoms of carbon, such as NH_3, H_2S, SiO and, of course, H_2O, as well as many 'deuterated' molecules (in which deuterium, D, has replaced hydrogen, H) are very commonly detected.

Isomers

Chemical compounds of different structure, but possessing the same overall chemical formula.

Isotope

Atoms possessing the same number of protons and electrons, but a different number of neutrons. Some isotopes may be stable, whereas others are radioactive.

Jeans Escape

A mechanism that enables the loss of atoms and molecules from planetary atmospheres. It comes into play when the thermal velocity of the atoms or molecules becomes greater than escape velocity. The lightest molecules and atoms (H, H_2, He) escape most easily.

Kuiper Belt

A wide ring, extending on both sides of the ecliptic plane, lying beyond Neptune, and thus more than 30 AU* away from the Sun. It consists of "dwarf planets", also known as "Trans Neptunian Objects", the prototype of which is Pluto. It probably contains also asteroids, but these are too far away to be detected individually. Some however can be deviated from their orbits and become detectable as they cross the inner Solar System. The Kuiper belt is a fossil from the very early stages of the formation of the Solar System.

Late Heavy Bombardment

A phase of intense cometary and meteoritic infall that peaked about 600 Ma after the formation of the solar system. It is estimated that it lasted from 50 to 150 Ma. The main evidence for this phase is provided by the abundance of impact craters on the surface of the Moon. It is understood nowadays as the result of a major, but temporary, gravitational resonance interaction between Jupiter and Saturn.

Lipids

Organic molecules carrying hydrophobic groups (most often hydrocarbons) that are of sufficient size to confer a hydrophobic* nature on the molecule. Synonym: fats.

Liquidus

Line wich, in composition *vs.* temperature or pressure *vs.* temperature diagrams, separates the domain where crystals and liquid coexist from the field where only liquid exist.

Lithophile

According to V.M. Goldschmidt's classification, a lithophile element is a chemical element that preferentially links with oxygen, most often in the form of silicates*. Examples of lithophile elements are Si, Al, K, Ca, Mg, etc.

Lithospheric Plate

A section of the Earth's rigid lithosphere that moves over the asthenosphere, which is ductile and subject to convective motion.

LUCA

See Cenancestor*.

Macromolecule

A large-sized molecule. Biological macromolecules consist of polymers (proteins, nucleic acids, lipids, polysaccharides).

Magma

Molten rock that may be completely liquid or may consist of a mixture of liquid and crystals. It arises from a high temperature melting (above 650 °C for a granite and 1200 °C for a basalt) of pre-existing rocks.

Magnetite

A mineral; iron oxide with the formula $Fe^{2+}(Fe^{3+})_2O_4$.

Magnetosphere, Stellar

A large-scale magnetic structure, rigidly linked to the star, to a first approximation analogous to a dipole, as is the case with the terrestrial magnetosphere. This structure thus follows the star's rotation and it extends to its accretion disk, at a distance known as the co-rotation radius (equal to several times the star's radius). It results in a central, magnetic cavity, separating the star from the circumstellar disk*. It serves both as a "bridge" between the star and the disk for the phenomenon of magnetospheric accretion* and as a "springboard" for the material that is ejected in the form of bipolar jets*.

Main Sequence

In the Hertzsprung-Russell Diagram*, a diagonal region in which stars, (as for example, the Sun) are the site of thermonuclear reactions taking place at the core, in which hydrogen is converted into helium. The preceding phase, where the central regions have not yet reached a temperature sufficient to initiate these reactions, is known as the pre-Main-Sequence stage.

Mantle

The portion of a planet lying between the crust* and the core*. The mantle represents 82% of the Earth's volume and 66% of its mass. It is sub-divided into the upper mantle (to a depth of about 700 km) and the lower mantle (down to 2900 km). It mainly consists of silicates.

Mass Function, Initial

The distribution of the mass of stars at the time when they are formed, with a range lying between 0.1 and 100 solar masses.

Mesophiles

Organisms whose optimum temperature for growth is moderate (15–35 °C).

Metabolism

The overall set of molecular and energy transformations that occur within a cell.

Metabolite

The product of the chemical transformation of a substance by an organism.

Metasomatism

Change in rock composition due to fluid circulation. For example, in a subduction zone, the fluids released by the dehydration of the subducted oceanic crust (and particularly water) rise and percolate through the overlying "mantle wedge" (peridotite). These fluids which also contain dissolved elements, not only rehydrate the mantle peridotite, but also modify its composition.

Methanogen

An organism that gains energy reducing CO_2 or acetate with hydrogen (H_2). All known methanogens are archaea belonging to the Euryarchaeota.

Methanogenesis

Formation of methane (CH_4) by the reduction of CO_2 or acetate by molecular hydrogen (H_2).

Mica

A family of minerals. Phyllosilicates (water-bearing sheet silicate). From a chemical point of view, they are divided into two main families: the biotite family (black mica) $[K(Fe,Mg)_3(Si_3AlO_{10})(OH)_2]$ and the muscovite family (white mica) $[KAl_2(Si_3AlO_{10})(OH)_2]$.

Micelle

A structure, often spherical and usually formed in aqueous solution, consisting of assemblies of amphiphilic molecules that have their hydrophilic heads on the exterior and their hydrophobic tails in the interior. Unlike vesicles*, micelles do not circumscribe an aqueous compartment within them.

Microfossil

The fossil of a microorganism.

Mid-ocean Ridge

A chain of submarine relief at the boundary of two divergent lithospheric plates. On Earth, the overall length of the mid-ocean ridges is 60 000 km.

Migration, Planetary

A gravitational phenomenon whereby a planet formed at a certain position in a protoplanetary disk reacts dynamically with it or with other massive planets and changes its orbit. There are several modes of planetary migration, both towards the central star or away from it, depending of the type of interaction taking place.

Mineral

Solid material defined both by its chemical composition and its crystalline structure.

Mitochondrion (Pl. Mitochondria)

An organelle typical of eukaryote cells where aerobic respiration* takes place. From an evolutionary point of view, mitochondria derive from ancient, endosymbiotic alpha-proteobacteria.

Molecular Cloud

An essential component of the cold interstellar medium, dense and very massive (10 000 to 100 000 solar masses). Most of its mass consists of molecular hydrogen (H_2), with traces of organic molecules, of which the most complex observed today contains a chain of 11 carbon atoms. Stars are formed within such clouds.

Nebula, Primitive (or Primordial)

A term used by planetologists to describe the environment that accompanied the formation of planets within the Solar System, as "seen from inside". With a young star of the solar type, this is analogous to its planetary disk (see Disks, circumstellar).

Nucleic Acid

A polymer of nucleotides* that carries genetic information.

Nucleoside

A molecule formed by the combination of a sugar and a nucleic base (nucleobase).

Nucleotide

A molecule formed by the combination of a nucleoside and a phosphate. Nucleotides constitute the building blocks for nucleic acids. There is a distinction between ribonucleotides (in RNA*; they are based on ribose) and deoxyribonucleotides (in DNA*; they are based on deoxyribose).

Nucleus, Eukaryote

A structure that is characteristic of eukaryote cells and which encloses the chromosomes*. The nucleus is surrounded by a double membrane that is continuous with the endoplasmic reticulum. It is in the nucleus that the transcription of RNA* takes place starting from the DNA*. In eukaryotes, this transcription is not coupled with translation, which takes place in the cytoplasm.

Oceanic Plateau

Thick pile of basaltic rocks covering extensive area, which erupted and emplaced above the oceanic crust. These magmas are generated in hot spot *environments.

Oort Cloud

A spherical reservoir of comets, orbiting the Sun beyond the Kuiper Belt*, up to an estimated distance of about 50 000 AU.

Organic (Chemistry)

Describes the chemistry of molecules based on carbon

Orion Nebula

A bright nebula that is considered the prototype of star-formation regions. It contains an estimated 2 000 young stars with ages less than 3 million years. There are certain indications that the Sun was born in a stellar association within a nebula of this type, but perhaps containing more massive stars, and a larger population (about 10 000 stars).

Oxidant

A component of a oxidation-reduction (redox) reaction capable of receiving electrons.

Paleomagnetism

Palaeomagnetism describes the ancient magnetic field, recorded by magnetic minerals within rocks (magnetite, etc.). From its measurements, it is possible to deduce the latitude at which the magnetic mineral-bearing rock was formed. Paleomagnetism allows us to retrace the movement of lithospheric plates over time.

Palaeosol

A fossil soil. Palaeosols are likely to have recorded the concentration of atmospheric O_2 and CO_2 at the time of their formation.

Partial Pressure

In a mixture of perfect gases, the partial pressure of one of the gases in the mixture is that exercised by the molecules of that gas if they alone occupied the whole of the volume given to the mixture, at the same temperature as the latter.

Peptide

An amino acid polymer.

Phenotype

The collection of traits of a living organism expressed from the genotype within a given environment.

Photosynthesis

A biochemical process that allows an organism to fix CO_2 and synthesize its own organic matter by exploiting light energy.

Phylogeny

The study of the evolutionary relationships (family connections) between different living organisms.

Phylogeny, Molecular

The study of the evolutionary relationships between different organisms by the comparison of the sequences of biological macromolecules (the genes of nucleic acids or proteins).

Pillow Lava

Lava extruded under water (ocean, lake) which produces its typical rounded pillow shape. Pillow lavas form the upper part of oceanic crust.

Planets

In the Solar System, and more generally in planetary systems, massive bodies orbiting a star, and which are generally divided into two categories: rocky or terrestrial planets (similar to the Earth), and giant, gaseous planets (similar to Jupiter). In contrast to stars, planets do not radiate their own energy. They are luminous because their radiate part of the energy they absorb or reflect from their parent star.

Planetary Nebula

The last stage in the evolution of stars with masses less than 8 solar masses (of which the Sun is one), which is characterized by an intense mass loss. Interstellar dust particles are formed within the ejected material.

Plasma Membrane

A membrane consisting of a bilayer of phospholipids that encloses the cell.

Prebiotic Chemistry

Abiotic organic chemistry having led to the origin of life. This term is therefore used, in general, for a sub-group of abiotic processes that took place before the origin of life and was directly linked to the emergence of life.

Platinoid (Metals of the Platinum Family)

A family of transition metals: ruthenium (Ru), rhodium (Rh), palladium (Pd), osmium (Os), iridium (Ir) and platinum (Pt). They have similar chemical characteristics (they are siderophile*). Their abundance in the terrestrial crust is very low, whereas they are abundant in undifferentiated meteorites.

Prokaryote

A generally unicellular organism which does not have a nucleus surrounded by an internal membrane separating the genetic material from the rest of the cell, and in which the transcription* and translation* are, as a result, linked. Bacteria and archaea are prokaryotes.

Protein

A polypeptide with a specific sequence of aminoacids – obtained through the process of translation* – and of a size that is most often sufficient to lead, through folding, to a specific three-dimensional structure that permits it to carry out a given function.

Protostar

The earliest stage in the formation of a star, which lasts about 10 000 years for a solar-type star. A protostar is a hybrid body, evolving very rapidly. Its outer layers consist of a rotating envelope of gas and dust that is in gravitational collapse and, at its centre, a growing stellar body that is in the process of accumulating mass at the expense of the envelope, through a circumstellar disk (via magnetospheric accretion, see "magnetosphere"). A portion of this mass is ejected in the form of bipolar jets*.

Protometabolism

A primitive metabolism in living organisms and, by extension, metabolic-like pathways related to the origin of life.

Pyrite

A mineral: sulfide (FeS_2).

Pyroxene

A family of minerals. Inosilicates (anhydrous, single-chain silicates). The pyroxenes are subdivided into two families: the orthopyroxenes (Fe Mg)$_2$Si$_2$O$_6$ such as enstatite, and the clinopyroxenes (Ca Fe Mg)$_2$Si$_2$O$_6$ such as augite and diopside.

Quartz

A mineral. A tectosilicate (a three-dimensional silicate) with formula SiO_2.

Radiation, Evolutionary

Great diversification of organismal lineages in a short period of time.

Radical

A chemical species with an unpaired electron.

Rare Earth Elements (REE)

Chemical elements with very similar chemical properties. This family (lanthanides) ranges from Lanthanum (Z = 57) to Lutetium (Z = 71). In geochemistry, they are commonly used as geological tracers of magmatic processes.

Reducer (Reductant or Reducing Agent)

A chemical species capable of acting as an electron donor in a redox reaction.

Replication

The process of copying a nucleic acid (or any other informative polymer), synthesizing a new strand by the assembly of monomers.

Reproduction

A process in which an organism perpetuates itself in producing new individuals. Binary division is the most simple form of reproduction.

Respiration

An oxidation-reduction reaction leading to the gain of free energy and based on the controlled oxidation of organic or inorganic molecules, in which the final acceptor of electrons is oxygen (aerobic respiration) or a whole range of organic or inorganic molecules (anaerobic respiration).

Ribozyme

RNA* possessing catalytic activity.

RNA (Ribonucleic Acid)

A nucleic* acid consisting of ribonucleotides (see Nucleotide).

mRNA (messenger RNA)

A ribonucleic acid obtained by transcription of genes and carrying series of *codons, which is translated by the ribosome to make proteins.

tRNA (transfer RNA)

A ribonucleic acid carrying the anticodon* and serving, in the synthesis of proteins, as adaptor for the activated amino acid corresponding to the anticodon.

Sagduction

Deformation of rocks under the effect of gravity. Sagduction arises when high-density rocks (for example, komatiites**) are emplaced above rocks of lower density (for example, TTG*). This then creates an inverse density gradient, the result of which is that the dense rocks sink into the less dense ones. Sagduction was widespread before 2.5 Ga.

Serpentine

A family of minerals. Phyllosilicates (water-bearing sheet silicate). For example, antigorite $Si_4O_{10}Mg_6(OH)_8$. These minerals, which are extremely rich in water, arise from the alteraton of olivines, and to a smaller degree, of pyroxenes.

Shield

A vast region of old, very stable, continental crust (often of Precambrian age). For example, the Baltic Shield that includes Norway, Sweden, Finland and the western portion of Russia. Synonym: craton.

Shocked Mineral

A mineral whose crystalline structure exhibits defects, characteristic of the very high dynamical pressures produced by a meteoritic impact.

Siderophile

According the V.M. Goldschmidt's classification, a siderophile element is a chemical element that has a tendency to bond preferentially with iron. Examples of siderophile elements are: Fe, Ir, Pt, Pd, Ni, etc.

Silicate

A vast family of minerals rich in silicon and in oxygen (silicon oxides). The basic structure is that of a tetrahedron $(SiO_4)^{4-}$.

Small Bodies (in the Solar System)

Unlike planets*, bodies with the lowest mass in the present-day Solar System (asteroids, comets, meteorites), witnesses to the very earliest stages of its evolution.

Snow Line

In a protoplanetary disk, the radius beyond which water is in the form of ice (see also: "ice line"). When planets are formed, and unless migration* has occurred, the rocky ones are closer to the parent star than the snow line, and gaseous or icy bodies lie farther away. But other effects may come into play: in the present-day Solar System, the Earth would be inside the snow line and should be completely frozen ("snowball Earth"), but it is not, and has never been, because of the greenhouse effect* associated with the presence of its atmosphere.

Solidus

Line which, in composition *vs.* temperature or pressure *vs.* temperature diagrams, separates the domain where crystals and liquid coexist from the field where only crystals (*solid*) exist.

Spherule

A small droplet of molten rock produced by a meteoritic impact and sometimes ejected very far from the crater.

Spinel

A mineral. An oxide with the formula $MgAl_2O_4$.

Sterane

A fossil lipid derived from a sterol. Sterols are aromatic organic molecules with a hydroxyl radical that confers stability to eukaryotic cell membranes. Cholesterol is a sterol.

Stromatolite

An organo-sedimentary structure showing laminated fabric formed by the activity of microbial communities.

Subduction

A plate-tectonic process where an oceanic lithospheric* plate sinks into the mantle, beneath another continental or oceanic, lithospheric plate.

Supernova

The final phase in the evolution of stars with masses greater than 8 solar masses, which involves a gigantic explosion, completely destroying the star, and leaving only a remnant in the form of a neutron star (a pulsar) or a black hole. The solar system is thought to have formed in the vicinity of a supernova, as indicated by the discovery of "extinct" radioactive nuclei (^{60}Fe) in primitive meteorites, because these nuclei can only be produced in supernova explosions.

Supracrustal

Describes rocks that have been emplaced at the surface of the Earth's crust and which consist either of sediments, or volcanic rocks.

Symbiosis

From the Greek "sym" (with) and "biosis" (living). A term suggested by Heinrich Anton de Bary in 1879 to refer to the tight interactive relationship established between two (or more) organisms that live together. Symbioses may be mutualistic, when both partners benefit from the association, commensal, when just one of the partners gains an advantage, without the other suffering any disadvantage, and parasitic, when one partner gains an advantage (increases its fitness) at the expense of the other. Endosymbiosis is the term used when one partner (the symbiont) lives within the other (the host).

Talc

A mineral. A phyllosilicate (water-bearing sheet silicate) with the formula $Mg_3(Si_4O_{10})(OH)_2$.

Tektite

Natural glass formed at very high temperature (>2000 °C) by meteoritic impacts. Fragments or droplets of molten rock are ejected from the crater and may be thrown at great distances.

T Tauri Stars

Young, solar-type stars that have a mass that is typically between 0.5 and 2 solar masses. T Tauri stars are at a stage of evolution known as the pre-Main-Sequence stage. They are a few million years old and resemble the Sun at the first stages of its evolution. The youngest are surrounded by circumstellar disks* and are the source of bipolar jets*.

Thermophile

An organism that can live and reproduce between 40 and 80 °C.

Translation

The synthesis of proteins* in the ribosome by the transformation of the message contained in the sequence of nucleotides on mRNA* in a sequence of amino acids.

Transcription

The synthesis of RNA* by copying the deoxyribonucleotide sequence of DNA* to a ribonucleotide sequence.

Traps

Thick pile of basaltic rocks covering extensive area, which erupted and emplaced above the continental crust. These magmas are generated in hot spot* environments.

TTG **

An acronym for tonalite, trondhjemite, granodiorite. Rock association typical of the continental crust generated during the first half of the Earth history (before 2.5 Ga).

Uraninite

A mineral. An oxide with the formula UO_2.

Vesicle

A bilayer consisting of amphilic* molecules which encloses within it an aqueous volume.

Wind, Solar/Stellar

The continuous flux of ionized material ejected at high velocity from a star (about 400 km.s^{-1}). The solar wind (currently about 10^{-13} solar masses per year), mainly consists of protons.

Zircon

A mineral. A nesosilicate (silicate made up of isolated SiO_4 tetrahedrons) with the formula $ZrSiO_4$.

Further Reading

Books and articles

ALBARÈDE, F., 2009. *Geochemistry: an introduction.* Cambridge University Press, Cambridge, 352 pp.

AMEISEN, J.-C., 2009. *Dans la Lumière et les ombres. Darwin et le bouleversement du monde.* Fayard/Seuil, 500 pp.

BALLY, J. et REIPURTH, B., 2006. *The birth of stars and planets.* Cambridge University Press, Cambridge, 306 pp.

BELL, E.A., HARRISON, T.M., McCULLOCH, M.T. and YOUNG, E.D., 2011. *Early Archean crustal evolution of the Jack Hills Zircon source terrane inferred from Lu-Hf, 207Pb/206Pb, and d18O systematics of Jack Hills zircons.* Geochimica et Cosmochimica Acta, 75(17): 4816-4829.

BERSINI, H. et REISSE, J. (eds.), 2007. *Comment définir la vie?* Vuibert, Paris, 126 pp.

BLICHERT-TOFT, J. and ALBARÈDE, F., 2008. *Hafnium isotopes in Jack Hills zircons and the formation of the Hadean crust.* Earth and Planetary Science Letters, 265(3-4): 686-702.

BRACK, A., 1998. *The molecular origins of Life – Assembling the pieces of the puzzle.* Cambridge University Press, Cambridge, 428 pp.

CANUP, R.M., 2004. "Dynamics of Lunar Formation", *Ann. Rev. Astr. Ap.,* 42, p. 441–475.

CHYBA, C.E. et HAND, K.P., 2005. "Astrobiology: The study of the living Universe", *Ann. Rev. Astr. Ap.,* 43, p. 31–74.

CONDIE, K. (ed.), 1994. *Archean crustal evolution.* Elsevier, Amsterdam, The Netherlands, 528 pp.

CONDIE, K., 2004. *Earth as an Evolving Planetary System.* Academic Press, 350 pp.

DANIEL, J.-Y., BRAHIC, A., HOFFERT, M., MAURY, R., SCHAFF, A. et TARDY, M., 2006. *Sciences de la Terre et de l'Univers.* Vuibert, 758 pp.

DAWKINS, R., 2004. *The ancestor's tale, a pilgrimage to the dawn of life.* Weidenfeld et Nicolson, 520 pp.

DE DUVE, C., 1991. *Blueprint for a cell: the nature and origin of life.* N. Patterson. Burlington, N.C., 275 pp.

DE DUVE, C., 1996. *Vital dust: life as a cosmic imperative.* Basic Books, New York, 362 pp.

EHRENFREUND, P. et CHARNLEY, S.B., 2000. "Organic Molecules in the Interstellar Medium", *Ann. Rev. Astr. Ap.,* 38, p. 427–483.

FENCHEL, T. and FINLAY, B.J. 1995. *Ecology and evolution in anoxic worlds. (Oxford Series in Ecology and Evolution),* Oxford University Press, 288 pp.

FENCHEL, T., KING, G., BLACKBURN, T.H. and FENCHEL, T. 2012. *Bacterial biogeochemistry: The ecophysiology of mineral cycling. 3rd edition.* Academic Press, London, 302 pp.

FOWLER, C.M.R., EBINGER, C.J. et HAWKESWORTH, C.J. (eds.), 2002. *The early Earth: Physical, chemical and biological development.* Geological Society of London, Londres, 352 pp.

FRY, I., 2000. *The emergence of Life on Earth: A historical and scientific overview.* Rutgers University Press, 344 pp.

FURNES, H., ROSING, M., DILEK, Y. and DE WIT, M., 2009. *Isua supracrustal belt (Greenland)-A vestige of a 3.8 Ga suprasubduction zone ophiolite, and the implications for Archean geology.* Lithos, 113(1-2): 115-132.

GARGAUD, M., CLAEYS, P., LÓPEZ GARCÍA, P., MARTIN, H., MONTMERLE, T., PASCAL, R. et REISSE, J. (eds.), 2006. *From Suns to Life: A chronological approach to the history of life on Earth.* Springer, Dordrecht, 370 pp.

GARGAUD, M., MUSTIN, C. et REISSE, J. (eds.), 2009. "Traces of past or present life: biosignatures and potential life indicators", *Comptes Rendus Palevol,* 8 n° 7, Académie des Sciences, Paris.

GARGAUD, M., BARBIER, B., MARTIN, H. et REISSE, J. (eds.), 2005. *Lectures in Astrobiology I. Advances in Astrobiology and Biogeophysics,* 1, Springer, Berlin Heidelberg, 792 pp.

GARGAUD, M., MARTIN, H. et CLAEYS, P. (eds.), 2007. *Lectures in Astrobiology II. Advances in Astrobiology and Biogeophysics, 2,* Springer, Berlin Heidelberg, 669 pp.

GARGAUD, M., LÓPEZ GARCÍA, P., MARTIN, H., (eds.), 2011. *Origins and evolution of Life. An astrobiological perspective,* Cambridge University Press, 526 pp

GILMOUR, I. et SEPHTON, M.A. (eds.), 2004. *An introduction to astrobiology.* Cambridge University Press, Cambridge, 364 pp.

GOULD, S.J., 1991. *La vie est belle: les surprises de l'évolution.* Points Sciences. Le Seuil, 480 pp.

GOULD, S.J., 2001. *L'éventail du vivant: Le mythe du progrès.* Points Sciences. Le Seuil, 299 pp.

HARRISON, T.M., SCHMITT, A.K., McCULLOCH, M.T. and LOVERA, O.M., 2008. *Early (>= 4.5 Ga) formation of terrestrial crust: Lu-Hf, [delta]18O, and Ti thermometry results for Hadean zircons.* Earth and Planetary Science Letters, 268(3-4): 476-486.

HAWKESWORTH, C.J. et al., 2010. *The generation and evolution of the continental crust.* Journal of the Geological Society, 167(2): 229-248.

HAWKESWORTH, C.J. and KEMP, A.I.S., 2006. *The differentiation and rates of generation of the continental crust.* Chemical Geology, 226: 134-143.

KASTING, J.F. et CATLING, D., 2003. "Evolution of a Habitable Planet", *Ann. Rev. Astr. Ap.,* 41, p. 429–463.

KLEIN, É. et SPIRO, M. (eds.), 1995. *Le temps et sa flèche.* Science et Culture. Editions Frontières, Gif sur Yvette, 281 pp.

KNOLL, A. H., 2003. *Life on a young planet; The first three billion years of evolution on Earth*. Princeton University Press, Princeton, Chichester, 277 pp.

KOLTER, R. and MALOY, S. (eds.). 2012. *Microbes and evolution. The world that Darwin never saw*. ASM Press, Washington, 325 pp.

KWOK, S., SANFORD, S. A. (eds.) 2008. *Organic Matter in Space, International Astronomical Union, Symposium 251,* Cambridge University Press, Cambridge, 490 pp.

LAZCANO, A (ed.), 2012 *Origin of life, special issue,* Evolution: Education and Outreach (2012) 5:334–336

LUISI, P. L., 2006. *The emergence of Life: From chemical origins to synthetic biology*. Cambridge University Press, Cambridge, 332 pp.

LUNINE, J. I., 1998. *Earth: Evolution of a habitable world*. Cambridge University Press, Cambridge, 348 pp.

MADIGAN, M.T., MARTINKO, J. M., DUNLAP, P.V. et CLARK, D. P., 2009. *Brock Biology of Microorganisms*. Benjamin Cummings Publishing Company, San Francisco, CA, 1168 pp.

MARGULIS, L. et FESTER, R., 1993. Symbiosis as a source of evolutionary innovation: Speciation and morphogenesis. The MIT Press Cambridge, MA, 470 pp.

MARTIN, E., MARTIN, H. and SIGMARSSON, O., 2008. *Could Iceland be a modern analogue for the Earth's early continental crust?* Terra Nova, 20: 463-468.

MAYR, E., 2004. *What makes biology unique? Considerations on the autonomy of a scientific discipline*. Cambridge University Press, Cambridge, 246 pp.

MCBRIDE, N. et GILMOUR, I., 2003. *An Introduction to the Solar System*. Open University/University Press, Cambridge, 418 pp.

MONOD, J., 1973. *Le Hasard et la Nécessité*. Points Essais. Seuil, 244 pp.

MONTMERLE, T., EHRENREICH, D., and LAGRANGE, A.-M. (eds.) 2010, *Physics and Astrophysics of Planetary Systems*, Les Houches Physics School, EAS Conf. Series, Vol. 41., EDP Sciences, Les Ulis, 522 pp.

MONTMERLE, T., EHRENREICH, D. et LAGRANGE, A.-M. (eds.), 2009. *Physics and astrophysics of planetary systems*. EAS Conference Series. EDP Sciences, Les Ulis, 534 pp.

MORANGE, M., 2003. *La vie expliquée? 50 ans après la double hélice*. Sciences. Odile Jacob 264 pp.

MOYEN, J.-F. and MARTIN, H., 2012. *Forty years of TTG research*. Lithos, 148: 312-336.

NORRIS, R. P. and STOOTMAN, F. H. (eds.), 2002. *Bioastronomy 2002. life among the stars: International Astronomical Union, Symposium 213*. Astronomical Society of the Pacific, San Francisco CA, 576 pp.

OLLIVIER, M., ENCRENAZ, T., ROQUES, F., SELSIS, F. and CASOLI,F., 2009. *Planetary Systems. Detection, Formation and Habitability of Extrasolar Planets*. Springer, 340pp

NORRIS, S. C., 2003. *Life's solution: Inevitable humans in a lonely Universe*. Cambridge University Press, Cambridge, 464 pp.

POPA, R., 2004. *Between Necessity and Probability: Searching for the Definition and Origin of Life*. Springer, Berlin Heidelberg, 258 pp.

PROSS, A., 2012, *What is Life? How chemistry becomes biology*. Oxford University Press, Oxford, 224 pp.

RASMUSSEN, S., BEDAU, M. A., CHEN, L., DEAMER, D.W., KRAKAUER, D. C., PACKARD, N. H. et STADLER, P. F. (eds.), 2008. *Protocells: Bridging nonliving and living matter*. The MIT Press, Cambridge, MA, 776 pp.

REIPURTH, B., JEWITT, D. et KEIL, K. (eds.), 2007. *Protostars and Planets V.* University of Arizona Press, 1024 pp.

REISSE, J., 2006. *La longue histoire de la matière: Une complexité croissante depuis des milliards d'années. L'interrogation philosophique*. Presses universitaires de France, 316 pp.

ROLLINSON, H. R., 2007. *Early Earth systems: a geochemical approach*. Blackwell publishing Ltd., Oxford, Malden, Carlton, 275 pp.

RUDNICK, R.L. and GAO, S., 2003. *The Composition of the Continental Crust*. In: R.L. Rudnick (ed.), The Crust. Treatise on Geochemistry. Elsevier-Pergamon, Oxford, p. 1-64.

SAPP, J., 2005. *The New Foundations of Evolution: On the Tree of Life*, Oxford University Press, 2009. 425 pp.

SARACENO, P., 2012, *Beyond the Stars - Our Origins and the Search for Life in the Universe*, World Scientific, Singapore, 377 pp.

SCHRÖDINGER, E., 1944. *What is life?* Cambridge University Press, Cambridge.

SCHULZE-MAKUCH, D. et IRWIN, L. N., 2004. *Life in the Universe: Expectations and constraints*. Advances in Astrobiology and Biogeophysics. Springer 172 pp.

STACEY, F. D. et DAVIS, P. M., 2008. *Physics of the Earth*. Cambridge University Press, Cambridge, 552 pp.

SULLIVAN, W.T. et BAROSS, J. A. (eds.), 2007. *Planets and life: the emerging science of astrobiology*. Cambridge University Press, Cambridge, 626 pp.

TAYLOR, S. R., 2001. *Solar system evolution: A new perspective*. University Press, Cambridge, 484 pp.

TAYLOR, S.R. and MCLENNAN, S.M., 2008. *Planetary crusts: Their composition, origin and evolution*. Cambridge Planetary Science Cambridge University Press, 378 pp.

VAN KRANENDONK, M., SMITHIES, R. H. et BENETT, V. (éditeurs), 2007. *Earth's oldest rocks. Developments in Precambrian Geology, 15.* Elsevier, Amsterdam, 1291 pp.

WOESE, C. R., KANDLER, O. et WHEELIS, M. L., 1990. „Towards a natural system of organisms: proposal for the domains Archaea, Bacteria, and Eucarya ", *Proc. Natl. Acad. Sci. USA*, 87, p. 4576-4579.

l = left; r = right; c = center; t = top; b = bottom.

Figure Credits

Cover:
Illustration: © 2009 Alain Bénéteau/www.paleospot.com
Photos: © Corbis – © PhotoDisc
p. 1: ESO/M. McCaughrean *et al.* (AIP).
p. 2l: ESO.
p. 2r: Nasa, ESA, M. Robberto (Space Telescope Science Institute/ESA) and the Hubble Space Telescope Orion Treasury Project Team.
p. 2c: Courtesy of Howard McCallon.
p. 3: R. Gendler/NOVAPIX.
p. 7r: Nasa, ESA, HEIC, and The Hubble Heritage Team (STScI/AURA). Acknowledgment: R. Corradi (Isaac Newton Group of Telescopes, Spain) and Z. Tsvetanov (Nasa).
p. 7r: ESO.
p. 8: ESO/M. McCaughrean *et al.* (AIP).
p. 12tl: Pr Mark J. McCaughrean.
p. 12tr: Hubble image © Nasa, ESA, N. Smith (University of California, Berkeley), and The Hubble Heritage Team (STScI/AURA) – CTIO Image © N. Smith (University of California, Berkeley) and NOAO/AURA/NSF.
p. 12bl: Nasa/JPL-Caltech/R. A. Gutermuth (Harvard-Smithsonian CfA).
p. 14: T. Pyle (SSC)/Nasa/JPL-Caltech.
p. 16l: ESO/M. McCaughrean *et al.* (AIP).
p. 16r: Nasa/CXC/Penn State/E. Feigelson & K. Getman *et al.*
p. 17: The solar X-ray images are from the Yohkoh mission of ISAS, Japan. The X-ray telescope was prepared by the Lockheed-Martin Solar and Astrophysics Laboratory, the National Astronomical Observatory of Japan, and the University of Tokyo with the support of Nasa and ISAS.
p. 21: © 2009 by Matthias Pfersdorff.
p. 24b: Nasa, ESA, P. Kalas, J. Graham, E. Chiang, and E. Kite (University of California, Berkeley), M. Clampin (Nasa Goddard Space Flight Center, Greenbelt, Md.), M. Fitzgerald (Lawrence Livermore National Laboratory, Livermore, Calif.), and K. Stapelfeldt and J. Krist (Nasa Jet Propulsion Laboratory, Pasadena, Calif.)
p. 33: Nasa/JPL-Caltech/T. Pyle (SSC).
p. 37: Ron Miller/Novapix.

p. 47: Nasa/JPL-Caltech/USGS.
p. 48t: Nasa/Johnson Space Center.
p. 48c: Nasa/Johnson Space Center (à g., scan courtesy of M. Gentry & S. Erskin, photo S71-43477, et à dr., scan courtesy of G. Lofgren & T. Bevill, photo S76-22598).
p. 48b: Nasa/Johnson Space Center.
p. 62: Nasa.
p. 81: T. Pyle (SSC)/Nasa/JPL-Caltech.
p. 87: OAR/National Undersea Research Progam (NURP), NOAA/P. Rona.
p. 156l: Nasa/JPL-Caltech /USGS.
p. 156r: Nasa/JPL-Caltech.
p. 165: Eros et Mathilde © Nasa/JPL/JHUAPL; Ida © Nasa/JPL.
p. 199: Nasa.
p. 225l: Image courtesy of the Image Science & Analysis Laboratory, Nasa/Johnson Space Center (image STS51I-33-56AA sur http://eol.jsc.nasa.gov).
p. 225tr: © Photo12.com/Alamy.
p. 225br: © Photo12.com/Alamy.
p. 244: Nasa/JPL/Space Science Institute.
p. 274: Nasa, Alan Watson (Universidad Nacional Autonoma de Mexico, Mexico), Karl Stapelfeldt (Jet Propulsion Laboratory), John Krist (Space Telescope Science Institute) and Chris Burrows (European Space Agency/Space Telescope Science Institute).
p. 275: image Nasa/JPL-Caltech.
p. 277: Mosaic of Clementine images. Image processing by Ben Bussey, Lunar and Planetary Institute.
p. 280: © 2009, Alain Bénéteau/www.paleospot.com

Printing: Ten Brink, Meppel, The Netherlands
Binding: Stürtz, Würzburg, Germany